The Moral Powers

The Moral Powers:
A Study of Human Nature

P. M. S. Hacker

WILEY Blackwell

Library of Congress Cataloging-in-Publication Data

Names: Hacker, P. M. S. (Peter Michael Stephan), author.
Title: The moral powers : a study of human nature / P. M. S. Hacker.
Description: First edition. | Hoboken, NJ, USA : Wiley-Blackwell, 2020. |
    Includes bibliographical references and index.
Identifiers: LCCN 2020015635 (print) | LCCN 2020015636 (ebook) | ISBN
    9781119657774 (paperback) | ISBN 9781119657804 (adobe pdf) | ISBN
    9781119657798 (epub)
Subjects: LCSH: Philosophical anthropology.
Classification: LCC BD450 .H235555 2020 (print) | LCC BD450 (ebook) | DDC
    128–dc23
LC record available at https://lccn.loc.gov/2020015635
LC ebook record available at https://lccn.loc.gov/2020015636

Cover design: Wiley

Set in 10.5/12.5pt Sabon by SPi Global, Pondicherry, India

For

Max Bennett

Contents

Prolegomenon   xi

1. Philosophical anthropology and the investigation
   of value   xi
2. The sinopia for a fresco   xvi

Acknowledgements   xxiv

Part I   Of Good and Evil   1

Chapter 1   The Roots of Value and the Nature
          of Morality   3

1. The place of values in a world of facts   3
2. Varieties of goodness   8
3. The framework of moral goodness   17
4. Morality   24
5. Individual critical morality   30

Chapter 2   The Roots of Morality and the Nature
          of Moral Goodness   33

1. Moral goodness   33
2. The roots of moral value   38
3. Respect   46
4. The relative permanence of the virtues   58
5. Constants in human nature   61

Chapter 3   The Roots of Evil   65

1.   The horror!   65
2.   The grammar of evil: preliminary clarification   76
3.   Philosophical problems: does evil exist?   83
4.   Philosophical problems: can evil be explained?   89

Chapter 4   Explanations of Evil   101

1.   The variety of explanations   101
2.   Reasons and motives for doing evil   103
3.   Can evil be a motive?   115
4.   Knowledge of good and evil   121
5.   Experimental psychology: Milgram's and Browning's
     explanations of evil-doing   125

Chapter 5   Evil and the Death of the Soul   129

1.   Body, mind, and soul   129
2.   The death of the soul   138
3.   Forgiveness and self-forgiveness   143
4.   Evil and the unforgivable   148
5.   From soul to soul: trisecting an angle with compass and rule   152

Part II   Of Freedom and Responsibility   155

Chapter 6   Fatalism and Determinism   157

1.   Of fate and fortune   157
2.   Fatalism   162
3.   Nomological determinism   169
4.   Flaws in reductive determinism   173
5.   The random and the determined   177

Chapter 7   Neuroscientific Determinism, Freedom,
             and Responsibility   179

1.   Neuroscientific determinism   179
2.   Explanations of human behaviour: a recapitulation   182
3.   Neuroscientific explanation and its limits   188
4.   How possible, not why necessary   192
5.   Varieties of responsibility   196

6. Elaboration   201
7. Irresistible impulse and temptation   203

Part III   Of Pleasure and Happiness   207

Chapter 8   Pleasure and Enjoyment   209

1. Varieties of hedonism   209
2. Pleasure, enjoyment, and being pleased   212
3. Pleasure, pain, and the pleasures of sensation   219
4. Enjoyment and the pleasures of activities   224
5. Pleasure, desire, and satisfaction   229
6. Comparability and quantification   231
7. First-person judgements of pleasure   235
8. The hedonic life   237

Chapter 9   Happiness   243

1. The linguistic terrain   243
2. A distinct idea of happiness   246
3. A clear idea of happiness   250
4. Preconditions of happiness   263
5. The epistemology of happiness   266
6. Two philosophical traditions   269
7. Happiness and morality   276

Chapter 10   The Science of Happiness   281

1. From eighteenth-century crudity and back again   281
2. How happiness is understood by happiness scientists   286
3. Psychological and epistemological presuppositions
    of the science of happiness   290
4. Measuring happiness   294
5. Some results of the science of happiness   298

Part IV   Of Meaning and Death   305

Chapter 11   The Need for Meaning   307

1. Meaning   307
2. The primacy of loss of meaning and the sense of
    meaninglessness   313

<table>
<tr><td>3.</td><td>The roots of meaninglessness</td><td>316</td></tr>
<tr><td>4.</td><td>Does life have a meaning?</td><td>326</td></tr>
<tr><td>5.</td><td>Finding meaning in human life</td><td>329</td></tr>
</table>

Chapter 12 The Place of Death in Human Life 334

<table>
<tr><td>1.</td><td>What is death?</td><td>334</td></tr>
<tr><td>2.</td><td>An afterlife</td><td>338</td></tr>
<tr><td>3.</td><td>The valuelessness of life</td><td>341</td></tr>
<tr><td>4.</td><td>The value of life</td><td>344</td></tr>
<tr><td>5.</td><td>Living for ever</td><td>349</td></tr>
<tr><td>6.</td><td>Thanatophobia – the fear of death</td><td>353</td></tr>
</table>

Appendices

Appendix 1: On Animal Beliefs and Animal Morality 361

<table>
<tr><td>1.</td><td>Animal morality</td><td>361</td></tr>
<tr><td>2.</td><td>Animal thinking, animal thoughts, and animal memory</td><td>364</td></tr>
<tr><td>3.</td><td>Counter-arguments and their rebuttal</td><td>367</td></tr>
<tr><td>4.</td><td>Animal knowledge of other animals' minds</td><td>378</td></tr>
<tr><td>5.</td><td>Animal emotions</td><td>384</td></tr>
</table>

Appendix 2: Diabology: Satan, Lucifer, and the Devil in Western Thought 390

Appendix 3: Hannah Arendt and the Banality of Evil 398

Appendix 4: The Pictorial Representation of Pleasure in Western Art 407

Index 412

Prolegomenon

1. Philosophical anthropology and the investigation of value

This volume is an essay in *philosophical anthropology*. Philosophical anthropology is not philosophy *of* anthropology. The absence of this term in anglophone philosophy is unfortunate, because we have to make do with 'philosophy of mind' on the one hand, and with 'ethics' or 'moral philosophy' on the other. But these are not coextensive with 'philosophical anthropology', which includes much more than philosophy of mind and does not incorporate the whole of moral philosophy. The expression *Philosophische Anthropologie* is common in the German philosophical tradition, as is patent in Kant's famous eponymous lectures. 'Philosophical anthropology', as I am using the expression, is the study of the conceptual framework in terms of which we think about, speak about, and investigate man (*Homo sapiens*) as a social and cultural animal.

It is a truism that human beings are extraordinarily complex creatures. But it is a truism worth bearing in mind when reflecting on the manifold concepts and multiple patterns of explanation that we invoke and employ in trying to describe and understand ourselves, and in trying to make ourselves intelligible to others. It is exceedingly improbable that any simple account or familiar 'ism' will render these transparent. The devil lies in the detail, and the gods in the overview. We are individual persons, each with a unique character and temperament, and each living an unrepeatable life. We differ widely in our thought and behaviour. Our relationships are often convoluted and indefinitely variable. Our ways of coping with stress are manifold. Our emotional lives are often opaque, not only to others, but also to ourselves. Self-knowledge and self-understanding of any depth are relatively rare. Moreover, our proneness to self-deception is both powerful and widespread. Over the millennia of recorded history, we have evolved a rich and subtle language both to express our thoughts, feelings, and will, and to describe our psychological and epistemic attributes and our emotional relationships. This is not a theory of anything, but a vocabulary – a set of instruments, as it were, to express ourselves, to understand and explain ourselves and others, and to enable a reasonable degree of mutual predictability. This vocabulary is very large indeed – English is doubtlessly blessed in being the richest among Western languages in this respect. The complexity of the logical and grammatical relationships between the words and phrases, the manifold common figures of speech, metaphors, and analogies we deploy, form an extensive conceptual network that is difficult to describe correctly, let alone to survey. To attempt to do so is a task for philosophical anthropology. When successful, it benefits the human sciences (psychology, sociology, economics, cognitive neuroscience) and humane studies (literature, history, and the arts). For it lays bare the conceptual framework for our thought and talk about ourselves, and makes clear the bounds of sense that we so commonly transgress in our describing and theorizing, in our theoretical descriptions and diagnoses, and in our prescriptions for ameliorating the human condition, both in the small and in the large.

Philosophical anthropology, one might say, is the philosophical study of human beings. As such it lies at the intersection of metaphysics, philosophy of biology, philosophy of mind, epistemology, and moral philosophy. More than any other branch of philosophy, it is synthetic – drawing together a multitude of threads to weave a many-coloured tapestry of the nature of humanity, of human life, and of the ways we think about our lives and about ourselves.

Metaphysics, understood as super-physics investigating the objective necessary features of all possible worlds, or perhaps just of this, the only actual world, is a pseudo-science and no more than a dream of philosophy. In this domain, the task of philosophy is to curb our perennial temptation to indulge in metaphysical reasoning by showing where and why it transgresses the bounds of sense. Critical analytic philosophy aims to dissolve such dreams and show that what seemed treasure is merely fools' gold. But metaphysics may also be understood as an analytic investigation of the most general concepts and concept types in any conceptual scheme capable of describing experience and its objects in an objective spatio-temporal framework. Philosophical anthropology overlaps with metaphysics thus conceived. For it too is concerned with general categorial concepts, for example with the concepts of substance, causation, agency, and power (potentialities, liabilities, and susceptibilities) – but only as they bear on the understanding of humanity and the concepts in terms of which we describe ourselves. For human beings are indeed substances (persistent individual things of a certain kind): they are agents that interfere in the course of events; they have a wide variety of causal powers and causal susceptibilities and liabilities, and numerous perceptual, cognitive, and cogitative abilities, abilities to act or refrain from acting, all of which present conceptual problems and generate numerous entanglements in the web of words.

Like philosophy of biology (properly practised), philosophical anthropology is concerned with teleology – goal directedness – in the biosphere in general and in human beings in particular. It is similarly concerned with the nature of teleological explanation of organs and their functions, for this bears directly on what is good for man: good health and the full exercise of native abilities. It is preoccupied with teleological explanation and understanding of the behaviour of animate creatures in general and of human beings in particular. For animals and humans alike act, as Aristotle put it, 'for the sake of an end', and a primary way of understanding what they do and why they do it is to explain their goals and to explain their behaviour and feelings (their frustrations and exasperations, their joys and satisfactions) by reference to the pursuit of their goals.

Unlike philosophy of biology, however, philosophical anthropology is also concerned with the nature of rationality and of reasonableness, subjects that lie at the heart of the study of man. Here philosophical anthropology provides conceptual analyses the understanding of which is fundamental for the social sciences and economics, and is pivotal for their constructive critique. The logical varieties of goodness: medical goodness, artefactual goodness, technical goodness, utile goodness, the beneficial, hedonic goodness, and moral goodness (see Chapter 1) undermine the idea of a simple relationship between preference and linear ordering. So too does the plurality of virtues. But the varieties of goodness and the multiplicity of virtues do not preclude rational choice. They exclude commensurability of value and the linear ordering of the plurality of values. But incommensurability does not imply the impossibility of preferences resting on good reasons. Conversely, the possibility of reasonable choice does not imply commensurability. The preferences and choices of individual human beings are personal, dynamic, dependent upon the exigencies of the situation, and unsystematic. These conceptual features constitute a radical challenge to received economic theory and social science.

Like philosophy of mind, philosophical anthropology is concerned with the investigation of the concept of a person, of the concepts of the mind and the body, and of the human faculties of sensation and perception, cognition and cogitation, memory and imagination, desire and will, the passions and the emotions. Unlike philosophy of mind, philosophical anthropology is not limited to the philosophical study of the nature of the human faculties, for it has to investigate the social nature of human beings and the manner in which that affects and moulds them. In particular, it has to investigate their most important defining characteristic: mastery of the social institution of a language and the way in which that determines what is possible for man. For a language is partly constitutive of a culture, and man, unlike all other animals, is a cultural being.

Mankind is distinctive, indeed unique, in the animal kingdom in valuing and pursuing so wide a range of things, and engaging in and enjoying such a wide range of activities. Human beings possess a very large range of skills in activities, and can accordingly aim to be, and are evaluated as being, excellent, good, or proficient. We are also unique in having a morality, possessing knowledge of good and evil, and having moral virtues such as compassion, honesty, and kindness. (The issue of whether animals can be said to have anything akin to a morality and moral virtues is examined in Appendix 1.) We are, above all, language-using, rule-following animals. Some of these rules are moral rules, partly constitutive of a morality, conformity with which is, other things being equal, partly constitutive of a morally good or admirable person. We cultivate, value, and admire certain human character traits, some of which – the moral virtues – are similarly constitutive of being a morally good person.

The investigation of human values and norms, and of human virtues and vices, is a province of moral philosophy that it shares with philosophical anthropology. It is, of course, not the whole of moral philosophy, for the latter incorporates the study of ethical theories, prescriptive ethics, and casuistry – applied ethics. But no study of the nature of man could be complete without extensive investigation of the nature of human morality (or moralities), of the concept and limits of morality, of axiology and normativity, of virtues and vices. However, the philosophical anthropological approach differs from that of moral philosophy. It is not concerned with constructing systems of morals (such as consequentialist and deontological ethics). Its primary task is to clarify the framework within which the existence of morality is intelligible, and to shed light on its scope and limits. In particular, it is to render the social phenomenon of a morality intelligible. (Here the enterprise is similar to David Hume's in his *Enquiry concerning the Principles of Morals*.) What is the phenomenon? How is it possible for such a phenomenon to exist? What is presupposed for its existence?

These are among the subjects that will be discussed in this book. It is the final volume in a tetralogy of books in philosophical anthropology. The first was *Human Nature: the Categorial Framework*, which provided the stage set. The second was *The Intellectual Powers: a Study of Human Nature*, which began the play with the presentation of the intellect and its courtiers. The third was *The Passions: a Study of Human Nature*, which introduced the drama of the passions and the emotions. This volume turns to the moral powers and the will, to good and evil, to pleasure and happiness, to what gives meaning to our lives, and to the place of death in our lives.

This tetralogy constitutes a *Summa Anthropologica* in as much as it presents a systematic categorial overview of our thought and talk of human nature, ranging from substance, power, and causation to good and evil and the meaning of life. A *sine qua non* of any philosophical investigation, according to Grice, is a synopsis of the relevant logico-linguistic grammar. It is surely unreasonable that each generation should have to amass afresh these grammatical norms of conceptual exclusion, implication, compatibility, and contextual presupposition, as well as tense and person anomalies and asymmetries. So I have attempted to provide a compendium of usage of the pertinent categories in philosophical anthropology to assist others in their travels through these landscapes.

Each of the previous volumes was designed, at the cost of a modest degree of repetition, to be self-contained and intelligible in its own right. This volume has the same goal, but since it does presuppose what has been achieved in the previous three volumes, the following section is an attempt to sketch out what I have taken for granted here – an attempt to provide as it were a sinopia – the underdrawing of a fresco. Everything adumbrated in the next section is argued for at length in one or other of the previous three books that constitute the fresco.

As in *The Passions*, I have in this volume made extensive use of quotations from fiction, drama, and poetry. For novelists, dramatists, and poets give concrete and vivid depictions of good and evil, of pleasure and happiness, and of the place of death in human life. Since the understanding of these themes is predominantly idiographic rather than nomothetic, such presentations anchor the intellect and imagination in the concrete character of human life.

2. The sinopia for a fresco[1]

We are animate, self-moving, relatively persistent, material substances – animals of the species *Homo sapiens* of the Hominidae family. So, like all other animals, we *are* bodies, consisting of matter of various kinds, variously classifiable, and composed of parts of various kinds, variously classifiable. Our organs, both internal and external, have functions essentially related to our health and characteristic activities.

Being bodies, we are space occupants, tracing a unique route through the world as we wend our way from birth to death. We speak of ourselves as 'having' or 'possessing' a body. It is not evident how a being that *is* a body can also *have* a body. But the body we have is not the same as the body we are. Or, to be more perspicuous, all talk of the body we have and of its features, of its being black or white, athletic or lithe, beautiful or dumpy, youthful or aged, is no more than talk of our *somatic characteristics*. If someone's body is painted blue all over (like the ancient Britons), then he is painted blue all over; if someone has a graceful body, then she is graceful; if someone has an unhealthy, fat body, then he is unhealthy and fat. *The body we have* is not a possession, but this definite description is a form of presentation of the various somatic characteristics that we wish to pick out from among the totality of our characteristics. The body one has is neither clever nor stupid, neither knowledgeable nor ignorant, neither cheerful nor miserable. These are not somatic features but rather psychological characteristics, not of the mind but of the living human being as a whole. Of course, when one dies, one leaves a body behind – but that is not the body one had (one's somatic characteristics) but the corpse that is left when one dies. (Note that when one dies one does not become a corpse: one ceases to exist. When we bury a corpse, we are not burying the person – who no longer exists – but their remains. Nor are we burying the body they inhabited, since people do not inhabit their bodies.)

We have perceptual powers, both passive and active, and powers of sensation, both passive and active. We sometimes take pleasure in

---

[1] Three quarters of the painted fresco itself is to be found in *Human Nature: the Categorial Framework* (Blackwell, Oxford, 2007; henceforth *HNCF*), *The Intellectual Powers: a Study of Human Nature* (Wiley-Blackwell, Oxford, 2013; henceforth *IP*), and *The Passions: a Study of Human Nature* (Wiley-Blackwell, Oxford, 2017; henceforth *TP*).

perceiving and feeling. At other times we may feel disgust and revulsion at what we perceive or feel, or shame at our perceiving or feeling it. We can build machines to perform perceptual tasks. However, they can be said to perceive (if at all) only in a secondary sense of the word – just as calculators can calculate, and computers compute, only in a secondary sense of the words. The robots we build may come up with the results of perceiving those things we build them to monitor, but they perceive them only in a secondary sense and feel neither pleasure nor revulsion at what they thus 'perceive', let alone shame at having 'perceived' it.

Our perceptual powers are cognitive faculties. By their voluntary exercise and through their non-voluntary cognitive receptivity in passively seeing what is evident to sight or passively hearing what is audible to the ear, we can acquire knowledge of our environment. We may achieve or attain knowledge as the upshot of our voluntary deliberately looking or searching, scrutinizing or examining, listening for and listening to. But we may also be *given* knowledge in recognizing or noticing something, and in awareness or consciousness of something or a feature of something in our perceptual field. Moreover, knowledge may dawn on us in realization. Receptive knowledge is non-voluntary: one can be ordered to look but not recognize, to voluntarily look but not voluntarily realize, to intentionally investigate but not intentionally become or be conscious of something, to look on purpose but not be aware of something on purpose.

Like developed animals and unlike plants, micro-organisms, and primitive creatures, we are conscious beings. Like most animals, we go through a diurnal cycle of wakefulness and sleep, and through injury, illness, or sedation we may lose consciousness and become unconscious. Following Norman Malcolm, these may be called 'forms of intransitive consciousness'.[2] The much cited question of what consciousness is *for*, if addressed to the phenomena of intransitive consciousness, is patently misconceived. The proper question is 'What is sleep for?' – and to that empirical question we have as yet no clear answer.

When conscious, we may enjoy or suffer various forms of transitive perceptual consciousness. We may become and then be conscious of the ticking of the clock in the background, of the smell of cooking wafting in from the kitchen, of Jack's embarrassment or Jill's amusement.

---

[2] Norman Malcolm, 'Consciousness and Causality', in D. M. Armstrong and Norman Malcolm, *Consciousness and Causality* (Blackwell, Oxford, 1984), pp. 1–101.

In such cases, our attention is *caught and held* by something within our perceptual field. Transitive perceptual consciousness is neither voluntary nor involuntary, but non-voluntary. One cannot intend to become or to be conscious of something, nor can one order another to be conscious of something – but only to pay attention to something perceptible. Transitive perceptual consciousness is at home with perceptibilia that lie on the periphery of our perceptual field. It is factive: one cannot be conscious of something that is not there or of what is not the case, and it involves reception of knowledge – what one is perceptually conscious of one knows to be so or to be there. Here too the questions 'What is consciousness for?' and 'What is the evolutionary warrant for consciousness?' are singularly inept. No animal could survive for long in the jungle or on the savanna without cognitive receptivity for peripheral perception. There are other forms of consciousness; they need not engage us now (see *IP*, ch. 1, for full elaboration of the varieties of consciousness).

We have cognitive powers. Our perceptual powers are sources of knowledge. They are cognitive faculties. But they are not infallible – one may mistakenly think that one perceives and hence that one knows things to be so. Being fallible, they are also sources of belief. For belief is the default when knowledge fails. One form belief may take is knowledge minus truth. It does not follow that knowledge is belief plus truth (*IP*, chs 5–6). By the active exercise of our perceptual powers, we may come to know how things are. But coming to know how things are is not the same as coming to know what things are true: knowledge does not 'aim (primarily) at truth' – at how things are said to be or might be said to be – but at reality.[3] Knowledge is not a state or mental state and knowing something is not a state of mind. To know something is akin to an ability – an ability to state what one knows, to answer relevant questions, to act or feel for the reason constituted by what one knows (namely that things are thus and so). Being ability-like, knowledge cannot in general exist only for a moment. A creature that can know things can also retain what it knows. Retention of knowledge is not storage (one cannot store abilities), and storage of knowledge (i.e. of what is known) is not retention. Our mnemonic powers consist of retention of

---

[3] Of course, strictly speaking, contrary to Dummett and his followers, knowledge does not aim at anything. What human beings primarily want to know is how things are. We want to know what is true when we are concerned with sayables, namely propositions, assertions, statements, announcements, declarations, and so forth, that have been expressed or made, or are envisaged as being expressed or made, all of which may be true or false.

those abilities constitutive of knowing what we know, not of storing what we know. What we know can, given our goals and projects, provide us with reasons for thinking, feeling, or acting.

We have cogitative powers: we can not only think (believe, hold) things to be so, we can also *think*, that is *reason*. To reason is to engage in reasoning. But reasoning is not an act or activity – it is to come to apprehend what is known or believed to be so *as a reason warranting* a conclusion. Things are as the conclusion states them as being *because* things are as the premises describe them. To reason is to come to apprehend a premise or array of premises as *justifying* the conclusion. The justifying relation between premises and conclusion may be deductive or evidential. If evidential, it may be inductive or non-inductive. Non-inductive reasoning may be from constitutive evidence – logical criteria – to empirical conclusion. It may also be from hearsay, established fact, or accepted principle.

Reasoning may be theoretical or practical. Sound theoretical reasoning is from premises known to be true to a conclusion consisting of a true judgement. Practical reasoning has been variously conceived. On some accounts, it is from assumptions that incorporate a goal, valuation or norm to intention or decision formation, or to action. On others, it is from such premises to judgements concerning how to act, what intention to form, what decision to make, or what act to perform. Only language-using creatures have full-blown cogitative powers. Only a being that has mastered a developed language can engage in *reasoning*. For only a language user can apprehend a premise or premises as *warranting* (or not warranting) a given conclusion – be the conclusion a judgement or an action. Only a language user can think this to be so *because* that is so. This 'because' is not causal.

It lies in the nature of reasons and of reasoning that a reason is characterized in terms of a degree of universality. If my needs provide me with a reason for acting, feeling, or thinking thus and so, then the needs of others provide them with reasons for doing likewise. This does not imply that if something is a reason for me it must also be a reason for others in like circumstances. Others will not have shared my peculiar history, they will not have my array of commitments, and they may not share my tastes, likings, and dislikings. Nevertheless, recognition of something as a reason for me requires acknowledgement that it *can* likewise be a reason for another person in similar circumstances, and if it isn't there must be some difference between us. So in learning that 'I need it' or 'I enjoy it' can be an explanatory or justifying reason for me to advance, one also learns that it *can be* an explanatory or justifying reason for another to advance.

Rationality, in one sense, is an ability to apprehend reasons and to engage in reasoning. Its absence is not irrationality but non-rationality. Only if one can be rational or reasonable can one be irrational or unreasonable. The intellect is the faculty of rationality. Only creatures with an intellect can be said to have a will, as opposed to mere appetites and wants. To have a will is to have the power to deliberate, form intentions and plans, make decisions, and choose reasoned goals by the use of the intellect. Creatures that have powers of thought and will can be said to have a mind. To have a mind is no more to own or possess something than to have a body. As Wittgenstein did not say, the mind is not a something but it is not a nothing either (cf. *Philosophical Investigations*, §304). A creature that has a mind is a creature with abilities of rational thought and will.[4] All idioms of mind, such as 'having something in mind or at the back of one's mind', 'calling something to mind', 'making up or changing one's mind', and so forth, can be paraphrased into descriptions of the exercise of powers of intellect and will. The idioms of mind cluster around the exercise of powers of thought, reasoning, and deliberation, and the formation of intentions and decisions, of attention and concentration, of memory and recollection, and of opinion. The mind is no more identical with the brain than the horsepower of a car is identical with its engine, or the purchasing power of a banknote is identical with the banknote.

It follows that the venerable problem of the relation between my mind and my body, or between the mind a human being has and the body a human being has, disintegrates. For there is *no* relation between the body one has (i.e. one's somatic characteristics) and one's intellectual powers (e.g. one's powers to know, think, believe, deliberate, plan, form intentions, understand, perceive[5]). Of course, one may be proud of one's figure, and one's physical weakness may make one

---

[4] This is to revert to the Aristotelian conception of the mind (the rational *psuchē*) and to reject the Cartesian conception of the mind as identified with the contents of consciousness (Cartesian thoughts, or *cogitationes*). The reasons for this are spelled out in *HNCF*, ch. 8.

[5] With us, but not with animals lacking a language, perception is a quasi-intellectual power. A dog perceives a cat in the tree; we perceive that a cat is in the tree. The dog knows a cat is in the tree: it can see it, and is barking at it. We know that a cat is in the tree, and hence that it is true that there is a cat in the tree and that the proposition that a cat is in the tree is true. We can deliberate on what we know, and invoke it in our reasoning. We can bear it in mind without taking action, and call it to mind subsequently ('the cat was in the tree'); a dog can do none of these. As Aquinas observed, animals have sense faculties and a *vis estimativa* (or instinct), but lack what we have, namely a *vis cogitativa* (*Sententia libri De anima*, II.13.221). For elaboration, see *IP*, 393–7, and Appendix 1.

depressed – but these statements do not describe relationships between one's mind and one's body.

Like all substances, we undergo changes, which may be internal, external, or both. Internally caused changes may be intrinsic (due to our natural species constitution) or contingent (due to our individual genetic constitution or to happenstance). Externally caused change is brought about by things acting on us. Like all living substances, we undergo growth, maturation, and decline in the course of a natural cycle from inception of life until death and dissolution. The concepts of neonate, child, youth, adult, middle-aged, old man or woman are phase-sortal concepts indicative thereof. Death is the termination of existence – a dead human being is no more a human being than a fake five pound note is a five pound note. A dead human being is a corpse. It is a question of some interest what living beings leave corpses on death (*HNCF*, 9.4). A copse of dead trees is not a copse of corpses of trees. The carcass of a slaughtered cow is not the corpse of a cow.

Like all substances, we have causal powers: we can *act on* things, thereby bringing about or preventing change in the world. But we not only act on things: we also act, take action, and engage in activities. To act or to take action presupposes the ability to do or to refrain from doing what one does. In the absence of such a two-way ability, there is no act but at most a so-called reflex action. But a reflex action is not an agential action at all. To act or take action presupposes not only a two-way ability but also an opportunity. (One-way powers require only an occasion, the occurrence of which triggers the power; but an opportunity does not trigger the two-way powers of sentient beings – whether they act or not depends upon their goals and decisions.) Opportunities are agent relative in so far as what counts as an opportunity for an agent to act depends upon the agent's skills. So what is an opportunity for the highly skilful is not one for the beginner. Skills are refinements of abilities through emulation, practice, training, and being taught. Being uniquely dextrous, we are not only thinkers and doers but also makers. Although apes, sea otters, and Corvidae make use of objects as tools or even fashion passing tools, few in their natural state keep tools for future use, or hone and improve them, and none of them designs new tools or makes tools for the making of tools. This 'technical' aspect of human powers (*techne* refers to what used to be called the 'mysteries' of an art or craft) was central to Marxist thought in the nineteenth century and to Heideggerian thought in the twentieth.

Like all living things, we have needs. Absolute needs are independent of our contingent goals, and their satisfaction is a condition of health and normal functioning. Relative needs are dependent on exigencies of our goals and wants. We have socially minimal needs, dependent on the received conception of the basic requirements of a tolerable social life. Socially minimal needs are historically conditioned. Our organs and faculties too have needs: exigencies of their normal functioning and exercise, the satisfaction of which will keep or make them well; or exigencies of their improvement, the satisfaction of which will make them better.

Like other animals we have appetites, both natural and acquired (addictions). Appetites are felt desires. So too are urges and cravings. We are subject to attraction and aversion, hence have likings and dislikings. So we have preferences, want certain things and are averse or indifferent to others, and choose from among the things we like to consume, to have, to do, or to enjoy. So we have inclinations. We have and pursue goals. So we have purposes, and do things on purpose. Our behaviour is commonly explained teleologically – by reference to that *for the sake of which* we do what we do. Teleological explanations of human acts, actions, and activities are compatible with causal explanations of muscular contractions, but they are not themselves causal. They answer the question 'Why?' by reference to reasons and motives, or by reference to chosen goals and intrinsically valuable activities, and by reference to abstract values such as truth or justice.

Because we are language users and possess cogitative powers of reasoning, we can take things to be reasons for us to think, feel, and do things. Reasons may be good, strong, or powerful as well as poor, weak, or bad. In deliberation we decide, prior to acting, which reason is, or which reasons are, decisive. In embracing a reason as decisive, we endorse the action it supports. In taking our action as done for a given reason, subsequent to acting, we make ourselves *answerable* for what we have done, thus conceived. For we then can answer the question 'Why?' by giving our reason. So, other things being equal, we are *responsible* for what we do or fail to do.

Unlike non-language-using animals, we have a *will*: the power to form *rational* or *reasonable* (as well as *irrational* or *unreasonable*) desires and wants in advance of action, on the basis of reasons that warrant or seem to us to warrant adopting a given goal, and hence provide or seem to provide reasons for wanting to pursue it, attain it, possess it, or engage in doing it. So we have the power to articulate our reasons for acting, thinking, and feeling, and to consider reasons pro and con. On that basis, we can come to a decision in advance of action. Everything that is reasonable is rational, but not everything rational is reasonable: the deliberations of 'rational economic man' are rational but not reasonable. We can form intentions, plans, and projects on the basis of reasons, and harbour them for future action (*HNCF*, ch. 7).

Like other animals, we are subject to the passions (*TP*, ch. 1). Among the passions are the emotions. Emotions may have causes and reasons. Our reasons for our emotions are what we take to warrant our feelings. Reasons for emotions may be good, poor, bad, irrational, or unreasonable. Emotions may be irrational or unreasonable in their object, their rationale, their intensity, or their duration. Irrational or unreasonable emotions need curbing by reflection. This capacity for rational control of emotions presupposes reflective powers that can be possessed only by language users. Some of our emotions, in particular emotions of self-evaluation such as pride, guilt and shame, regret, embarrassment, and humiliation can be ascribed only to self-conscious language users.

I have harped here, as in previous volumes, on our mastery of a language as the main key to our nature as rational animals. Linnaeus, in the Age of Enlightenment, optimistically characterized us as *Homo sapiens* – but wisdom is a scarce commodity among mankind. We are above all *Homo loquens* – talking man. It has become fashionable in the last three decades to emphasize our continuity with the rest of the animal kingdom, in particular with the great apes. It is true that, in our animality, we share much with them. But most, if not all, that makes us distinct from them is attributable to our being language users. This includes our knowledge of good and evil and our susceptibility to guilt, shame, and remorse; our possession of an autobiography – a story of our life; our beliefs and myths about the history of our people; our technical and theoretical knowledge; our productive activities and the products that flow from them; our mathematical knowledge and employment of it in empirical judgements; our creative capacities in literature and the arts, and our aesthetic sense; our sense of humour and irony; and our knowledge of our mortality.

So far the sinopia of the fresco has been painted in *Human Nature: the Categorial Framework*, *The Intellectual Powers*, and *The Passions*. It should be evident that the painting is very large indeed, and that what I have sketched here only hints at the complexities and refinements that have been introduced on the huge, many-coloured fresco. One large array of items that is conspicuous by its absence thus far is the place of value in the scheme of things. In this volume, *The Moral Powers*, I fill in the axiological gaps.

Acknowledgements

Friends, colleagues, acquaintances, and ex-students have aided and encouraged me in writing this book. Correspondence with them was always helpful and illuminating. Conversations with them were a great delight, with much laughter and merriment. I am indebted to them all. They pulled up weeds, sowed seeds, and showed me how to train stubborn bushes to grow into pleasing shapes and to produce fruit.

I am grateful to Robert Cooter, David Ellis, Andrew English, Hanjo Glock, Edward Greenwood, Jonathan Hacker, Bernard Harrison, Lassi Jakola, Edward Kanterian, Anselm Mueller, Parashkev Nachev, Anthony O'Hear, Juan Pascual, Dennis Patterson, Amit Saad, Constantine Sandis, Harry Smit, Cosmin Vaduva, Christian Wenzel, and Aryeh Younger, all of whom read one or more, often many more, chapters of the book. Their comments, corrections, and advice were invariably helpful.

Edward Greenwood generously permitted me to print two of his philosophical poems that powerfully illustrated points I wished to make.

Hanoch Ben-Yami, Anthony Kenny, Iddo Landau, Hans Oberdiek, Herman Philipse, and Gabriele Taylor read the whole book as it was being written and went over each chapter with a fine tooth-comb. Their detailed and painstaking criticisms were invaluable, saving me from error again and again. Their sapient advice helped me find my way through the conceptual jungles and philosophical mountain ranges that I traversed. Their encouragement strengthened my will and raised my spirits. I am deeply indebted to them all and grateful for their friendship.

P.M.S.H.
Oxford
November 2019

The Valuers

This weary world's weighed down with woe,
Or so the pessimists will say,
And some so hate its being so
They cannot face another day.
Optimists bless the world's delight,
Savouring the sweetness of each hour,
Rejoicing in each blissful sight,
When each day opens like a flower.
For worthlessness is twinned with worth,
And hope can counterpoise despair,
If some can find their hell on earth,
Others can find their heaven there.
The optimist or pessimist
May be one self on different days,
The deepest moods do not persist:
The vagaries of mind amaze.
And change of time or change of place,
May do what lies beyond intent,
And with strange potency efface
Our misery or merriment.
Such is the state of humankind,
Sole valuers in a world of facts,
For they alone have powers of mind
To give them words to judge their acts.
When humankind is at an end,
No reckoning will keep a score,
Then light and dark will not contend,
And praise and blame will be no more.

Edward Greenwood

PART I

Of Good and Evil

1

The Roots of Value and the Nature of Morality

1. The place of values in a world of facts

No fact–value gap In worlds that lack life, there is no value. For all that, there is no mystery about 'the existence of values in a world of facts'.[1] The world does not consist of facts; rather, true descriptions of the world consist of statements of fact. It is as much a fact concerning the world that there are things that are of value *to* living things, that human beings value things and possess valuable characteristics, perform valuable deeds, and stand in valuable relationships to others, as it is a fact that there is life on earth. There is no 'gap' between fact and value, and we don't 'jump' across a logical gulf when we judge some things to be good or bad for the roses, some artefacts good or poor, some artisans good or incompetent at doing things, and some people virtuous or wicked. But, to apprehend this, we have to take into account the needs of living things and the preferences of sentient animals, human abilities and their cultivation, human relationships and activities, human societies and their histories. And these have to be seen in the presupposed framework of the nature of the world we live in, on the one hand, and of our nature, on the other.

---

[1] According to Christine Korsgaard, 'The fact of value is a mystery, and philosophers have been trying to solve it ever since [Plato]' (*The Sources of Normativity*, Cambridge University Press, Cambridge, 1996, 2).

The Moral Powers: A Study of Human Nature, First Edition. P. M. S. Hacker.
© 2021 John Wiley & Sons Ltd. Published 2021 by John Wiley & Sons Ltd.

Human biological nature   It is important to realize that human beings, like all animals, have a biological nature that was determined by evolution and is transmitted genetically. It is not a social construction. We have a large variety of basic physiological needs for certain types of foodstuffs, without which we sicken and die. Our basic perceptual capacities are determined by our neurophysiology, even though they admit of some degree of training and refinement through experience and culture. We have a large variety of reflexes: numerous eye reflexes, such as the corneal reflex, pupillary light reflex, and pupillary accommodation reflex; various limb reflexes, such as the triceps reflex, the biceps reflex, and the patellar reflex; numerous reflexes of internal organs; coughing, yawning, and sneezing reflexes, which can be partially inhibited. Some reflexes are distinctive of neonates and disappear later, such as the sucking reflex and the Palmar grasp reflex. We have an orienting reflex to attend to anomalies, which is the root of perceptual consciousness (in which our attention is caught and held by some anomaly in the periphery of our perceptual field) as well as of our curiosity and tendency to investigate anomalies. We have a large variety of social interactive instincts manifest in body language: sexually attracting instinctual behaviour, facial expressions in response to stimuli, eye contact instincts, voice modulation, and so on. There are gender differences that are predetermined and can be measured statistically, behavioural age differences, intelligence differences, and so on. Some of these are amenable to modification and sometimes improvement through training and teaching; some are not.

Autonomy of morals misconceived   It was a persistent feature of much analytic philosophy in the twentieth century to restrict reflection on the place of value in the world and the role of axiological concepts in our conceptual scheme to moral values. Non-moral values were commonly assumed to be limited to instrumentality, which was interpreted as an unproblematic causal concept: something is *good as a means* if it causally contributes to the attainment of some contingently desired goal of an agent in a given circumstance. Moral values were held to be 'intrinsically good'. This dichotomy was also characterized as being between *absolute* and *relative value*, and it was assumed that if any account could be given of absolute value, it would be independent of any consideration of other forms of non-moral value. This might be characterized as a degenerate residue of the salient Kantian thesis of the autonomy of morals: the thesis that the nature of moral value is to be elucidated independently of any contingent facts about the empirical world and about the nature of human beings as creatures with a vegetative *psuchē* and a sensitive *psuchē* in addition to their rational powers.[2] Moral value was argued to be intelligible only in terms of the dictates of the rational will. It was left to Nietzsche, and more generally to the twentieth century, to abandon the role of reason and to endeavour to explain the nature of moral value in terms of the will alone. Moral values were accordingly viewed as the creation of the untrammelled individual will. This conception gave rise to existentialism, on the one exaggeratedly dramatic hand, and to emotivism and prescriptivism, on the other rather feeble hand.[3]

Von Wright's varieties of goodness

It was above all Georg Henrik von Wright, in his book *The Varieties of Goodness* (1963), who challenged this approach to the philosophical study of value in general and of moral value in particular. He reverted to the great tradition of Aristotelian naturalism. His salient negative claim was to deny the thesis of the autonomy of moral value. His great achievement was to present a brilliant overview of what he called 'the varieties of goodness' in which the morally good is presented as a derivative or secondary form of goodness dependent upon the primary forms of non-moral goodness. The following discussion is much indebted to his insights. Although I shall have reason to depart from his account of moral value and to suggest minor modifications to his analyses of some of the varieties, I shall be standing firmly on the staircase he built. However, I am less concerned than he was to investigate a reductive ordering of the varieties of goodness, that is to pursue the question of whether there is a primary form of goodness in terms of which all others can be explained. I am more

---

[2] To translate Aristotle's use of *psuchē* as 'soul' does no service either to his thought or to clarity. The *psuchē* is a set of powers of living things, not an ethereal object. It is a biological concept, not a theological or metaphysical one. The vegetative *psuchē* consists of the distinctive powers of plants, and the sensitive *psuchē* of the additional distinctive powers of self-moving, sentient creatures that act in pursuit of goals. The rational *psuchē* is, as far as we know, unique to mankind. It consists of the power to reason and sensitivities to reasons. Note that 'rational' in this context incorporates what is reasonable. For discussion of the differences between the rational and the reasonable, see *HNCF*, 7.1.

[3] The substantive alternatives, as they presented themselves to Western thinkers of the mid-twentieth century, were primarily utilitarianism (correctly associated with capitalism), Marxism (which viewed existing morality as an instrument of the bourgeoisie in the class war with the workers), monotheist ethics, and fascism.

concerned to arrive at a connective analysis (as in the previous three volumes of this tetralogy) and to investigate what natural phenomena and what human needs link the various forms and render their existence intelligible.

Grammar of 'good'
As a first step in the investigation of the place of values in the lives of living beings in general and of human beings in particular, and of the point and purpose of our axiological concepts and their position in our conceptual scheme, a brief overview of the use of 'good' is needed. 'Good' is the most general axiological concept. This expression can occur as an adjective, a noun, an adverb, and an exclamation. Its comparative is 'better' and its superlative is 'best'. It has a variety of opposites (contraries and contradictories) depending on the noun the adjective qualifies, for example 'poor', 'weak', 'bad', 'wicked', and 'evil'. One can be *well*, that is in good health, or be 'unwell' or 'ill' – in 'poor' health. 'Good' is the most general adjective of commendation. But not every assertoric predication or attribution of goodness is a commendation, let alone a recommendation. Nor are occurrences of 'good' in the antecedents of hypotheses, in interrogatives or imperatives. Goodness is intrinsically connected with preferential choice, but not all things that can be said to be good in one way or another are objects of choice, nor is what is chosen always good or even thought to be good. Some things may be said to 'be good' or to be 'a good' such-and-such. The range of the subjects of such attributions or predications determines, together with the context, the nature of the goodness attributed or predicated. Notoriously, 'good' (unlike, say, 'yellow'), used attributively (as an operator on a noun) is not detachable (otherwise one could infer that a good soldier is a good man). Something may *be good for* a being, or *do good to* a being. Something may *be good for* a purpose or *for* an artefact. A creature, animal or human, may be *good at* an activity. Someone may *be* or *do good to* another, or be *good with* other beings (e.g. children or horses) or with an instrument (musical or otherwise), implement, or weapon. Things may *be good of their kind* and may be *good as* such-and-such a kind. A human being may be *good as* a tinker, a tailor, or a candlestick maker. He or she may also be *a good person* or *a good human being* – which are not social roles. It may be *good to do* such-and-such or *good to be* a so-and-so. One may *make good* or *make something good*, and something may *come good*.

The concept of a variety
It takes great acumen and uncommon insight to be able to find order and put order into this welter of detail. For this we are much indebted to von Wright.

He pointed out that these manifold forms of goodness are not a matter of lexical ambiguity – if it were, one would not find it trans-linguistically ubiquitous, since lexical ambiguity is a local linguistic accident not preserved in translation. The multiple forms of goodness are patently not related to each other as species of a genus sharing a common feature and distinguished by specific differentia. Nor is the concept of a form of goodness a family resemblance concept, since it is not open to the addition of new members of the family (as are the concepts of art or of number) and is not explained by means of a series of examples together with a similarity rider 'and other things like that'. It is not a focal concept, like health (as in 'healthy creature', which is the focal point of healthy food, healthy exercise, and a healthy environment), although some of its forms are (obviously the goodness of health is). Von Wright characterized the different forms as 'varieties', which are related to each other in a complex web of logical connections. Other concepts displaying similar patterns of a conceptual network, he suggested, may be *truth* and *existence*.[4]

Value and living beings
The key to a perspicuous overview of axiology is the realization that all values arise from life. I do not mean that the only things that are valuable are living things, which is patently false, since human artefacts are obviously valuable. Nor do I mean that the only things for which something has value are living things, since humus is patently good for the soil, as is wax for antique furniture. What I mean is that no form of value can be rendered intelligible save by reference to living beings. Before examining this claim, we must bear in mind the familiar salient features of living things in general and mammals in particular. All living things are created sexually or asexually. Sexual animals are either male or female. They have a life cycle that can be demarcated by phases – in the case of mammals, by being a neonate, a cub (pup, kitten, etc.), being a young animal of its kind, being mature, and being an old animal of its kind; in the case of human beings, by infancy, childhood, youth, adulthood, maturity, and senescence. All living things have a natural pattern of growth and a variety of needs. All animals have natural appetites, and some can acquire appetites. They

---

[4] Suggested to me tentatively in a private conversation. So, the truth of an observation statement, the truth of empirical generalizations, the truth of a scientific theory, the truth of a mathematical equation, the truth of an apparently metaphysical statement (e.g. 'Nothing can be red and green all over'), the truth of a moral principle (e.g. Democritus' principle), and the truth of an aesthetic judgement are all different varieties of truth.

care about things, are attracted by some things and averse to others, and have powerful emotions. They are susceptible to pain. They take pleasure in sensation, perception, activity, and achievement. They have natural and acquired abilities that may be enhanced by emulation and practice. They pursue goals, and may succeed or fail in their endeavours. Mammals may be solitary by nature, or social, which in turn may involve small groups (packs, pods, prides, tribes) or large ones (herds, flocks). Parenthood may be solitary or shared.

Set against this familiar context, there is no puzzle about the existence of value in a world of fact, only a puzzle as to how so many philosophers could have persuaded themselves otherwise. As we shall see below, similar considerations apply to the existence of moral values. Given these biological and zoological facts, it is patently unintelligible that there should be no values and no things that are valued.

2. Varieties of goodness

The second step of this investigation is to give a brief overview of von Wright's categories, or 'varieties', of goodness. This discloses the diverse forms of goodness and the complex patterns of conceptual relations between them.

2.1 *Medical goodness*

Health and value   Medical goodness is the most elemental variety of natural value and disvalue. Its forms are firmly linked to the notions of the health of a being, that is, to thriving and flourishing somatically or mentally, to declining and deteriorating, to being well or ill, or to being damaged, injured, or maimed.[5] Only living things have a welfare or ill-fare. The good health of a living being is a constitutive component of its welfare. Bad, poor, or ill health detrimentally affects the life and welfare of its subject. It may disfigure. It may stunt natural growth and development. Among sentient beings it may involve pain and suffering. It may prevent or impede people from pursuing, or effectively pursuing, their projects and from engaging in their characteristic activities. The notion of good health – of being

____

[5] I shall bypass vegetal forms of life in what follows, even though, of course, they may be healthy or diseased, may flourish or wilt, have needs, and be beneficiaries of what is good for them.

well – primarily signifies the *absence* of disease, injury, or weakness, and is in that sense privative. But it has a positive counterpart in the notion of somatic flourishing, as when one is said to 'radiate' good health and vitality. A creature enjoys good health when its organs perform their function well, that is without impairment, and its superficies (skin, feathers, fur) are not diseased or defective, so that the animal can lead a normal life, free from the evils of ill health. A creature may be disfigured or maimed from birth, from past illness or injury. These defects need not involve ill health, but they may be included in the category of medical disvalue as somatic (or psychological) conditions profoundly detrimental to the good of the being, affecting its ability to lead a normal life. The idea of normalcy here is not, or is not only, statistical but axiological (dyslexia was no significant deficiency or abnormality in preliterate or innumerate societies). Normalcy requires the satisfaction of one's needs (*HNCF*, 5.1–5.3), and absence of somatic constraints of ill health, disfigurement, and other forms of deleterious physical or mental condition. The goodness of organs and faculties, eyesight, hearing, memory, and imagination is characterized as medical goodness. Organs and faculties are characterized as good, bad, poor, or weak according to whether they perform their function well. Failure to do so deleteriously affects the welfare of their owner inasmuch as it prevents or impedes normal activities.

Value-free science The concepts of medical goodness of organs and of faculties, and of the goodness of health of a being, are both biological and axiological. Neither medicine nor biology can dispense with these notions. That is why theoretical attempts by philosophers of biology to eliminate axiology from biology in the name of value-free science are incoherent. Value-free science is rightly a goal of scientists, for they should not be biased by prejudice or ideology, and should have no axes to grind. But that does not mean that all axiological descriptions and all evaluative language should be eliminated from true science. They are ineliminable from the sciences of life.

2.2 The technical goodness of skills

Valuation of skills Acquisition of the skill of V-ing implies mastery of the techniques that define a V-er (e.g. a tinker, a tailor, or a candlestick maker). Acquisition of a skill involves training, learning, and practice. It is altogether natural that we should evaluate those who are acquiring skills (pupils, apprentices, trainees, students), in order to judge their progress. We submit them to tests which they must pass as a qualification for admission into the class of V-ers. We may grade them as excellent, good, weak, or bad according to the degree to which they meet the standards demanded. It is likewise natural to evaluate those who have acquired a skill in order to select the best among them for our purposes or for the purposes of society. Possession of skills, like possession of the abilities of which they are refinements, admits of axiological gradation. These are degrees of *technical goodness* – the goodness at an art or craft (*techne*). To become a V-er one must not only be able to engage in the activities of V-ing; one must be able to V well – be a reasonably good V-er. A good V-er is *good at* V-ing. Being merely able to run does not make one a runner, and being able to play bridge does not make one a bridge player. The opposite of a good V-er is a poor V-er – someone who can V but not very well. A poor V-er or a bad V-er is someone who has learnt to V, but has not achieved the standards of excellence demanded. One would not choose a poor V-er to do a task that involves V-ing unless there is no better. A bad V-er should be shunned in order to avoid collateral damage.

It is evident that in a social group the pursuit of valued things will produce *differential skills* among its members. Given shortages and scarcities of desiderata, this will unavoidably generate competition. In human societies it cannot but produce inequalities of distribution. Males and females alike tend to select the most successful specimen available for breeding. The combination of differential skills and inequality of distribution of desiderata will generate hierarchies within the social group. The central problem for human societies and for political and economic theory is how to balance competitiveness with the good of society. So we unavoidably need the concept of a good (just, decent) society as well as the concept of the social good. These have been essentially contested throughout recorded history.

2.3 *Instrumental and artefactual goodness*

Goodness of artefacts   Because we pursue goals the achievement of which may be facilitated or made possible by the utilization of objects, because we, unlike other animals, are sophisticated tool-makers and instrument inventors, and because we can master techniques of using tools and instruments, there is a natural need, in any mature language, for concepts tailored for the form of goodness von Wright called 'instrumental'. Unlike von Wright, I shall treat instrumental goodness narrowly as the goodness of instruments and tools. As such it may be deemed to be a subform of *artefactual goodness*. The goodness of human or animal artefacts that are not tools or instruments is exemplified by such things as houses, cars, roads, harbours, on the one hand, and by beavers' dams, birds' nests, spiders' webs, on the other. The boundary between instrumental and non-instrumental artefactual goodness is not sharp. Both are related to the idea of purpose: artefacts are *made for* a purpose, are *used for* a purpose, and are *good for* a purpose. Good and bad, better and worse, *ways of doing things* are linked to, but not identical with, technical goodness. They are not good for a purpose, but facilitate the achievement of a purpose.[6]

Goodness of tools　　It is inconceivable that we should not evaluate and value tools and instruments, for they enlarge the horizons of our powers and opportunities. Instrumental goodness is essentially connected with reasoned preferential choice. Hence also with commendation and recommendation. Tools and instruments may be good, poor, or bad, or better or worse than alternative instruments. A poor tool or instrument may be better than none, but a bad one may do more harm than good. The very best tools and instruments may be flawless of their kind. It is not always rational to prefer the best available instrument for one's purposes – it may be much too valuable and one's competence in using it may be deficient.

Goodness of techniques and custodial goodness　　In using tools and instruments and in making use of artefacts that are not instruments, we do things and exercise our skills – we act, take action, and engage in activities. These acts and activities are characteristically purposive and intentional. There are commonly different *ways* of using or misusing, and abusing, such artefacts. It is obvious that creatures that make, use, and make use of instruments of any degree of

---

[6] Von Wright invoked the category of *utilitarian goodness* here, by which he meant the useful (including the favourable and advantageous), and considered the beneficial to be a subform of utilitarian goodness (*Varieties of Goodness* (Routledge & Kegan Paul, London, 1963), ch. 3). Given the association of the term 'utilitarian' with hedonic consequentialism, I prefer the term 'utile goodness'. I am not convinced that the beneficial is a subcategory of the useful. A liberal education leading to knowledge and understanding of the canon of Western literature and music is beneficial and serves the good of human beings, but not, or not necessarily, because it is useful. If anything, the useful, though not a subcategory of the beneficial, is often beneficial.

complexity, and possess skills for using them effectively, should find it necessary to evaluate the manner of their use. There are right and wrong ways of using tools and instruments, correct and incorrect ways of manipulating or handling them, better and worse ways of achieving our ends by using them or making use of them in pursuit of those ends. In the case of non-instrumental artefacts too one may use them well or badly. Cars are not instruments but artefacts that we use for purposes (getting from place to place, transporting goods or passengers). We may take good care of the car we drive (not straining the engine, not grinding the gears, making sure that the engine is well oiled and greased), or we may abuse it. Similarly, there are better and worse ways of using one's house (responsibly and with care, or neglectfully – allowing it to go to rack and ruin). The correct and proficient manner of using instruments may be called the *goodness of techniques* (as opposed to technical goodness). Taking good care of artefacts, instrumental and non-instrumental alike, when using them and storing them well may be termed *custodial goodness*. In both cases it is easy to see the rationale of the evaluation.

There is nothing strange or mysterious, let alone supernatural about these forms of goodness. There is here no gulf between fact and value. The attributions of these forms of goodness, poorness, weakness, badness, as well as forms of being better or worse are truth evaluable even though the concepts are vague.

2.4 *The goodness of the beneficial*

The beneficial Things may be *beneficial*, that is *be good for a being*, and they may *do good to a being*. One might suppose that such predicative phrases apply only to living things, but that would be precipitate. It is true that only living things *have* a good and a welfare. But it is false that things can be good for, and only for, living beings. Inanimate things, both natural and artefactual, have needs the non-satisfaction of which causes deterioration. Manure and rainfall are good for the soil, grease and oil are good for engines, and some degree of humidity is good for panel paintings and antique furniture. But these are *derivatively* and *preventatively* good. Manure is derivatively beneficial for the soil, since it enables *us* to grow good crops on it and obtain a good yield. It *makes the soil better*, but it does not serve the good of the soil, since the soil does not have a good. Grease is preventatively good for engines because it stops them from deteriorating. The beneficial for artefacts is preventative, whereas for living things it may *also* be augmentative and productive.

The *beneficial* for human beings may be divided into three subforms:

(i)   that which causally restores, maintains, or augments their *wel-fare*: welfare is concerned with human health and the satisfaction of human needs (see *HNCF*, 130–4);

(ii) that which causally contributes to their *doing well*, that is their *prospering* in their work and its remuneration, in their receiving recognition for their skills and achievements, and so forth;

(iii) that which contributes constitutively to their good and to their *flourishing* – to their living a good life.[7] This includes those things that contribute to favourable moral development, to character formation, to sharper understanding of good and evil, to greater moral sensitivity and better moral judgement.

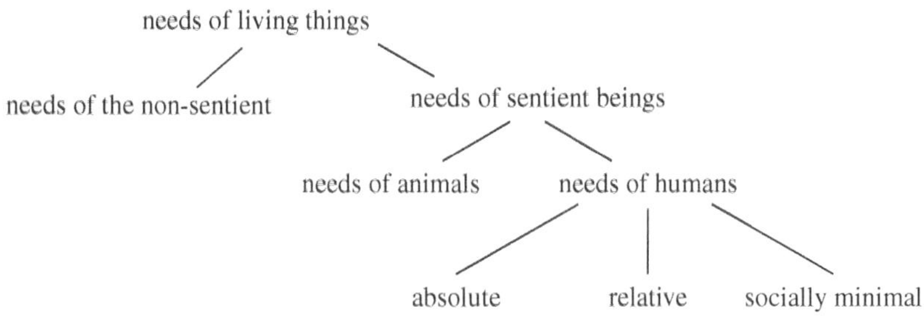

Figure 1.1   *Needs*

Human needs   Human needs may be absolute (or basic) natural needs common to all human beings. Their satisfaction is necessary for health and survival. They may be socially minimal needs the satisfaction of which enables a person to engage in the pursuit of projects that are normally necessary for a tolerable life in a given society at a given period. These are historically and socially conditioned. And they may be relative needs dependent on the contingent goals of individuals or required for the pursuit of individual life-projects (see fig. 1.1). Provision for people's welfare ensures that they are 'all right'. It does not imply that they will fare well or that they will do well. But, for the most part, it is a prerequisite for what von

---

[7] I am here drawing sharp lines where there are none, for example in taking welfare narrowly or in distinguishing thus between prospering and flourishing. So my remarks are explicative (in Carnap's sense).

Wright characterizes as *the good of man* (a notion, which, in the absence of an English equivalent of *das Wohl*, is useful).

2.5 *The goodness of the useful; a good life*

Doing well and living well
Someone *does well* if they prosper. Prospering is a form of success in life. This includes, at the upper end of the social scale, achieving fame, honour, power, and wealth. These are not given to many. At the lesser heights, someone can be said to be doing well, or *doing well for themselves*, if they receive recognition for their skills and work in terms of pay, promotion, and praise. Anything that contributes to a person's doing well or doing well for themselves can be said to *benefit* them or to be *advantageous* for them in their pursuit of the contingent goals they have. *Utile goodness* – the useful – is not a subform of the beneficial, but what is useful is commonly beneficial. Something that has utile goodness may be useful for a purpose, may be advantageous to have, and may serve someone's interest. The goal or end to which the useful is relevant may be the prosperity of a person. It may contribute to a person's flourishing (which is not the same as prospering). But it need not (that is why the utile is not a form of the beneficial). Something may be beneficial without being useful, and useful without being beneficial. The useful is good for a purpose; the beneficial is good for a being. The purpose that the useful serves need not be beneficial to anyone. A good plan is useful and its fulfilment, if its goal is wisely chosen, may be beneficial.

We must distinguish between *doing well* and *living well*. Doing well is compatible with unhappiness and lack of virtue. Wealth, fame, and power may notoriously be accompanied by loneliness, lack of genuine friendships, little pleasure, and no joy (vividly illustrated in Orson Welles's film *Citizen Kane* or by Karenin in *Anna Karenina*). The life of an artist or engineer, of an author, scientist, or philosopher, may be lonely and unhappy even though their achievements may be great and admirable, and their lives well spent. But one may live well – live a good life with the exigencies of welfare satisfied – without prospering or doing well but nevertheless flourish (e.g. Joe Gargery in *Great Expectations* or Caleb Garth in *Middlemarch*). The good of human beings involves the taking of pleasure in things and the enjoyment of activities, engagement in subjectively worthwhile projects, the attachments of friendships and the good of reciprocal love, the goods of liberty, and the maintenance of dignity (fig. 1.2). It is essentially connected with happiness (see

Chapter 9). The notion of a *good life for a human being* is linked problematically with that of the *life of a good human being*. The nature of *this* link has disturbed and preoccupied thinkers since the dawn of thought and will preoccupy us in subsequent chapters.

2.6 *Hedonic goodness*

The hedonic   Given that we are creatures with senses, given that by nature we may enjoy good health or suffer illness and disease, that we are doers and makers, that we have abilities and possess skills, that we have projects and pursue goals, and harbour attitudes such as liking and disliking, it is inconceivable that we should not make room in our conceptual scheme for the form of goodness von Wright called *the hedonic*. The hedonic includes the pleasant, pleasing, and pleasurable; the satisfying, gratifying, and enjoyable; the delightful, enchanting, and luxurious. Taking pleasure, delight, and joy in things; enjoying activities; and finding satisfaction, gratification, and fulfilment are constitutive elements of *a good life* for a human being. We ascribe hedonic goodness to things that contribute to the pleasure, enjoyment, satisfaction, or gratification of those who possess, consume, perceive, apprehend, or engage in or with them. We shall examine hedonic value in detail in Chapter 8.

Need for axiological   So, let us take stock. Human nature is the *concepts*   source of many kinds of value. The need for axiological concepts in general is readily intelligible. The concepts of good or poor health and of medical goodness are altogether natural for the characterization of the state of health of living beings in general and the identification of what is conducive to ameliorating, protecting, or advancing it. It is also needed by language users to ask for help and advice for themselves, for other people, or for other living things to explain why they can't do something or to excuse another who can't, and to express sympathy for other suffering creatures, human or animal, to whom they are attached or whose suffering touches their heart. The concept of the technical goodness of skills is, as it were, thrust upon any language-using social beings who can acquire and practise skills. For where there is division of labour, there is a need to select the best from among a group of people with a certain skill for a given task at hand. Any language-using creature that has the skills to make and to use tools, instruments, and other artefacts is going to need the concepts of artefactual goodness and its subcategory of instrumental goodness. For instrumental goodness too is essentially connected with preferential choice. A salient use of the concept is in commending and recommending tools or instruments. All beings with perceptual sense and sensibility take pleasure in feeling and perceiving things from time to time, and enjoy activities of various kinds. The concepts of the hedonic are expressively and descriptively indispensable. They are needed in making choices, in advising others to make choices, and in practical reasoning and reflection. Clearly, since we are goal-pursuing creatures, the categories of the useful – of utile goodness – are indispensable for considering means to our ends, for planning the way of achieving our goals, both in the small and in the large. Since prospering and flourishing are goals of human beings, the concepts of the beneficial will readily find a place in the conceptual scheme of any user of a well-developed language.

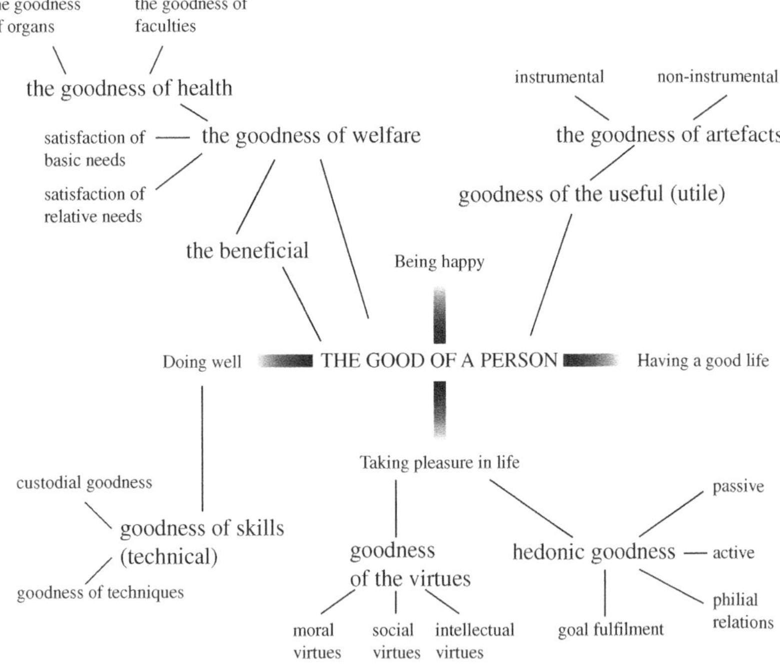

Figure 1.2   *Varieties of goodness that contribute to the good of a person*

What is conspicuous so far by its absence in our survey of the sources and varieties of value is reflection on our familial and social nature and the forms of value that spring from it – in particular, moral goodness in its various forms. Here, I shall argue, von Wright was mistaken to think that one can give an adequate account of moral goodness in terms of its relationship and dependency on the primary forms of goodness. That there is such dependency is undeniable, but that one can give a general account of moral goodness solely in terms of its relationship to the forms of goodness is contestable.

3. The framework of moral goodness

Moral goodness   We speak of good – morally good – men and women. A good child is more commonly a well-behaved child than a morally admirable one, whereas a good man or woman need not be especially well behaved (which is partly a matter of mores or 'small morals', as Hobbes put it, or of not making a nuisance of oneself). A good person is, roughly speaking, considerate, benevolent, and righteous. We speak of morally good deeds and of the morally good reasons and intentions that inform them, and of knowing the difference between right and wrong. We describe the morality of a society and the moral standards endorsed in a past or present society. Sometimes we talk of a moral system as we talk of a social, political, or economic system. So the term comes easily to our lips when reflecting on ourselves, on others, on social issues, and on history. But we would be hard put to answer the question 'What is morality?'.

Where does morality come from?   The question of where morality or moral value *comes from* is often raised. It is not a felicitous question, as it admits of too many different interpretations. It may be a question about the evolutionary origins of morality: can sexual selection and survival of the fittest explain how such complex patterns of norms and values arise, and if so how? It may be a question concerning social genealogy of the morals and values of a society (e.g. Plato (Adeimantus, Glaucon, and Thrasymachus in the *Republic*), Nietzsche in *The Genealogy of Morals*). How does it come about that we have the morality we have? It may be a question of the authority of morality as such. For, whatever moral considerations are, they are non-trivial and not merely a matter of individual likings and dislikings. What gives moral considerations their weight? It may be a question about the social functions of a morality: what is its point and purpose? And it may be a question of justification: what validates moral norms and values?

These are all respectable questions. Some are empirical, and have at best only indirect relevance to philosophy. Attempts to provide evolutionary explanations of the emergence of morality or diverse moralities among primitive human communities are conjectural, and will remain so. The sole task of philosophical enquiry here is to clarify what a morality is, and to advance conceptual criticisms of evolutionary theory when it transgresses the bounds of sense or when it overreaches the limits of evolutionary explanation. An investigation that runs parallel to evolutionary speculation investigates the pre-human seeds of moral behaviour among animals, in particular among chimpanzees, which are evolutionarily our closest relatives. Here too philosophy can make only a critical contribution, for example in explaining why non-language-using animals can have no morality (see Appendix 1). Interesting though ethology and primatology are in revealing the animal roots of the possibility of moral judgement and moral values, their bearing on philosophical questions concerning morality and moral values is indirect. Their results belong to the presupposed framework for the existence of morality.

Naturalism   This book, as has been made clear in the Prolegomenon, is an essay in philosophical anthropology. Our concern now is, on the one hand, with the place of morality in human lives and, on the other, with the roles of moral concepts in our thought and talk. It is with disclosing the roots of moral values in our nature as rational animals (capacity rationality) and in the nature of the world we live in. The first philosophical questions to be addressed are: What human needs and tendencies generate moral values, on the one hand, and call forth moral concepts, on the other? What forms of act, action, and activity, of attitude and emotion, are required for the fulfilment of human nature, in so far as there is such a thing? In this respect the following reflections begin by following in the footsteps of Hume's enquiries: how are the phenomena of morality to be accounted for? Some philosophers have expressed puzzlement at the very existence of moral values (see n. 1). Like Hume, I shall try to show that there is nothing to be puzzled about. Like Aristotle and Hume, I shall argue that moral values (moral goodness, compassion, kindness, etc.) and moral norms (duty and obligation imposing rules) are not supernatural either in origin or in their authority. They are not divine ordinations. Nor are moral values metaphysical; in particular they are not Platonic Ideas or Ideals. They are not the product of evolution and sexual selection, even though these provide part of the framework in which they have developed. But they can be rendered fully intelligible in natural terms by reference to the nature of the world we live in and to our nature as language-using social beings with sensitivity to reasons and the ability to reason.

The account I shall advance is subject to a number of self-imposed constraints.

(i)   It must allow for a diversity of moralities in different times and places. For this is a historical fact that may not be disregarded. Nevertheless, I shall argue, not anything goes. Evil societies that crush morality can exist.

(ii) It must tolerate a degree of relativity without tolerating ethical relativism. For what can be counted as a morality is to be judged by reference to what can be counted as 'the good of a human being'.

(iii) It must explain the intelligibility of critical assessment of defective moralities that pay homage to false ideals of the good, and of evil societies inimical to morality.

I shall accordingly begin the investigation with *reminders* of very general empirical facts about the world we live in and about ourselves in it. For it is against these background facts that the needs and tendencies that give rise to moral norms and values exist. The statements of these facts are truisms known to all, but readily overlooked.[8] They are contingent.

Contingencies of nature

The world might have been friendlier: the rivers might have flowed with milk; the trees might have produced honey; the climate might have been warm but never too hot or too cold; there might have been no disease-producing viruses and bacteria – and the lion might have lain down with the lamb, and the wolf with the sheep. But that is not how things are. In most parts of the earth in which human beings live, the climate imposes extremes of heat and cold, rains and storms, against which we seek to protect ourselves by means of clothing and dwellings. Nature, for the most part, is not over-generous in supplying us with the means of livelihood. Food is commonly difficult to come by and has to be wrested from nature. Disease and injury are common. The decrepitude of old age is well-nigh unavoidable save by early death.

Contingencies of mankind

As for the human world – we might all have been benevolent creatures, with no will to power, without possessive drives, with no aggressive instincts and no

---

[8] H. L. A. Hart's discussion of the minimum content of natural law in *The Concept of Law* (Clarendon Press, Oxford, 1961, ch. IX, §2) was a salutary reminder of such facts. I am also indebted to Anthony O'Hear's 'Morality, Reasoning and Upbringing', *Ratio* (2020).

fierce competitive desires. But we aren't like that. We are sexually driven,[9] predatory and possessive, fiercely competitive, and given to sexual jealousy, envy, anger, aggression, and vengeance. We have deeply tribal instincts (which no doubt once served a purpose) and are intolerant of strangers, of deviant behaviour, and of deviant beliefs. We have a limited innate proneness to sympathize, a limited ability to empathize, a tendency to form friendships and group allegiances, an innate second-order ability to acquire a language, and a language possession of which has endowed us with the ability to reason and with sensitivity to reasons.

It is an intrinsic feature of our biological nature that we are born virtually helpless and require a long period of childhood and youth prior to attaining biological maturity. We are, both in childhood and in adulthood, relatively vulnerable to disease and to accidental injury. We are also susceptible to injury or death through physical aggression. So we need protection – had we carapaces like turtles there would have been less need for constraints on physical violence. We differ in our facial appearance (imagine the consequences of our all looking exactly the same), so we can easily identify other human beings and relatively easily recognize and reidentify those we have encountered before. Indeed, we not only recognize other human beings, but we intuitively recognize them *as* human beings – not merely as other animals, a point emphasized nicely by Simone Weil.[10] In such encounters, from a rather early age, there is a reciprocal urge to understand the other and an awareness of that urge in the other – we try to make sense of each other.[11] In this sense, we have an intuitive (fragile) ability to recognize the humanity of others. We share similar basic needs. Our mammalian nature normally generates instinctual mother–child

---

[9] We might have been neuter beings that go into a sexual phase for three days a month, during which period we might, arbitrarily, be male or female; if the latter, we may conceive and, subsequent to giving birth and lactating for the neonate, revert to neuter status, after which, in a subsequent cycle, we may father a child. This fantasy was brilliantly worked out by Ursula Le Guin in *The Left Hand of Darkness* (1969).

[10] Simone Weil, 'The *Iliad*, Poem of Might' (1939), in *Intimations of Christianity among the Ancient Greeks* (Routledge, London, 1987), 24–55, in which she pointed out that we do not react to human beings as we respond to things. To step aside to avoid a street sign is altogether unlike stepping aside for a person in our path.

[11] Philippa Foot, 'Morality, Action, and Outcome', in T. Honderich (ed.), *Morality and Objectivity* (Routledge, London, 1985), and David Wiggins, *Solidarity and the Root of the Ethical: The Lindley Lecture* (University of Kansas, 2008), argue that this natural propensity is the root of the notion of human solidarity.

bonding both before and after birth, and instinctual child–mother bonding through lactation, cuddling, handling, and associated facial features and voice. Such bonding is facilitated by sympathy and empathy alike. The human capacity for sympathy is an innate emotional susceptibility, the capacity for empathy is an innate cognitive power (*TP*, ch. 12) that involves neither theory nor projection.

What we care about   We are highly emotional creatures, with a range of animal emotions such as rage, fear, affection, and curiosity, and a wider range of emotions characteristic only of language users, such as pride, humility, shame, humiliation, guilt, and remorse. The emotions essentially involve caring about something or someone (*TP*, ch. 1). What we care about is what matters to us. Caring about something involves valuing it. All animals care about avoiding physical danger, and about satisfying their appetites – were that not so, the species would not have survived. Animals that have to rear their young care for them and care about them – were that not so, the species would not have survived either.[12] Most animals that construct dwelling places (nests, warrens, tunnels, or just appropriate holes) jealously defend them, and territorial predators patrol, demarcate, and protect their territory. We are social animals that need the company of others of our kind. We care not only about our own projects and interests, but also about the joint projects we undertake with others, and the interests and projects of those to whom we are attached.

Rule followers   We have powerful instinctive emulative propensities and the ability to reproduce behaviour, to recognize regular behaviour, and to apprehend it in given contexts as a norm – a standard to follow, deviations from which are reprehensible. In short, we are rule-following creatures with the ability to differentiate correct from incorrect, right from wrong, forms of behaviour. Were that not so, mastery of a language would not have been possible. Nor would the duty-imposing norms of conduct essential to the existence of any

---

[12] This is compatible with varying degrees and limits: with some animals only the female cares about and for her offspring (e.g. tigers, bears); with others (e.g. some kinds of bird) only the male. Some will fight to the death to protect their young; others, when under persistent threat, will actually eat their neonates (e.g. mice). Birds will labour all summer long to feed their chicks, but if a second clutch is late in the season and migration is imminent they will abandon their chicks without hesitation and *without looking back* (Charles Darwin, *The Descent of Man* (John Murray, London, 1871)).

society, and to the control of any social group, be possible. Without mastery of a language there can be no knowledge of the difference between right and wrong, or between good and evil.

Homo loquens   Being language users is part of our social and cultural nature as human beings – better deemed *Homo loquens* than *Homo sapiens*. We are born into linguistic communities. A linguistic community is a culture. We require a long period of training and learning to become fully functioning, rule-following members of the cultural group into which we are born. We need to be *brought up*, for we have to learn *how to behave*. We also have to learn to speak and understand the speech of others, to learn how to perform a wide variety of speech-acts and how to engage in a multitude of language games characteristic of our culture. Being language users brings with it substantial normative pressure, for it is virtually unavoidable that value should be placed on truth telling and that lying should be considered reprehensible. For it is developed linguistic communication that makes it possible for us to be eyes and ears to each other, for knowledge to become communal and communicable, for teaching to replace subjective personal experience, and for information to be transmitted from one generation to another by word of mouth. For this truthfulness is requisite and the value of truth unavoidable. The more slippage that occurs here – and unfortunately it does – the more corrupt the society.

Homo ludens   We have cooperative instincts manifest in our early playing propensities (Huizinga called us *Homo ludens* in his eponymous book) and exhibited in our friendships. Play is a feature of most mammals, with an obvious evolutionary warrant. It serves both as preparation for adult life and as a means for partially establishing early social hierarchies without the dangers of serious combat. It also provides important foundations for the possibility of rule following, for among advanced animals it involves regularities and recognition of regularities. For human beings games evolved on this foundation. Playing games brings in its wake a variety of norms (not cheating) and values (fairness), as well as virtues (fortitude, discipline, self-control). To be sure, it is not only games that involve both cooperation and competition. In general, our cooperative instincts clash with our competitive instincts and regulation is indispensable. Some forms of regulation are more conducive than others to the good of human beings and to a decent, good, or just society.

Plasticity of second nature   While there are natural constants to our nature, it is a distinctive feature of human beings that we also have a nature that is, in a wide variety of respects, plastic. We are highly adaptable animals, both biologically and culturally. We can readily adapt to very different climatic conditions, to very different kinds of food. We are all born with a propensity to evolve a capacity to acquire a language, but we are capable of learning any one of a large multitude of languages, and are able to learn more than one language. Our 'second nature', as Aristotle called it, is highly malleable. It is moulded, initially, in childhood and youth as our character is formed. Virtues, which are traits of character, and excellences are acquired in a moral community through training, teaching, emulating, and practice.

> Intellectual virtue in the main owes both its birth and its growth to teaching (for which reason it requires experience and time), while moral virtue comes about as a result of habit. ... From this it is also plain that none of the moral virtues arises in us by nature; ... Neither by nature nor contrary to nature do the virtues arise in us; rather we are adapted by nature to receive them, and are made perfect by habit. ... of all the things that come to us by nature we first acquire the potentiality and later exhibit the activity (this is plain in the case of the senses ...); but the virtues we get by first exercising them ... For the things we have to learn before we can do them, we learn by doing them ... We become just by doing just acts, temperate by doing temperate acts, brave by doing brave acts. ... Thus, in one word, states of character arise out of like activities. ... It makes no small difference then, whether we form habits of one kind or of another from our very youth; it makes a very great difference, or rather *all* the difference.[13]

It is against the background of such familiar and largely contingent facts that we must come to understand the nature of moral goodness, its relation to other forms of goodness, and its roots in natural human sentiments and in the mastery of a language and the rational powers that flow from it (see list 1.1). There would be no morality without animality, and likewise no morality without capacity rationality.

---

[13] *The Nicomachean Ethics of Aristotle*, trans. Sir David Ross (Oxford University Press, London, 1925), II. 1.

- Social animals
- Approximate equality of strength and vulnerability
- Similar basic (absolute) needs
- Powerful sexual appetite and propensity to procreate
- Prolonged infancy and childhood, needing care, protection, training
- Similar range of emotions
- Limited sympathy
- Natural recognition of the humanity of others
- Empathetic cognitive abilities
- Emulative powers essential for learning the difference between right and wrong
- Rule-following abilities
- Mastery of a language
- Limited rationality (ability to reason and sensitivity to reasons)
- Playing propensity
- Tendency to form friendships
- Cooperative abilities
- Competitive instincts
- Plasticity and malleable second nature
- Character formation by action and habit

List 1.1   *The framework of human powers and propensities presupposed by morality*

4. Morality

Morality is essentially a social phenomenon – it is an aspect of the form of life of a community. In evolved societies, it overlaps with and fades into law, into social mores or small morals, and into individual critical morality. Virtue, or moral goodness, as well as the moral virtues, which are moral character traits, are attributes of social beings who are, and who identify themselves as being, members of a community, people, or nation.[14] This does not imply that there are no self-directed moral principles or duties to oneself. Nor does it imply that the unfortunate Alexander Selkirks (or Robinson Crusoes) may not live virtuously in their enforced solitude, exercising the self-regarding virtues.

Composition of morality A morality, a social morality, can be viewed as consisting of the overlap between the following:

(i) *Social norms or rules that specify what one must or must not do in recurrent situations*: These are duty- and obligation-imposing rules, primarily concerned with protective prohibitions and with agreements. If communal life is to be possible, there must be rule-governed control of violence (hence the duty not to kill or to assault others), of communication (hence an obligation to tell the truth or at any rate not to tell lies), and of cooperation (an obligation to keep one's promises and to comply with voluntary contractual agreements, as well as obligations of reciprocity). Given the strength of human sexual drive, on the one hand, and of the long period of pregnancy and the needs of neonates and young, on the other, there must be norms of sexual behaviour, of parenthood, and of responsibility for the care and upbringing of children. Given the human life cycle, the limited powers of the ageing, and the decrepitude of senescence, there must be obligations upon children or the community to care for the aged.

The domain of the deontic is normative (norm or rule involving). It deals with *recurrent* situations in personal and social life in a community. So, as von Wright sapiently observed, *the path of duty is laid out in advance*. But, given the complexities of human relationships and circumstances, the dictates of duty are for the most part defeasible. Apprehension of their defeat requires experience and good judgement.

(ii) *Ideal norms specifying what one ought to be*: The virtues are intrinsically valuable character traits the citing of which specifies not what *one ought to do*, but what one *ought to be*. They are exhibited in action that is intended to serve the good of human beings – of oneself and/or others (hence the virtues of courage, industry, temperance, patience, fortitude), and of others (e.g. generosity, compassion, kindness, being just and fair). The path of virtue, and hence the exercise of

---

[14] There are, of course, non-moral virtues or excellences, such as wit, congeniality, and charm.

the virtues, is not laid out in advance, but is determined from occasion to occasion by good judgement concerning what is good and right. The judgement and behaviour of the virtuous and experienced provide one with reliable guidance as to what one ought to do in the particular case. (There is no vicious circularity here.)

(iii) *Values*: These are ideals or abstract goals, both moral and non-moral, that are respected and that provide final justifications of action. They can be cited, as Aristotle put it, as that 'for the sake of which' one acts. But, unlike the citation of contingent goals, the justification given by abstract values are termini of chains of justification. Such distinctively *moral* values are justice and fairness, honesty and trustworthiness, generosity and compassion. Acting for the sake of such moral values *requires no explanation*.

(iv) *Moral conscience and social sanctions*: The phenomena of morality in a community essentially involve a moral psychology, without which there could be no moral community. Given that human beings are, for the most part, endowed with limited sympathy, with a proneness to cooperate for mutual advantage, and have a need for friends and a procreative urge, coupled with protective instincts towards partner and offspring, it is only to be expected that the phenomenon of a moral conscience should be widespread. Most members of a flourishing community care about fulfilling the deontic requirements in their society. They know the difference between right and wrong – which is a form of practical, emotion-involving knowledge that essentially involves caring about the former and condemnation of the latter. Hence it carries in its wake both shame and guilt as subjective responses to their own failure to live up to the requisite standards of duty and obligation. In short, they have a moral conscience consequent upon the internalization of moral norms. The pangs of conscience, contrary to what was suggested by Bentham and Mill, are not sanctions.[15] Sanctions are intentionally applied on the grounds of a misdemeanour, but pangs of conscience are not intentional and are not 'applied' with a purpose. They are a corollary of internalization, a mark of caring about conformity to the norm in question, and constitutive of knowing what is right. However, precisely because those who internalize deontic requirements care about nonconformity there are social sanctions, which

---

[15] See P. M. S. Hacker, 'Sanctions Theories of Duty', in A. W. B. Simpson (ed.), *Oxford Essays in Jurisprudence*, 2nd series (Oxford University Press, Oxford, 1973), 131–70.

take the form of manifest social disapproval, social criticism, and severance of social relationships.

Variations and constants   This characterization of morality allows for a wide degree of variation in social norms, both as regards norm subjects and as regards beneficiendiaries (to use Bentham's term) – those intended to benefit. While there must be regulation of violence, there may be much variation in the class of people who are permitted to kill others and the class of people who are protected from arbitrary violence (e.g. Spartans and samurai were permitted to kill helots and peasants respectively at will). While there must be rules of property, this allows for any number of variations in distribution of property. And so on. This is not of direct concern to our axiological investigations. The natural roots of morality also allow for variation in the ranking of the virtues proportional to social needs (the need for a warrior class to defend society against other social groups or peoples, the need to regulate legitimacy of children and their rearing, etc.). Four historical features, however, are striking:

(i)   The fact that there can be other forms of 'social glue' than a shared morality; it is rare, but as we shall see below, it is possible for a society to exist that crushes morality.

(ii) The relative constancy of most of the virtues in societies that have a morality despite substantial changes in the rank ordering of the virtues.

(iii) The fact of periodic fundamental change in moral ideas in the known history of humanity.

(iv) The slow emergence of moral enquiry and criticism of traditional social morality. This is marked, in the ancient world in the West, by the iconic figure of Socrates and by some of the sophists of his time. It is marked, in the modern era, first in the Age of Reason, coincident with the European discovery of the Americas and India, and subsequently by the Enlightenment.

Must every society have a morality?   It is sometimes said that every society must have a morality. So, it has been claimed by some, Nazism was a morality. It might be suggested that even Aztec society, at the peak of its intoxication with blood and human sacrifice, had a morality – a set of rules constitutive of the form of life of a society and a corresponding set of character traits encouraged and cultivated within the social order.

No doubt the concept of a morality, and indeed the concept of the moral, are not sharply defined. It does not follow from this that anything goes – that whatever system of rules a society has constitutes a morality. It seems to me quite wrong to aver that Nazism was or had a morality. It involved a demonic ideology (including genocidal racism, lethal eugenics, and murderous homophobia) incorporating evil beliefs, an elaborate system of wicked rules (e.g. the Nuremberg laws) and a perversion of moral virtues (e.g. of loyalty and courage). This was imposed upon a population that had a Western, by and large Christian, morality, and yet it was accepted by a large proportion of the population, by some enthusiastically, by others reluctantly and fearfully. It was, to a degree, a superstructure imposed upon a pre-existing society with a traditional morality, which it duly corrupted wherever it could (systematically so in the Hitler-Jugend). But the Nazi ideology itself, with its worship of the Führer, with its Nazi party cadres, its depraved judiciary, its sanctification of the *Volk* and suppression of individualism, its fanatical militarism, its compliant if not indeed enthusiastic bureaucracy, and its rabid antisemitism cannot be deemed a morality. Here residues of common morality coexist with an evil system of norms and practices superimposed upon it.

Unlike Nazism, the Aztec Mesoamerican societies lasted for centuries. Their sacrificial and cannibal practices were part of a religious ethos sustained by an overwhelmingly powerful priestly caste, requiring constant appeasement of the gods by human sacrifice. A society that lauds the torture and sacrifice of children, that celebrates the bloody sacrifice of prisoners of war, which at the peak of the imperial power of Aztec society was said to have reached thousands of sacrifices in one four-day festival, is not a society with a morality that we find distasteful, but an evil society. Its obsession with blood and sacrifice, with horror and terror, is patent in much of its art. Its wickedness is not mitigated, or transformed into a morality, by the fact that the bloodbaths dedicated to the gods were held to appease them and to ensure rainfall and agricultural prosperity. On the contrary, to hold superstitious religious beliefs that demand the infliction and public celebration of hideous suffering, and the suppression of human sympathy and compassion, is itself depraved.[16] Nevertheless, one might think, *some* common morality of reciprocity, of keeping one's

---

[16] Such hideous practices of human sacrifice seem to be common to very many societies in early stages of human social and urban development (as in the worship of Moloch in the Middle East and Carthage, and Druid practices in ancient Britain).

word, of respecting the property of others, of reciprocal parent–child relations must obtain side by side with the depravity for the sake of daily coexistence. That is surely correct.

Limits of the licit   One thing seems clear: no forms of faith, no systems of ideology, no forms of entertainment that sanction great human suffering should command greater allegiance than our common humanity and natural sympathy.[17] To endorse the torture and immolation of human beings because of their religious beliefs or lack of religious beliefs, even if this is sincerely held, to further the Second Coming, or to ensure collective salvation in this life or an afterlife is evil (a profound moral insight we owe above all to Sebastian Castellio: see Chapter 5). Similarly, to endorse the slaughter of millions of people for the sake of an ideologically sanctioned millennium when the perfect society will be achieved and a transformed humanity will inherit the earth, is evil (as the twentieth century painfully learnt). Such systems of beliefs and ways of living are not new forms of morality, but ancient forms of wickedness. For a society to have a morality at all, a minimally effective concern with the good of man must be manifest in social practices and form of life, as well as approbation of other-regarding virtues that serve the good of man, and a publicly acknowledged conception of the qualities of a morally good person. That is inextricably interwoven with public recognition of common values of truth and honesty, of justice and fairness, and of generosity and compassion, even though there is common failure to live up to these standards.

Of course, one may stipulatively define the notion of a morality so that it severs any connection with the good and the beneficial, and signifies merely a set of rules and values that is the cement of a society. But there is no good reason for doing so and many reasons for not doing so. Among other things, it is likely to diminish our sense of evil and to enhance our modern propensity to suppose that morality is wholly relative and a function of the will alone or of the interests of the ruling classes. It is equally prone to delude us into justifying present evil for the sake of a glorious future and to hold that the goal of

---

[17] The Roman practice of the public *entertainment* of gladiatorial games was made all the worse by the fact that they did have a well-entrenched morality and no priestly caste to impose savagery to appease the gods. The Romans just found it entertaining to watch men fight to the death or wild animals tearing men and women to pieces. The pleasure taken today in the violence of Hollywood films and computer games shows how close we still are to this mentality.

a perfect society justifies the means – as was done in communist regimes, or to condone present evil for the sake of other-worldly recompense after death. And it facilitates the self-deception involved in advancing mono-valued consequentialism as a form of morality.

5. Individual critical morality

Functions of social morality   More or less unquestioned social morality characterizes traditional societies. It constitutes an array of values, norms, and ideals that are constitutive of the identity of a society and that impart to its members a part of their own sense of identity. It lies at the foundation of the moral upbringing of children – of their learning how to behave and acquiring the habit of good behaviour in the course of which they learn the difference between right and wrong. It provides the framework for agreements and disagreements, and for the determination of socially acceptable reasons and patterns of informal reasoning and justification within a traditional society. Given division of labour, the existence of social hierarchies, and the unequal distribution of wealth and power, it allows for a variety of modes of life and ideals of life, and of various different ways in which a member of society may live a virtuous life. What it does not license is critical evaluation of the modes of life and of the nature of the virtues within the society.

To be sure, a traditional society may allow for great public moralists. Ancient Israel was remarkable for the public institution of prophets whose fierce castigation of moral turpitude was respected and for the most part tolerated. Their criticisms of the kings of Israel and of the moral state of the nation had no parallel in the ancient world for its merciless moral preaching. However, they were not criticizing or questioning the moral order; they were striving to restore it. They were fearless critics of the state of society, not of the moral standards of the society that the rulers and aristocracy failed to meet.

Challenging social morality   It seems that it was above all Socrates in Athens and the sophists throughout ancient Greek city-states who challenged the moral rules and values of traditional society. It was Socrates, more than any other, who taught Western civilization to question rather than to accept unquestioningly, to demand clarification, to insist upon justification, to draw distinctions, and to show differences. It is perhaps difficult for us moderns in the West to imagine how threatening to the social order the interminable questioning encouraged by Socrates appeared to be.[18] It is hardly surprising that he was, in due course, condemned to death. But it was his example, and the example of those Greeks, and later Romans, who followed in his footsteps, that produced the idea of critical evaluation of the morality of the society of which one is a member. This presupposes a clear apprehension of the moral values and standards of the day as well as reflective powers to challenge them and probe for their rationale. It unavoidably spills over into criticisms of current social and political policies and attitudes. To this extent it is a challenge to the existing social order, which may be, and be perceived to be, beneficial or destructive. With the overthrow of Roman religion and the elevation of Christianity to the official religion of the declining empire, moral questioning and confrontations with the new Christian moral order were not welcome for the next millennium and more. Challenges to official religious beliefs and to doctrinal justifications of moral demands imposed by the church were dangerous in the extreme. It was not until the scientific revolution and the Thirty Years War in the seventeenth century that systematic reflection upon the social and moral order was again cultivated, and freedom of thought and speech gradually became a cherished right of any member of an enlightened community. It can be said, without gross exaggeration, that the Enlightenment created a new kind of human being with a new configuration of interrogative ideals and ideals of toleration. It was, to be sure, in many ways flawed and in other ways shallow. It sorely lacked a robust and rich philosophy of psychology. Unsurprisingly, therefore, it little appreciated the darkness in the human soul and the powers of unreason. But it strove mightily for the good of humanity, it widened human horizons, and it broadened moral, social, and political self-consciousness. Despite dreadful vicissitudes in the course of nineteenth- and twentieth-century autocracies, dictatorships, theocracies (both Catholic and Muslim), fascism and communism, the ideas and ideals of the Enlightenment have survived so far. But it was, and seems likely to continue to be, a close-run thing.

It is asking questions, demanding justifications, challenging received views that make it possible for members of a non-traditional society in which freedoms of thought and speech are not crushed to superimpose upon the social morality of their peers their own reflective moral

---

[18] It is well depicted in Roberto Rossellini's film *Socrates* (1971).

standards and values. These may often override the received social morality with critical moral attitudes towards warfare, the exploitation of nature, social welfare, received sexual relationships, and the subordination of women. How successful the liberal endorsement of the individual critical moralist living within a social community sharing a broad moral consensus is likely to be remains to be seen. In an age that is increasingly rejecting the ideals of the Enlightenment and endorsing populism and fanaticism its prospects are unclear.

2

The Roots of Morality
and the Nature of
Moral Goodness

1. Moral goodness

Concept of
a person

Human beings are good or bad, virtuous or vicious, inasmuch as they are persons.[1] The concept of being a person, I have argued (*HNCF*, ch. 10), is not itself a substance concept. It is a qualification upon the substance concept of an animal of some kind or other *with the powers of rational agency.* Persons are free beings of one kind or another, with two-way abilities to act or refrain from acting, with powers of reason and rational will, with reflexive and reflective powers of self-consciousness, with knowledge of good and evil, who are able to explain and justify their acts by reference to goals and reasons, and hence are answerable for their deeds. In origin a Greek theatrical concept (*prosopon* – the mask of an actor indicating his role in the play), it was transformed in Rome into a legal concept (*persona*) signifying a legitimate legal role in the agonistic procedures of the Roman law courts. So the concept of a person was a status concept. It has remained so, although the character of the associated status has been detached from the law courts and transformed over the ages. Persons not only have moral virtues and

---

[1] If neonates, sufferers from dementia or Alzheimer's disease, or those who suffer devastating brain damage are not counted as being persons, then an explanation is needed for why they must nevertheless be treated as persons. See below, p. 53.

vices; they are also subject to moral duties and obligations, and are bearers of rights. (Virtue ethics and deontological ethics are not alternatives between which we must choose. Each misleadingly presents only one facet of a complex whole.) This is an aspect of their constitutive nature.

Goodness of a person   I doubt whether one can usefully *define* a (morally) good person and whether any such definition will shed light on the philosophical problems associated with this notion. There are multiple different criteria for being a good person. Satisfaction of different subsets of the set of criteria for being a good person corresponds to the different ways in which a person may lead a morally good life. In static societies with rigid social hierarchies, such as medieval Europe, there is little social mobility. Nevertheless, even in such a socially immobile society there were different modes of life (lords, ladies, knights, men-at-arms, clergy, craftsmen, yeomen, peasants), each of which admitted different sets of virtues and different criteria for leading a good (virtuous) life. In modern, socially mobile societies there is commonly a choice between different modes of life. In some cases, the choice is existential, for one is in effect choosing which of two different kinds of human being to become (as did Sartre's pupil, faced with the decision of whether to go to fight with the Free French or to stay at home to protect and care for his ageing widowed mother).

The multiplicity and diversity of criteria for being a good person do not imply that there are not common formal features of, as well as substantive commonalities to, the goodness of human beings. It is evident that a good person, for the most part, fulfils their moral obligations (e.g. to keep their word, to reciprocate favours and benefits, to take care of their family) and duties (of role and office), and possesses the virtues associated with benevolence (e.g. kindness, compassion, considerateness, generosity) as well as self-regarding virtues (courage, industry, tenacity, and the like) willingly exercised in the service of others or of their community. A balanced judgement as to whether someone is a good human being is often difficult because we do not know enough about each other. Sometimes it is impossible, because there are no scales on which to measure a person who has done both good and bad things, or to weigh wickedness against remorse, or ways of drawing a sharp dividing line between wickedness and evil. It is commonly systematically unclear whether and when acting wrongly defeats the satisfying criteria for being a good or even merely a half-decent person. But it is evident that one who *merely* fulfils duties and obligations for their own sake is reliable and trustworthy. Such a person does what is right, but without the virtues of benevolence they will *not* be held to be a good person.

Goodness of deeds   It is not only people, their character and personality, that we characterize as good or benevolent (or bad, wicked, and malevolent, or as an average sort of fellow, neither impressively good nor particularly bad). We speak of *good deeds*, of acts or activities that it is (morally) *good to do* (by contrast with being *hedonically* good or *useful* to do) and which it is *good* (praiseworthy) *of* the agent to have done. We must distinguish between the goodness of an act and the rightness of an act. The rightness of an act is often a matter of its conformity with an obligation- or duty-imposing norm. Then its rightness is independent of the intention in the act. One may do the right thing for the wrong reason, but that does not derogate from its rightness – although it may derogate from its merit. Fulfilling one's humdrum moral obligations, for example doing what one agreed or promised to do, and abiding by one's duties, for example not to kill, steal, or lie, are not usually morally good or praiseworthy, although their nonfulfilment is morally bad and blameworthy and in some cases wicked. Unlike the obligations and duties arising out of personal relationships, it is merely granting to others what is their due as fellow human beings.

For the most part, it is good to assist others, because one thereby benefits them – but not always, for, as we have seen, there are different forms of benefiting a person. One may serve their (minimal) welfare, their prospering, or their flourishing (which includes their leading a morally good life). It is not generally good to assist the wicked by helping them evade justice (as members of the Vatican helped Nazi war criminals escape to Argentina). But, in some cases, helping someone evade justice may be an act of profound wisdom, as when Monseigneur Myriel saves Jean Valjean from arrest for theft (in Hugo's *Les Misérables*). In the former case, one would merely be helping them to prosper, and in no way affecting for the better the wickedness of their lives. In the latter case, the good bishop opened to Valjean a life of outstanding virtue.

There is no merit in accidentally benefiting someone. There are morally good (admirable, praiseworthy) acts that do not benefit anyone. They may be purely expressive acts (e.g. commemorative acts), either collective or individual. They may be self-sacrificial acts done out of solidarity or fraternity, even though no one benefits from them.

It is not only deeds that may be morally good. One's *intentions* may likewise be morally good – the goodness of intentions being determined by the goodness or rightness of the planned act. According to Kant, intentions are the only thing capable of being morally good without qualification. I see no reason for accepting this view (see below).

Altruism or selfishness: a false dichotomy   It is a common error to suppose that human behaviour is either moral or self-interested, either altruistic or selfish (as many evolutionary socio-biologists assert), either free or causally determined by desires and inclinations (as Kant argued). These are false dichotomies and misleading categories. There is much that we do that is neither morally right nor morally good to do, nor selfish. Very often, it is good (e.g. enjoyable, beneficial, or useful) to perform such acts or to engage in such activities. There is nothing selfish in spending a Sunday afternoon reading a book, going for a walk with a friend, gardening, doing crossword puzzles, listening to music, enjoying a discussion, or a thousand other ways of spending one's leisure time. It is selfish only if one ought to be giving a helping hand to another, doing a duty of one's role or fulfilling an obligation previously undertaken. The utilitarian thought that at every single moment there is something that is the best thing to do, and the Kantian thought that there is always a duty (if not to others, then to oneself) that has to be fulfilled, are deep distortions of our lives, of morality, and of our freedom to pursue our freely chosen projects or harmlessly to amuse ourselves as we please. There is nothing self-interested in pursuing one's interests in bird-watching, studying philosophy, or writing poetry, for *being* interested in something does not imply *having* an interest in it (as when one has a vested interest or a financial interest in something). Nor is doing something out of self-interest (a bachelor earning his living, for example) necessarily selfish. It is selfish only if there are other pressing claims upon one that override one's self-interest. It is not true that all morally admirable deeds are altruistic: acting fairly or justly in adjudication is not. Nor is preserving one's dignity and integrity at the cost of one's life (e.g. John Proctor in Arthur Miller's *The Crucible*). Moreover, acting well may often benefit one (as indeed it benefited Mr Darcy in *Pride and Prejudice*), and that need not derogate from the goodness of one's deed or intention – it depends upon one's reason for doing what one did. One may adhere to Democritus' principle, and suffer evil rather than do it. Finally, it was a grievous fault for Kant to suppose that doing something to satisfy a desire, want, or inclination is to be causally determined and unfree. To act in order to satisfy a desire or to obtain something one wants, to do something one wants to do, or to do something because one feels so inclined are only in the exceptional case *being caused* to do what one does. That is so primarily in cases of satisfying or aiming to satisfy irresistible appetites, urges, and addictions, or succumbing to uncontrollable phobias. More typically, one has reasons for one's desires, wants, and inclinations, and one is free in acting for those reasons, and answerable for what one does in so acting. (Freedom of the individual will be explored in Chapter 6.)

Moral goodness and the good of man   Von Wright argued that moral goodness is a derivative form of goodness. Indeed, he averred that it is not a form of the good on a level with other forms, but 'If it be called a form of goodness at all, it is this in a *secondary* sense'. By this he meant that there is no special moral sense of the word 'good', but that its conceptual nature has to be explained in terms of some non-moral form of the good. He proceeded to give an account of the moral goodness of an act (which he averred to be 'perhaps the most important category of things, which are judged good or bad "in a moral sense" or "from a moral point of view"'[2]), in terms of the good of man. For an act to be morally good depends upon its being intentionally beneficial – it must benefit and be intended to benefit another. As is evident from previous remarks, this is not quite right. Although morality is logically bound up with the good of man, there are acts that are morally praiseworthy even though they neither benefit anyone nor are intended to benefit anyone. Among these are acts whereby the subjectively beneficial is repudiated for the sake of solidarity with others, or where, in accordance with Democritus' principle, one chooses to suffer evil rather than to do evil. Moral goodness is exhibited in one's attitudes towards other people (and other living things), and in the respect one accords them as free, autonomous beings. One's attitude, as Wittgenstein observed, is 'an attitude towards a soul'.[3]

It is quite mistaken to present morality as a device to 'ameliorate the human condition' (as G. J. Warnock once argued[4]) or to present the moral virtues as merely instrumental in attaining the good of man (as von Wright argued) and so instrumentally or externally related to being a good person. Contrary to Warnock, morality is not politics

---

[2] Von Wright, *Varieties of Good*, 119.

[3] Wittgenstein, 'Philosophy of Psychology: a Fragment', in *Philosophical Investigations*, rev. 4th edn by P. M. S. Hacker and Joachim Schulte (Wiley-Blackwell, Oxford, 2009), §22.

[4] G. J. Warnock, *The Object of Morality* (Methuen, London, 1971), ch. 1.

writ small. Contrary to von Wright, possessing the virtues, on the one hand, and fulfilling one's duties and obligations in the right spirit, on the other, are *constitutive* of being a good human being. To be a good human being is intrinsically, not instrumentally, valuable. It is something one ought to be and ought to strive to become.

Given

(i)   our animal nature, with inherent attractions and aversions, likings and dislikings;

(ii)   our social nature and our inherent conflicting propensities for cooperation and competition;

(iii)   our inherent goal-directed propensities essentially linked to (i);

(iv)   our possession of two way abilities (consequent on (i));

(v)   our mastery of a language;

(vi)   our rational powers to reason and to respond to reasons (consequent on (v));

(vii)   our consequent freedom to act for reasons and to be answerable for what we do (consequent on (i)–(vi));

(viii) our natural reactive attitudes that manifest intuitive acknowledgement of (vii);

(ix)   our emotions (and our natural propensity for caring) in general and proneness to feel sympathy in particular;

(x)   our need and capacity for friendship (consequent on our social nature (ii))

– given all this, it is neither mysterious nor surprising that we should value a wide disjunctive range of character traits (virtues) as *constitutive* of being a good person. That we do so is manifest in our having ideals of character – a moral subset of *things we ought to be*. Of course, it is equally unsurprising that any human society will have a wide range of conjunctive duty- and obligation-imposing norms (a moral subset of *things we ought to do*) to control violence, ensure mutual commitments, assure communication, and underpin reliance on communication, regulate property and possession, both to protect members of society and also to ensure a degree of social predictability.

2. The roots of moral value

Natural roots of morality   Philosophical anthropology must render the phenomenon of morality intelligible. In this respect, our endeavour resembles Hume's. Like Hume's, the approach here adopted is naturalist. There is no need to appeal to anything supernatural, such as the commandments of God, to render moral values, moral norms, or moral virtues intelligible. Nor need one invoke anything metaphysical, such as a noumenal self, a willing self, or an immortal soul. It is naturalist in a further sense: the aim is to clarify the nature of morality by reference to human nature and the natural world in which we live (see list 1.1). Like Hume and Adam Smith, and contrary to Kant, I shall suggest that the *roots* of moral value lie in human sympathy, in maternal love, in intuitive recognition of the humanity of others, and in the nature of loving friendship – not in doing one's duty for duty's sake (see fig. 2.1). They also lie in a sense of fairness that is itself rooted in natural reactive attitudes and is the spring of moral resentment and indignation – not in the idea of a universal moral law self-imposed by pure practical reason. But, more or less like Kant, I shall argue that moral value presupposes the form of freedom (sometimes mischaracterized as freedom of the will) possible only for beings capable of reasoning and acting for reasons. That in turn is one root of the formal respect for others as rational agents and recognition of the formal dignity of rational agents that have explicitly become a mark of morality. I shall clarify this historical claim below.

We are subject to passions (*TP*, ch. 1). We like certain things and dislike others. We are attracted by some things and are averse to others. We have goals and engage in projects about the achievement of which and about the engagement in which we care. We are subject to powerful emotions that exhibit what we care about and whom we care about. We naturally feel sympathy for others, and we have empathetic powers to understand other people – their affective states (emotions and moods) and desires, their behaviour, and their sufferings. Were that not so, there would be no such thing as moral concern, moral reasons for action, or moral commitment. (It does not follow that 'reason is the slave of the passions' (Hume[5]), but only that practical reason presupposes caring.)

Caring   The idea of agents who have appetites, can perceive things, and can suffer physical pain, yet *care about nothing*, including their own well-being, their own pleasures and suffering, is intelligible in pathological cases of persistent profound depression, but

---

[5] An unfortunate and quite mistaken remark in *A Treatise of Human Nature*, II. iii. 3, not repeated in the *Enquiry concerning Human Understanding*. Among other things, there are reasons for one's passions.

is necessarily an exception to the rule. For it is a constituent element of the idea of appetites that their subjects want to satisfy them and derive relief or gratification from so doing. Similarly, it is a constitutive feature of physical suffering that, other things being equal, one wants it to cease and is relieved and glad when it does.[6] So too it is part of what is meant by having a goal or purpose that one aims to achieve it or live up to, and is typically pleased to do so or fulfilled in doing so.

Perhaps we can imagine intelligent social creatures who care nothing for the good of others save when it benefits themselves. They would, to be sure, require no less elaborate social norms than we do, since forms of purely self-interested cooperation within a competitive framework require regulation. But such creatures would enjoy no reciprocal relations of affection within the family, no loving friendships, no feelings of comradeship – but only 'business' partnerships; they would feel no compassion, no remorse or guilt, but only regret. This, I think, is conceivable, but perhaps only as a limiting case exhibited, for example, by young gamblers on the floor of the Bourse. In general, the capacity to care for others is a presupposition of the possibility of morality. (It is grievous that it often goes hand in hand with the disposition to hate others ('them', not 'us'), which it is a task of morality to curb.)

Sympathy   Normal human beings have an innate susceptibility to feeling sympathy for others (as well as an innate cognitive capacity for empathetic understanding of others (*TP*, ch. 12)). This is no mystery, since it is a corollary, common to many mammals, of bonding during lactation, of the exigencies of maternal care during prolonged infancy and childhood, and of the requirements of paternal care and protection. However, unlike other, non-language-using mammals, human children are normally also *taught* to care about others, to join in celebrations for others, to exchange gifts, to show gratitude for gifts or favours done, and to be sympathetic to and show consideration for the sufferings of peers and elders. Moral education is built upon the natural recognition of others. Children normally learn that the suffering of others in the family circle or peer group and beyond is *a reason* for commiseration, for offering assistance, for foregoing some things they want. They learn that manifestations of care, concern, and sympathy are proper objects of approval. Alas, they also learn who is customarily *excluded* from the circle of beneficiendiaries: the stranger

---

[6] Of course, other things are not equal if one believes that the physical suffering leads to a further good (e.g. to good health for the sake of which it is reasonable to put up with it, or to felicity in an afterlife that will follow martyrdom) or is a constituent part of a good activity (such as mountain climbing or an endurance race).

in one's midst who is not 'one of us', the heretics or disbelievers who threaten our system of beliefs, the abject poor and crippled, the lower orders, the racially distinct. To be sure, the seeds of sympathy need to be watered and protected from subjective rapacity, aggression, and a desire to dominate and from social bigotry and prejudice in the family and society in which the child matures. Natural sympathy is all too easily destroyed, or curtailed to 'us' while excluding 'them'.

Friendship　　The need for friends is deeply rooted in our social nature. Forging friendships comes naturally to most of us. Aristotle remarked that 'We consider having a friend to be one of the greatest goods, and friendlessness and solitude to be quite terrible'. Indeed, he added: 'without friends no one would choose to live, though he had all other goods.'[7] Friendship no doubt serves the good of the friends and is useful; but it is also of intrinsic value – good in itself. For true friends are dear to us. (The Greek *philia* – roughly speaking 'loving friendship for those we cherish' – is better suited to capture this form of attachment than the weaker and more general English expression (*TP*, 327).) This natural relation essentially involves a range of moral virtues and excellences: fidelity, trustworthiness, honesty, benevolence, and a willingness to make sacrifices for the sake of one's friend (*TP*, 352–3). This relation between people carries its value on its face, so to speak. Enjoying a few such relationships (a few *philoi* is the most we are capable of sustaining) is partly constitutive of a good life for a human being. They are not to be confused with business friendships that are primarily utile, or hedonic friendships (with those with whom one goes on the razzle in one's youth) that tend to be brief and superficial.

Lack of *any* proneness to feel sympathy or having only a minimal tendency to do so is pathological among human beings. So is lack of recognition of the humanity of others. The sentiment of sympathy is virtually ubiquitous, but sympathetic propensities vary from person to person, being partly a function of their imagination (*TP*, ch. 12). Some people, women in particular, are sensitive to the joys and sufferings of their fellow human beings more markedly than others. Furthermore, human sympathies are limited, malleable, and destructible. They are limited by kinship, acquaintanceship, community, common language, distance, and also by our attention span to the woes of others and our preoccupation with our own projects and concerns. They are malleable by ideological fanaticism, religious antagonism, racialism, xenophobia, and the widespread 'tribal' need to ensure our own identity by finding the 'other' to hate or despise. They are selec-

---

[7] Aristotle, *Eudemian Ethics*, 1234^{b}32–3.

tively susceptible to being destroyed with respect to their objects – one of which is the suffering and humiliation of others – by the demonization and dehumanization of the 'other' (as is patent in warfare, religious bigotry, and slavery). They are destructible with regard to their subject through obliterating the subject's own capacity to feel sympathy by their religious and ideological fervour and fanaticism (powerfully depicted in *The Crucible*), and by conditions of extreme suffering and degradation (as experienced by prisoners in Nazi concentration camps, in the death marches of the Armenian Genocide, in the Russian gulag, in Japanese prisoner-of-war camps, in Iranian prisons today) in which feelings of humanity may be eroded in the victims of evil deeds, in all but moral saints and heroes (see Chapters 4 and 5).

Nonetheless, sympathy is an emotional tap-root of moral reasons and a presupposition of moral reasoning. It is an immediate motive for action inasmuch as we may *act out of sympathy* when we aim benevolently to alleviate the suffering or to share in the joys of others. Sympathy is a spring of action for a wide range of acts that are done in goodwill (benevolence). It is manifest in the exercise of many virtues, such as compassion, kindness, generosity, charity, considerateness, and tenderness, that serve the good of man. As Hume emphasized, with us it is natural to approve, commend, and encourage such manifestations of sympathy. No explanation of why we do so is called for.

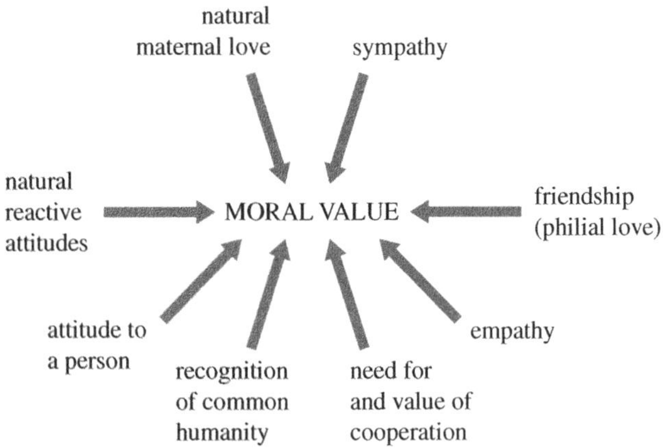

Figure 2.1   *The roots of moral value*

Respect   One of the perennial roots of fundamental prohibitory moral norms (e.g. not to kill or assault others, not to steal or defraud others, not to tell lies or slander, not to cheat or bear false witness, not to break one's promises) is that, without them, society would not survive but would deteriorate into a war of all against all. But another, much more recently developed root of morality is that respect for human beings, for rational free agents, requires one to recognize their liberty to pursue their goals within the framework of what is licit. Hence each person must enjoy the equal protection of the laws irrespective of social status or merit, and each person must be equal before the law irrespective of lineage, wealth, and power. In respecting these norms of conduct, one is also recognizing other human beings *as* persons. This Foot and Wiggins characterized as a form of solidarity with humanity.[8]

Sympathy for others is also manifest in the manner in which we fulfil many of our duties and obligations to others. For one may fulfil one's duties reluctantly, or fulfil them merely for duty's sake (as did Aline Solness in Ibsen's *The Master Builder*) and not out of goodwill. A doctor may treat his patient conscientiously, but without sympathy for the patient's fear or suffering, just as a teacher may lecture painstakingly, without sympathy for his students' misunderstandings or lack of understanding, uncertainties, and lack of self-confidence. These are faults. According to Kant, one must indeed do one's duty for duty's sake, but the temperament of so doing, he insisted, is loving obedience to the demands of the moral law.[9]

Duty for duty's sake It is, however, moot whether acting out of a sense of duty, even with loving obedience to the moral law, is a necessary condition for acting morally. Acting

---

[8] Philippa Foot, *Natural Goodness* (Clarendon Press, Oxford, 2001); Wiggins, 'Solidarity and the Root of the Ethical', The Lindley Lecture, The University of Kansas, March 27, 2008.

[9] Friedrich Schiller mocked Kant for allegedly holding that the merit of doing one's duty for duty's sake was proportional to the reluctance one must overcome. So the more one detests doing one's duty, the better one is (*Xenian*, a book of satirical epigrams by Schiller and Goethe (1796)). Schiller's considered criticisms were in his earlier essay 'On Grace and Dignity' (1793). There he had asserted that Kant presented acting for the sake of duty in 'the frame of mind of a Carthusian', thus robbing it of all gracefulness. Kant replied to Schiller in a long footnote in *Religion within the Bounds of Mere Reason* (6:23), where he declared that the temperament of duty is, as it were, courageous and hence joyous, not weighed down by fear and dejected: 'The latter slavish frame of mind can never be found without a hidden *hatred* of the law, whereas a heart joyous in the *compliance* with its duty (not just complacency in the *recognition* of it) is the sign of genuineness in virtuous disposition ... [without a joyous frame of mind] one is never certain of having *gained* also a *love* of the good.' The idea of loving obedience to the moral law is a secularization of the Old Testament's insistence that the commandments of God must be followed out of loving obedience (see *TP*, appendix, 397–8). Nevertheless, as we shall see, this is not a wholly adequate defence.

merely out of sympathy, Kant averred, has no moral merit at all. It may be attractive, but its dependence on the contingency of the agent's subjective character and disposition and on fleeting desires and inclinations that cause the agent to act or make him act deprives it of moral value. For were these features, which are not under the control of the agent and his will, different, he would not do the right thing at all. If an agent does what is morally right out of 'pathological' love (i.e. an emotion one is passive in feeling and cannot feel at will), rather than willing it for duty's sake, then, Kant averred, the deed has no moral worth. It may be beautiful but it is not morally praiseworthy. According to this conception a moral act has to be the product of the will – the rational will, to be sure – not of the sentiments.[10]

My objections to this are threefold – two in the small and one in the large. First, the naturalist contention that I am advancing is not that sympathy provides the motivating force of all moral action. Nor am I claiming that one's sympathetic inclinations may not lead one into acting wrongly – as when one's sympathy for a charming rogue or *femme fatale* leads one to protect them from their just deserts. My claim is that without there being, by and large, a proneness to sympathy among human beings there could be no moral reasoning at all, and no recognition of moral reasons. That we are in general prone to sympathy, I suggest, belongs to the presupposed framework of morality. (That a minority of human beings lack it is irrelevant; it is sufficient if most are prone to sympathy.)

Secondly, sympathy *can* be a moral motive, and acting out of sympathy may be morally praiseworthy. A person who unreflectively comes to the aid of others in appropriate circumstances, unhesitatingly assists those who need assistance, and has natural sympathy for the distress or suffering of another is *morally* admirable. That one's natural sympathy can also lead one astray in other circumstances does not show that it is therefore morally worthless when it does not. Moreover, in many cases, the intrusion of principle in one's reasoning, as demanded by Kant, would be, as Bernard Williams observed, 'one thought too many'.[11]

---

[10] It was left to the existentialists to discard the rational will in favour of the naked will and to the emotivists to disregard the will altogether.

[11] Bernard Williams, 'Persons, Character, and Morality', reprinted in *Moral Luck* (Cambridge University Press, Cambridge, 1981), 18. Even if the principle is one which one does or would embrace, it plays no role in one's action or in one's justification for one's action: 'She is my wife (daughter, son)', 'He is my friend', 'She was in pain', or 'He was dying', in certain circumstances, require no supporting principle to carry overriding weight, and are fully intelligible as such. (That is not to say that they are not defeasible.) Were that not so, our lives would be colourless affairs.

Natural goodness Thirdly, there is such a thing as *natural* moral goodness in action and in personality that is quite independent of considerations of the moral law and of the motivation of duty. Tolstoy depicted such a character in *War and Peace* in the person of Platon Karataev, Dostoyevsky in *The Brothers Karamazov* in the person of Alyosha Karamazov. It would not occur to such a person to act out of selfishness or self-interest when someone else needed help or solace. Benevolence is part of their nature, part of their character and personality. Such a person, it seems to me, is good without qualification and needs neither the spur of a categorical imperative valid for all nor the accolade of doing one's duty for duty's sake. It is indeed rare to encounter such goodness, but if one does, it is this, and not the moral law, that strikes awe in one's heart. Very, very few of us are like that. Nevertheless, most of us, from time to time, do something right or good spontaneously and without reflection on principles or duties, and would not, if asked why we acted as we did, cite universal maxims of action. This is morally admirable, not merely attractive.

Cooperation There are many forms of cooperation that are not, or not merely, instrumental and not, or not merely, reciprocal agreements. Many of them are characteristic of friendship and are valuable and valued. Cooperation in a joint enterprise involves caring for the collectively beneficial outcome. Fruitful cooperation over time in a joint endeavour often yields a sense of *camaraderie*, and the goodness of camaraderie is not merely utile goodness. Shared adventure in a common enterprise, shared danger in war or natural disaster, and shared adversity commonly lead to a sense of comradeship. These are among the seeds from which fraternity and solidarity spring, watered by reason and imagination.

Reactive attitudes Positive and negative reactive attitudes are natural to mankind.[12] It is intelligible that character traits such as truthfulness, honesty, reliability, and trustworthiness should be valued and praised. It is equally obvious why lying, dishonesty, unreliability, and untrustworthiness should naturally call forth resentment, indignation, and disapprobation. Such vices are harmful to individuals and to society. Their manifestation shows lack of solidarity with 'the party of humanity', as Hume put it. Their proliferation is a mark of a corrupt society. For social creatures with the powers of reason and

---

[12] See P. F. Strawson, 'Freedom and Resentment', in *Freedom and Resentment and Other Essays* (Methuen, London, 1974), for elaboration of the idea of natural reactive attitudes.

subject to the passions, it would be unreasonable *not* to respond with resentment and indignation at unwarranted injury, humiliation, unfairness and injustice, free riding, and failures in reciprocity.

To be sure, a great moralist advocated turning the other cheek. This is a matter for reasoned debate. But it is clear that both positive and negative reactive attitudes are highly sensitive to the intentions and motives of the offender. These conjunctively constitute an *attitude to a person*, or recognition of the humanity of others (Foot, Wiggins, Vlastos[13]). Other things being equal, we consider each other able to act or refrain from action at will, to know what we are doing, to have reasons for doing what we are doing that are themselves subject to evaluation, and hence to be answerable for our deeds and responsible for what we do. This is part of our rational nature as human beings, although the realization and clear articulation of personal act responsibility was evidently slow to emerge in Western civilization, as is evident from the *Iliad* (*TP*, 254–5). We shall examine the varieties of responsibility in Chapter 6.

3. Respect

Formal respect   This fundamental attitude – an attitude towards a person – is no less elemental than natural sympathy and empathy. It is patent even in childhood in natural reactive attitudes and sensitivity to the intentions and purposes of others with whom we interact. For the sensitivity to the intentions of others, the ready recognition that if the harm done to us was not meant and was not negligent then resentment is not warranted, and that benefits bestowed on us for the sake of the interest of the bestower require little gratitude, carries within it the inarticulate recognition of others as free agents. These patterns of natural recognition of the humanity of others and the sensitivity to their intentions, are the roots of the moral idea of *formal respect for persons*.[14] However, the full and *explicit*

---

[13] G. Vlastos, 'Human Worth, Merit and Equality', reprinted in Joel Feinberg (ed.), *Moral Concepts* (Oxford University Press, Oxford, 1969).

[14] I use the term 'formal' here because this kind of respect is indifferent to the character and deeds of a person. No matter how contemptible and wicked he may be, and hence how undeserving of any substantive or 'material' respect, we owe him formal respect simply in virtue of his being a human being a rational person. The importance of formal respect was made patent in the Nuremberg trials and the Eichmann trial. These were landmarks in our slowly evolving moral and legal thought.

emergence of that idea and its integration into moral reflection was even later in the history of Western civilization than introduction of the concept of a person.

Respect for status　　The notion of respect as such is as ancient as the existence of social hierarchies. It is a paradigmatic status concept and bound up with the notion of the dignity of one's station. One's position in the social hierarchy demands recognition from one's inferiors in the form of respect, manifest in their deferential attitudes towards one. Non-recognition is an offence against one's station and against the social order. One's status, save in the case of those at the bottom of the pecking order, demands acknowledgement not only from one's inferiors, but also from one's equals and superiors. Respect is also associated with valued achievements and with admired skills. Respect for social status is linked to the idea of dignity. Tasks that are *beneath one's dignity* are those that should be performed by one's social inferiors. To force someone to do or undergo what is beneath their dignity is to humiliate them and to undermine their self-respect (*TP*, 148). Respect for one's parents, prominent in most societies and in monotheist religions, is not determined by social status but by familial order and parental (typically primarily paternal) authority. Respect for God, prominent in monotheistic religions, is not social-status respect, but reverence (see fig. 2.2).

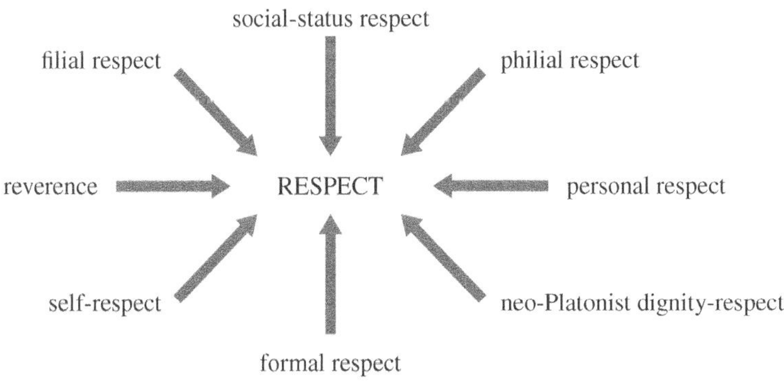

Figure 2.2　*Different kinds of respect*

Although the concept of a person is a status concept, the conception of respect bound up with the mere idea of a person has nothing to do with position in a social hierarchy. Indeed, formal respect for a person is owed to all human beings, no matter what their status in society is. The concept of a person is essentially linked to the notion of freedom, but that is not the freedom before the law that is contrasted with being a slave (and hence a mere chattel) and that was made prominent in Justinian's *Institutes* (AD 533). It is the notion of freedom intrinsically associated with the idea of a *rational being* as such, that is with possession of two-way powers, with the ability to deliberate on and to choose goals and projects, with forming intentions, and with being answerable and responsible for one's deeds and beliefs. Formal respect for a person *qua* rational being is quite distinct from the *personal respect* that one may have for someone on the grounds of their moral stature, the quality of their leadership, or their great achievements, and it is equally distinct from the notion of *respect for social status and office*. The formal respect due to a person is owed independently of any contingent characteristic a human being may have.

Formal dignity   It is surprising that the conception of *formal respect* for a person as such and the associated idea of *the formal dignity* of a person emerges so late in the history of human thought and morals in the West. To be sure, it has ancient roots. Psalm 8 in the Old Testament celebrates the uniqueness and glory of mankind in a passage that deliberately echoes the creation myth of Genesis:

> What is man,[a] that thou art mindful of him?
> And a human being[b] that thou care for him?
> For thou hast made him but little lower than the gods[c]
> And hast crowned him with glory and honour
> Thou hast given him dominion over the works of thy hands
> Thou hast put all things under his feet
> All sheep and oxen, yea, and all the beasts of the field;
> The fowl of the air, and the fish of the sea
> That pass through the seaways.
>
> [a] *enosh*   [b] *ben adam* (son of Adam, human being)   [c] *elohim*

What corresponds to this sentiment in ancient Greek culture is the elevation of rationality to the defining feature of man. It is, according to Plato, rationality that allows us to penetrate the veils of illusion to apprehend the ultimate perfection of the world of Ideas. According to Aristotle, it is possession of a rational *psuchē* that differentiates man from the rest of the animal kingdom. Although the Roman aristocracy were much preoccupied with their *dignitas*, the extension and generalization of that notion to mankind in general was, on the whole, alien to a slave-owning society and to an empire dominated by Rome and Roman citizens[15] (until the universalization of citizenship to all freemen by the Edict of Caracalla in AD 212). It may be thought to be but a short step from Boethius's (AD 480–525) definition of a person as an individual substance of rational nature (*HNCF*, 287) to the idea of respect for and dignity of human beings *qua persons* – *qua* moral beings. But it took until the Renaissance for the idea of the dignity of man to emerge forcefully.

That idea is prominent in the writings of the Renaissance *umanisti* in Florence, such as Gianozzo Manetti (1396–1459),[16] Marsilio Ficino (1433–99), and his pupil Giovanni Pico della Mirandola (1463–94) in his so-called *Oration on the Dignity of Man*[17] (written in 1486, but published posthumously in 1496). This was written as a prefatory oration for a disputation on his *900 Theses*, which was subsequently prohibited by the pope. The theses are a curious syncretic amalgam of Christianity, Platonism, neo-Platonism, the magic of Hermes Trismegistus, and the mysticism of the Jewish Kabbalah. Like many others, Pico presented mankind as midway in the Great Chain of Being (*scala naturae*) between inanimate matter, at the lowest level, and the ranks of the hierarchy of angels, culminating in God, at the highest.[18] However, according to Pico, man is unique among animals in being able to *choose of his own free will* whether so to conduct his life as to ascend in the scale or to descend.[19] A person's dignity is a

---

[15] It is unsurprising, given that Cicero was a Stoic, to find him stating that man's nature is vastly superior to that of beasts: 'their only thought is for bodily satisfactions … Man's mind, on the contrary, is developed by study and reflection … From this we may learn that sensual pleasure is wholly unworthy of the dignity of the human race' (*De Officiis*, I. 30). The notion of dignity here is extended from the dignity of social status to the dignity of man as such. But the idea was evidently premature and did not, as far as I know, take root.

[16] Manetti's *On the Excellency and Dignity of Man* (1452) was written as a response to Pope Innocent III's *On the Misery of Man*.

[17] The title was a later editorial addition. The original title was simply *Oration*.

[18] The idea is said to have originated in Plotinus (albeit with anticipations in Plato's *Timaeus*). For detailed discussion see A. O. Lovejoy, *The Great Chain of Being* (Harvard University Press, Cambridge, MA, 1936), and E. M. W. Tillyard, *The Elizabethan World Picture* (Chatto & Windus, London, 1943). It is altogether unlike the Jewish conception of man as the acme of creation.

[19] Here he disagreed with with his teacher Ficino, according to whom man, like everything else, has a fixed place in the metaphysical ladder of beings. Pico, by contrast, held that, in virtue of his freedom, man's place was not fixed but depended on his free choice: he could ascend or descend, depending on whether he lived a virtuous life or not. In this sense, Pico took the first steps in delivering humanity from the shackles of the Chain of Being.

function of freedom of choice and rationality. It is fully realized only in a life of perfect virtue. So the virtuous have greater dignity than others. The dignity of a person depends upon the freedom to ascend or descend in the divinely ordained Chain of Being, and upon the immortality of the soul. The virtuous achieve greater dignity and the wicked lesser. In these respects Pico's conception is unlike what I have been calling formal dignity and formal respect.

The idea of celebrating the dignity of man as conceived by Pico was swiftly crushed, first, by the papal prohibition on the publication of Pico's *900 Theses*, secondly by Savonarola in Florence to whose impassioned admonitions Pico succumbed, and more generally by the subsequent Counter-Reformation.[20] The idea of human dignity made no impact on Lutherans and Calvinists, who (like the later Augustine) were obsessed by predestiny rather than free will, by dependency on grace rather than autonomy, by the depravity of man rather than his perfectibility, and by the infinite gulf between God and man that is to be emphasized by self-abasement and a pervasive sense of sin.

Kant on formal respect   It is only with Kant in the late eighteenth century that the notions of the formal respect for man as such, irrespective of his social status, and of the essential dignity of man as a rational agent, irrespective of his moral standing, emerge fully into the light and become a major theme in moral, legal, and political reflection.[21] Why that should have been so is a matter for speculation. I am inclined to think that it was a consequence of the dominance of Christianity from the fourth century until the deism and subsequent agnosticism or atheism of the Enlightenment. In Christian thought, respect, that is reverence, for God the father was paramount, and within the family order respect for the father and (to a much lesser extent) the mother are prominent. The family metaphor dominated Catholic religious thought, and the notion of dependency rather than that of autonomy dominated the Christian conception of man's place in the world. We are all children of God, who is our father in heaven,

---

[20] Psalm 8 finds an echo in Hamlet's great speech 'What a piece of work is man! How noble in reason, how infinite in faculty, in form and moving how express and admirable, in action how like an angel, in apprehension how like a god – the beauty of the world, the paragon of animals' (II. i). However, it ends with the characterization of man as the 'quintessence of dust'.

[21] Respect for life as such in all its forms is a prominent theme in Buddhism and a dominant one among the Jains. This is quite different from formal respect for rational beings.

hence we are all brothers and sisters in the shared faith; we are all born sinners since we inherit original sin through the concupiscent form of our conception, and we can achieve felicity in the afterlife only through the loving grace of God and the redemption of our sins through Christ. The notion of the dignity of man *qua* rational autonomous being has little place in this framework of thought.[22]

Kant was a secularizer of the Pietism in which he was brought up, residues of which still cling to his moral philosophy (extreme individualism, a sharp and unambiguous distinction between right and wrong, the demand for guidance by universal principle, the centrality of the idea of duty in the moral scheme of things, emphasis on the state of the individual will) and to the Ideas of Reason he invokes (hope that a benevolent creator exists and hope for an afterlife in which virtue is rewarded). He, more than any other thinker, put the idea of respect and dignity on the philosophical, moral, and political agenda in the modern era. However, Kant's argument in support of a requirement of equal formal respect for human beings as moral agents is problematic. It is linked to his distinction between ourselves as noumenal beings subject to laws of practical reason and ourselves as phenomenal beings subject to causal law. But that distinction is not viable. Moreover, he offers two apparently distinct reasons for respect. The first is that the moral law as such is the object of respect, and that respect for a person is parasitic on that:

> The *object* of respect is therefore simply the *law*, and indeed the law we impose upon *ourselves* and yet as necessary in itself. As a law, we are subject to it without consulting self-love; as imposed upon us by ourselves it is nevertheless the result of our will; ... Any respect for a person is properly only respect for the law. (*Groundwork of the Metaphysics of Morals*, 4. 402n.)

But this is in effect to *deny* that a human being is worthy of respect simply as a person, a free rational being with intellect and will, answerable for his deeds. Moreover, there is surely something awry in suggesting that what is wrong with torturing and slaughtering innocent people or committing genocide is that it is 'against the moral law' or that it offends the principles of practical reason (see below).

---

[22] For a diametrically opposed understanding of the development of ideas of individuality and autonomy, see Larry Siedentop, *Inventing the Individual: the Origins of Western Liberalism* (Allen Lane, London, 2014).

Ends in themselves   The second explication depends on the idea of treating persons also as *ends in themselves*. 'Act so that you treat humanity, whether in your own person or in that of another always as an end and never as a means only' (4. 429, cf. 4. 420). To speak of persons, human beings, as 'ends in themselves' is anomalous. An end or goal 'in itself' is something for the sake of which one acts, something that is not a means to some further goal. Items that are said to be *ends in themselves* are signified by abstract nouns, such as 'the true', 'the good', and 'the beautiful' (the three traditional transcendentalia). But it was an innovation to speak of persons or human beings as 'ends in themselves', and it is clear enough that to say that one stayed at home for Jack's sake or went for a walk for the dog's sake has nothing to do with treating Jack or the dog as ends in themselves, unless that means no more than merely *not* treating them solely as means. But if that is all it amounts to, then it is not only persons that are ends in themselves but also dogs. Similarly, it can be argued that the sublime in nature deserves both respect and awe, and is not to be treated as a means (for fracking, or laying oil pipelines and electricity pylons). But that does not explain the sense in which human beings merit formal respect and dignity simply because of their status as rational persons.

Rationale for formal respect   The niceties of Kantian exegesis are not our concern here. The fundamental thought that is of concern to us is that human beings as persons deserve respect in virtue of;

  (i)   being free – possessing two-way abilities to act or refrain from action at will;

  (ii) belonging to a moral community, a society woven together out of a common language, common morals and mores, knit together by common narratives, which presupposes

 (iii) being rational – able to reason and make judgements, and sensitive to reasons for thinking, feeling, or acting – which is presupposed by

 (iv) having powers of deliberation and decision – being able to consider reasons for acting and for believing, to weigh reasons, and to decide on future action on the basis of reasons, which is presupposed by

  (v) being able to form intentions in advance of acting, which is presupposed by

 (vi) having the power to set ends for themselves on the basis of deliberation on values, on beliefs, and on their past histories or 'autobiographies';

(vii)   being responsible for their deeds (act responsibility) and consequently having liability responsibility for their moral misdeeds, subject to justifying or excusing conditions (non-culpable ignorance, accident, mistake, lack of somatic control, insanity), and being responsible for their beliefs.

To possess these powers is to be a person, to have a mind and a soul (the secular conception of a soul will be investigated in Chapter 5).

Such powers are characteristic of all human beings, save small children, some mentally ill people, and those suffering from dementia. But such human beings either *will* grow to maturity and acquire these powers (children), or *would* possess these powers but for brutal accidents of fate. The latter are defective, damaged, human beings, not non-human beings, and must be so treated. Respect is due to them too (contrast Bedlam – a place to visit for entertainment in the eighteenth century – with a mental hospital in a civilized society today).

Formal respect and human rights
It is noteworthy that *respect for persons*, for human beings with intellect and will, and knowledge of good and evil, is a far deeper and more fundamental notion than the recognition of human rights. To suggest that what was evil about the Armenian Genocide, the Holocaust, the gulags, the terrors of Pol Pot, and the horrors of Mao's Cultural Revolution was that human rights or natural rights were being violated is, as Simone Weil wrote in a different context, 'ludicrously inadequate'.[23] Nor can one decently just say that the millions of victims were treated *unjustly*. Rather, they were not treated as human beings. They were tortured, and subjected to unimaginable physical and mental suffering. They were humiliated and degraded, and subjected to unlimited terror. And they were murdered.[24] What was being violated was not so much human rights as the intrinsic dignity of a human being – the respect that is due to a person. (Reflect on how indignant one might be to find someone urinating on a great work of

---

[23] 'If someone tries to browbeat a farmer to sell his eggs at a moderate price, the farmer can say: "I have the right to keep my eggs if I don't get a good enough price." But if a young girl is being forced into a brothel she will not talk about her rights. In such a situation the word would sound ludicrously inadequate' (Simone Weil, 'Human Personality', in G. A. Panichas (ed.), *The Simone Weil Reader* (David McKay, New York, 1977), 325; see Rai Gaita, *Good and Evil: an Absolute Conception* (Macmillan, Basingstoke, 1991), 5–6).

[24] This will be examined in detail in Chapter 3.

art. How much more indignant one must be when seeing someone urinating on a human being.) Inflicting such evil is a deliberate attack on the very idea of a common humanity as such, as well as an assault on the moral order.

Formal respect for persons is owed to all human beings irrespective of their social status, their merits, and their iniquities. In granting others the formal respect due to them, one is not exhibiting *admiration* for the fact that they are free and autonomous beings, but *recognition*. The notion of formal respect is manifest in what we are obliged *not to do* to our fellow human beings. We must not humiliate other human beings as such, no matter how wicked they are – although we may deny them recognition of their socio-political status and dignity; we must not inflict degradation upon them by treating them as subhuman or non-human beings even if they have done evil – although we may deny them social-status respect; we are under an obligation to recognize them as human beings, as free and responsible agents with rational abilities, and we may not destroy their humanity in their own or in other eyes.[25]

All human beings are owed *equal formal respect*. They must enjoy equality before the law, with an equal right to a fair trial and due process, and with equal rights to the protection of the law. They must not be rejected as human beings. This places substantive restrictions upon the way they may be treated. To grant people the respect due to them is to recognize their dignity as human beings. Again, the notion of dignity here involved is formal. It is detached from the scalar dignity of social roles, since people all have equal dignity. This imposes an irrevocable prohibition on humiliating them and on assaults on their dignity as human beings. So it is also linked to preservation of their self-respect as human beings (but not to their self-esteem, for it is altogether appropriate that those who have done wrong should be brought to feel remorse, to be ashamed of themselves and to repent).

It is important to note that failure to live up to this standard of behaviour not only inflicts profound suffering on one's victim; it also inflicts harm on oneself – since in so doing one falls beneath the standards of behaviour that qualify one for membership of the moral community of mankind. In denying the humanity of others one derogates from one's own. In humiliating others, in destroying their self-respect,

---

[25] See Avishai Margalit, *The Ethics of Memory* (Harvard University Press, Cambridge, MA, 2002).

one besmirches oneself whether one realizes it or not. One becomes morally corrupt.

Evil doers and formal respect   One may well ask why *evil people* merit respect in any form whatever. Why should the formal dignity of Nazis, of the organizers and implementers of the Soviet gulag, of the Young Turks who instigated the Armenian Genocide, of the slaughterers of Pol Pot, or of the mass murderers in Rwanda be upheld and their self-respect protected? Is the answer 'Because they are human beings, persons' adequate? Is this answer a terminus of justification, akin to 'Because it is just (fair, kind, compassionate, etc.)'? Or is there a further rationale behind it? Religious believers in monotheist traditions may appeal here to the idea that human beings are made in the image of God. Respect for human beings is required because mankind is God-like. It is, therefore, a form of reverence.

Another religious claim is that human beings, unlike animals, deserve respect because they have the ability to repent of their sins or misdeeds, can come to recognize the evil they have done, feel remorse, and wish to make amends or expiate their crime. This, in some views, is a condition for redemption in an afterlife.

Neither of these warrants is available to a naturalist philosophy or to a humanist one. One can, of course, detach the possibility of repentance from religion and from belief in transcendent redemption. But then it becomes unclear why the mere possibility of repentance should justify equal formal respect for all human beings. It is true that such evil people as Hitler, Stalin, Enver and Talaat, Mao, Pol Pot, Khomeini, and their henchmen were free, rational, capable of deliberation, able to set themselves ends, and responsible for their deeds. But they abused these distinctively human powers in the most appalling ways imaginable. Have they not forfeited any claims that might be made on the basis of their being human, rational, and free? Have they not freely chosen moral self-degradation as Pico conceived it? Have they not voluntarily descended so far in the Great Chain of Being that they deserve unlimited contempt rather than formal respect?

On one interpretation of Kant's claim that human beings are ends in themselves, this principle is tantamount to saying that each human being is unique and irreplaceable, and therefore intrinsically valuable. But, while it is true that each human being is unique, it is not obvious why that should imply that each human being deserves equal respect. They may simply be uniquely evil and contemptible. They may indeed not be fungible, but who would want to replace such monsters as Hitler, Stalin, Mao, and their like with their equals?

Humiliation   It has been suggested that the negative ideal of non-humiliation can be justified by reference to the general prohibition on cruelty, which is said to be the most fundamental of all evils.[26] Humiliation is a form of mental suffering, the memory of which often long outlasts that of physical suffering. But, while it is true that humiliation is a form of mental suffering and that physical torture is often inextricably interwoven with humiliation, there is more to humiliation than mental suffering and more to its prohibition than the prohibition on inflicting mental suffering. That is because, unlike, say, grief or jealousy, which may likewise involve great suffering, intentional humiliation aims to destroy self-respect. It is an assault upon one's status as a person, as a free moral agent. One may, indeed one should, punish wrong-doers in general, and evil-doers in particular – they deserve punishment. But one may not humiliate them.

It seems to me that there is no single underlying reason for the principle that all human beings deserve formal respect. Rather there is a small circle of linked ideas that all stand together. Those who embrace the principle see it as inextricably bound up with acknowledgement of the equal dignity of moral agents. That in turn is internally related to the prohibition on humiliation. Respect for human dignity is duly essentially connected with the idea of self-respect, inasmuch as humiliating a person, destroying their dignity, is crushing their self-respect as human beings too. And that is evil. (Many punishments throughout history were, and many punishments to this day are, not merely cruel but also deliberately humiliating.) It is this small circle of interwoven ideas, now bound up with the conception of a human being as a person, that make the response 'Because they are human beings – persons' a *terminating* answer to the questions 'Why should their dignity be preserved?', 'Why do they merit respect?', and 'Why should they not be humiliated?'

Respect and moral progress   The explicit emergence of the idea of formal respect is, I believe, one of the great advances in our evolving understanding of human nature and our conception of ourselves. It is one of the fruits of Enlightenment thought and it is as fragile as all Enlightenment ideas. It can easily be destroyed. It is deeply disturbing that, despite the notion's being familiar today, it is

---

[26] Avishai Margalit, *The Decent Society* (Harvard University Press, Cambridge, MA, 1996), chs 4 and 5.

so readily disregarded to a greater or lesser degree, virtually everywhere, by state organs, by bureaucracies, and by law enforcement agencies and penal systems. It is mindlessly swept aside by religious intolerance, xenophobia, racialism, and bigotry. And it is trampled on in warfare, in the maltreatment of enemy civilians and of prisoners of war.

The requirement to respect others, not to humiliate them, not to derogate from their dignity *qua* human beings, reaches far deeper than the social and political domains. It permeates personal relationships. I cannot compete with the following reflections, in a novel by Amos Oz, of an ageing kibbutznik Srulik, who, having written in his diary on the hopes of mankind to put an end to hunger, bloodshed, and the grosser forms of cruelty, and having concluded that the battle is not hopeless, reflects:

So far, so good. But that is where the trouble begins.

I say 'the battle' – yet as soon as I say it, I sense, staring down at me through the thin curtain of ideology, the savage peaks of a suffering far more primeval. The very suffering that drives us to look constantly for battlefields, for 'challenges', to fight, to defeat, to win. How shall we tame the ancient instinct to seize … a spear or sword and run after an antelope and stalk it, hunt it, kill it, and then celebrate the killing? What can we do against the weariness of heart, the subtle, cunning cruelty which is not openly sadistic and can even masquerade as the most reasonable and 'constructive' of behaviours? What shall we say in the face of the secret brutishness lurking within each of us, what our forefathers called the uncircumcised heart and what even a logical-minded, self-disciplined, monkish, musical village priest like myself sometimes discovers in his own soul? With what weapons can we repel this interior wilderness? How can we overcome our dark desires to dominate others, to humiliate them, to subjugate them, to make them dependent on us, to chain and enslave them with the gossamer threads of guilt, shame and even gratitude?

I have just reread my last few lines. 'With what weapons can we repel?' 'How can we overcome?' Even as I seek to avoid the horror, the horror is infecting my own way of speaking. To repel. To overcome.

I am seized with fear and trembling.[27]

---

[27] Amos Oz, *A Perfect Peace* (Fontana, London, [1982] 1990), 363. Srulik is not a village priest but calls himself one out of self-mockery.

4. The relative permanence of the virtues

It is striking that, although different virtues have dominated Western societies over the past 3,000 years and the rank ordering of the virtues has changed (partly in response to social change), the list of virtues, as far as I can see, has changed little since the days of Socrates, even though the interpretation of the virtues has. Christianity, to be sure, reordered the virtues and vices, placing more emphasis on compassion, pity, mercy, chastity, and humility than before, but these had been recognized as virtues long before. It made much of faith – not of mere belief in God, but rather doctrinal faith. But it is not at all clear that this is a virtue at all, and faith in an ideology is surely not one.

Intrinsic value of virtues   Be that as it may, we have barely discovered or invented a *new* virtue in the modern era, despite very considerable changes in our conceptions of the right and the good.[28] The character traits that are essentially at the service of the good of others, such as kindness, considerateness, compassion, charity, generosity, trustworthiness, reliability, truthfulness, and honesty are, it seems, unlikely to become obsolete no matter how much they may conflict with the desire to gain and maintain power or wealth (unless they are forced underground and destroyed by a ruthless police state, as in Orwell's *1984*). Similarly, virtues such as courage, industry, tenacity, prudence, and temperance, which serve the good of their subject and *can* serve the good of others, are not likely to sink into desuetude with the advance of science, technology, and the ever-increasing powers of government. Even when the welfare state takes over the functions of charity in the form of welfare payments and social services, or of prudence in the form of compulsory national insurance, that does not render charity and prudence obsolete or make them lesser virtues. Even when technology takes over the functions of fortitude in facing physical pain (by analgesics and anaesthetics), that does not diminish the value of fortitude in the face of the manifold misfortunes of life. It is true that the invention of effective means of contraception in the last decades has displaced one of the functions of the virtue of chastity; nevertheless it has not transformed lust or *luxuria* into a virtue or even into a neutral character trait. Finally, the master

---

[28] It might be suggested that toleration is a new virtue unknown in antiquity and in the medieval Christian world prior to the great writings of Sebastian Castellio. Castellio was rapidly followed in his views on religious toleration by Jean Bodin and Montaigne.

virtues of self-control, integrity, and practical wisdom are not likely to be eroded by time. The virtues have intrinsic value, and they are constitutive of what it is to be a good human being.[29]

Nietzsche's challenge   To be sure, this conception was challenged by Nietzsche. He held that Homeric Greece had a splendid array of worldly virtues in which the Apollonian and Dionysian qualities were held in perfect balance in an ethos of good and bad. (He gives a similar characterization of Norse culture.) This was a civilization of masters and slaves in which the excellences of the masters were happiness, beauty, vitality, courage, power, success, ruthlessness, and defiance. This 'yea-saying' ethos that glorified life in the world with all its dangers and tragedies, he held, was displaced by the Judaeo-Christian morality rooted in the slaves' *ressentiment* of the masters. This was an unworldly and other-worldly morality of good and evil, in which the (false) virtues of meekness, humility, chastity, submission, charity, pity, compassion, guilt, and piety were inculcated by promises of rewards in an afterlife. Declaring that 'God is dead', he aimed to bring about a future 'transvaluation of values' in which the supposed ethos of noble, vital, yea-saying 'blond beasts' will be revived, Christian slave morality will be trampled underfoot, and the enlightened will pave the way for a new kind of humanity, the supermen.

I don't believe that this paean to a mythical Mycenaean past is of any value. As a contribution to history, it is unreliable. Its overview of the social and moral history of the West is defective. Among other things, the poetic tales of the *Iliad* composed in the eighth century BC were fantasies of a vanished world of mythical heroes that had existed three centuries earlier. Nietzsche's glorification of the supposed morality of early Greek slave-owning warlords is as irresponsible as the romantics' glorification of the alleged chivalry of medieval serf-owning barons enmeshed in interminable internecine strife. Their lives, in so far as we can get a glimpse of them behind the verses of the *Iliad* or the *chansons de geste* were nasty, brutish, and short. The ethos of the Homeric heroes was remarkably similar to the Pashtunwali code of honour of the Pashtuns to this day (see *TP*, 154–5). Nietzsche's supposition that Christianity emerged from the oppressed underclass and slaves of Rome is now known to be false – conversions to Christianity in Rome came largely from the urban middle classes.

---

[29] See A. J. P. Kenny, 'Mental Health in Plato's *Republic*', reprinted in *The Anatomy of the Soul* (Blackwell, Oxford, 1973), 26–7.

(Similarly, the Norse conversion to Christianity was top-down rather than bottom-up – led by superstitious Norse chieftains rather than by common warriors, let alone by slaves.) He is oblivious to the significance of the economic base for the flourishing of Christianity after Constantine made it the official religion of the empire, relieved the church and priesthood of taxes, and permitted the church to own wealth and land. It is true that Christianity then brought about a transvaluation of values in imperial Rome. This involved a fundamental change in the position of women in society, restoring them to something akin to the inferior role of women in archaic and classical Greece (as opposed to Rome), a fundamental doctrinal change in sexual morality and attitudes towards sexuality (the impact of which is unclear[30]), and a shift in the ordering of the virtues and vices.

Putting aside the historical claims, the delineation of an alternative morality amounts to a glorification of the vigour, vitality, and spontaneity of arrogant and exuberant young barbarians (Homeric and Celtic 'blond beasts'). This academic fantasy does not deserve to be taken seriously today, no matter how appealing it may have been in the stultifying social atmosphere of Wilhelmine Germany and late Hapsburg Austria. This is patent when one reflects on how such blond beasts are going to cope with post-industrial, globalized capitalist societies. If there is anything of value in Nietzsche's vision, it lies in:

(i)   the rejection of the martyrology of early Christianity and the conception of a good human life that Christianity then advanced;
(ii) the rejection of any religious foundation for morality and of any promises of an afterlife;
(iii) the repudiation of the hypocrisies of nineteenth-century Europe, with its often superficial pieties, its complacency in the face of rising nationalism (which Nietzsche hated) and imperialism, its philistinism, and its extensive humbug;
(iv) the not very well argued rejection of Kantian Pietism, British utilitarianism, and Schopenhauerian pessimism.

None of these is original.

---

[30] Emmanuel Le Roy Ladurie's *Montaillou* is a fascinating reconstruction of the mores and sexual morals of villagers in the Pyrenees in the early fourteenth century. The gulf between church doctrine and actual practice is striking.

5. Constants in human nature

Moral insight in history   There are constants in certain aspects of human nature – constants that may not be noticed because they are taken for granted. They may be suppressed by a wicked ideology and they may be corrupted within the form of life of a ruthlessly competitive society. An overview of Western civilization over the last 3,000 years does suggest that from time to time radical insights into the nature of the good of man and into the nature of a good person emerge as different aspects of human nature come into the focus of thinkers and moralists. Such, it seems to me, is the idea of loving one's neighbour and the stranger in one's midst in the Torah; and the idea of the Golden Rule advanced by the House of Hillel in the first century BC (*TP*, appendix).[31] So too is the extension of certain moral rules and principles (natural law) from the Hellenes to the whole of mankind, as was advocated by the Stoics. The gradual development of the ideas of individual moral responsibility and the principles of excuse and justification too display deep insight into human beings, their deeds, and the limits of rational responsibility. The idea that human beings should not be hounded and burnt at the stake for their beliefs on matters of opinion and interpretation (advanced by Sebastian Castellio[32] (1515–63) against Calvin[33]) was as deep as it was revolutionary. It was a passionate

---

[31] It does not seem to me at all obvious that Jesus was an original moral thinker. He was a prophet in the mould of the prophets of the Old Testament – fiercely insisting on a return to faithful and loving obedience to the commandments of God. His views, I am inclined to think, are largely derived from the Pharisees, the House of Hillel, and the Essenes. He was a reformer, not a revolutionary. He deviated from Jewish thought primarily in his prohibition of divorce, in his belief that he was the son of God, and in his belief in 'turning the other cheek'. But that had been anticipated centuries before in Greece by Democritus.

[32] Especially in his books *De haereticis* (1554) and *Contra libellum Calvini* (written in 1554 but published only in 1612), in which he writes: 'To seek truth and to utter what one believes to be true, can never be a crime. No one must be forced to accept a conviction. Conviction is free'; 'We can only live together peacefully when we control our intolerance'; 'Let us be tolerant towards one another, and let no man condemn another's belief. For a fine description of the Calvin–Castellio conflict, see Stefan Zweig, *The Right to Heresy* (Cassell, London, 1936).

[33] Calvin, like Luther, Melanchthon, and Zwingli, had himself earlier advocated religious tolerance within Christianity, but (like the others) only until he had achieved power. It is noteworthy that both Luther and Melanchthon gave their approval to Calvin's horrific immolation of Michael Servetus in 1553.

appeal for tolerance that anticipated the Enlightenment. But the idea of tolerance of beliefs is a form of respect for human dignity and freedom of thought. The struggle for freedom of thought and for toleration of disagreement is unlikely ever to cease. It goes against the herd instinct, the sense of identity derived from consensus, and hence the fear of new ideas. So it wars with populist and demotic rule. It also wars with autocracies, theocracies, totalitarian regimes, and dictatorships that fear criticism and opposition. But the principles are now evident and perspicuously articulated and defended.[34] The painfully slow recognition in the West of the *rights of man* and the *rights of women* has gathered pace since the days of Thomas Paine (1737–1809) and Mary Wollstonecraft (1759–97) and is an advance in moral understanding of the human condition and the nature of mankind. The struggle for the idea of *human rights* began with the Nuremberg trials and continues to this day. The very late emergence of the ideas of formal respect for and the formal dignity of human beings are fundamental insights into human nature, on the one hand, and into the nature of the good and the right, on the other.

Moral progress in history   Does this mean that despite the horrors of the twentieth century there is progress in morality and in moral standards? Yes and no. The twentieth century unleashed evil on a scale never hitherto dreamt of, and human beings sank to levels of bestiality that give the most savage of animals an excellent name. This continues to this day in one part of the world or another. The human capacity for wickedness is now empowered to an ever greater degree by science and technology, on the one hand, and by state and bureaucracy, on the other. In this sense, to be sure, contrary to the beliefs of the Victorians, there is little sign of individual moral betterment. Despite the huge increase in wealth and prosperity, in health care,

---

[34] But tolerance of thought was rejected in the twentieth century by communists (who believed that their ends justified any means), by Nazis (who held the individual to be but a grain on the seashores of the *Volk*), by Maoists (who believed themselves to be custodians of the truth), and at times by the USA (e.g. Senator McCarthy and his followers). Today similar rejection of toleration is evident among Islamists who, like the Christians of earlier times, believe themselves to be doing the will of a merciful God while burning, beheading, and torturing. It is also repudiated by populists who seek to undermine representative liberal democracy.

in provision for unemployment, and in education, our individual and collective capacity for evil is undiminished.[35]

At the same time, there has, I think, been progress of sorts in the painfully slow growth in our understanding of *ourselves as moral beings*, and of the *place of morality* and of the *good* in human life. I do not mean scientific progress – even though we have learnt much about human beings in medicine, psychopathology, neurology, and the social sciences. Nor do I mean that this painfully slow increase in understanding of ourselves *as moral beings* is either linear or cumulative. Whatever advances in psychology, neurology, and cognitive neuroscience have occurred, these sciences cannot rationally be expected to shed light on what is morally right, what the good of man is, let alone on what a good man or woman is. Indeed, they are more likely to corrupt our meagre understanding.[36] I should rather compare what I see as gains in moral insight and understanding to gains in philosophical understanding over the past 2,500 years. There is no linear progress over generations, advancing ever forward to greater heights and building on the insights of our predecessors. On the contrary, there are advances and regresses. There is deeper understanding of virtue and the good of man in Aristotle and Aquinas than in later neo-Platonists of antiquity, Augustine, and the Renaissance neo-Platonists. There are greater insights into virtue and the roots of morality in Hume than in the later Bentham and the Mills. Despite the unpredictable tides of fashion and fortune, the great works survive (as long as libraries do) and can be read and studied. But each generation of thinkers, in philosophy and in reflection on the good and the right, must 'roll their own'.

Moral relativism   One cannot but ask what grounds there are for favouring some moral principles and conceptions

---

[35] The Victorian belief in moral progress was neither naive nor self-deceptive. The nineteenth century did indeed see significant moral and social advances in ameliorating the lot of the oppressed (with the abolition of slavery and legislation on child labour), in the law and the judicial systems, in improving the health of cities and city dwellers (clean water and efficient sewerage), and in the penal codes. That vision of moral progress, however, was destroyed by the frenzy of killing in the First World War. That spelled the beginning of the end of European civilization in much the same way as the Peloponnesian wars spelled the beginning of the end of Greek civilization.

[36] See M. R. Bennett and P. M. S. Hacker, *Philosophical Foundations of Neuroscience* (Blackwell, Oxford, 2003).

over others. I have applauded the moral insight of the commandment to protect the stranger in our midst. I have expressed admiration for Hillel's formulation of the Golden Rule – which was a great step forward in moral enlightenment. I have lauded ideas of human responsibility and its limits that moralists and jurists slowly struggled (and continue to struggle) to articulate. I have praised the emergence of the ideas of the formal dignity and respect for persons, the notion of the rights of men and women, and the recent struggle for recognition of human rights. I have commended principles of toleration and of benevolence in the service of the good of man. But with what right can one justify these in the face of moral relativism? Over most of human history, and in very large parts of the world today, these values and principles are not respected in practice, and often not acknowledged at all. Must one not agree that if the heavens are empty then 'man is the measure of all things', and that human beings in different ages and in different places measure differently?

To this, it seems to me, there are three complementary answers. First, that given the constants in human nature such conceptions of the good, if recognized in a society and pursued with the limited powers we possess, are more conducive to human felicity than any others (the notion of happiness will be discussed in Chapter 9). Secondly, in a society in which such general values and principles are recognized, the opportunities for each person to realize their positive potentialities (including potentialities for valuable human relationships) and to pursue life projects of their choice are much higher than if they are not. Accordingly each person is more likely to find fulfilment and to live a meaningful life (see Chapter 11). Finally, these values are not merely instrumental in the pursuit of the good of man but are constitutive of human goodness. In their recognition, and in a life that is guided by them, one may perhaps be able to overcome one's self-centredness, self-interests, and self-concern, and to find one's own soul.

3

The Roots of Evil

Humans are caught – in their lives, in their thoughts, in their hungers and ambitions, in their avarice and cruelty, and in their kindness and generosity too – in a net of good and evil. I think this is the only story we have and that it occurs on all levels of feeling and intelligence. Virtue and vice were warp and woof of our first consciousness, and they will be the fabric of our last, and this despite any changes we may impose on field and river and mountain, on economy and manners. There is no other story. A man, after he has brushed off the dust and chips of his life, will have left only the hard, clean question: Was it good or was it evil? Have I done well – or ill?

John Steinbeck, *East of Eden*, ch. 34

1. The horror!

The varieties of badness Just as 'good' is the most general adjective of commendation, so too its contrary, 'bad', is the most general adjective of condemnation. Both are scalar modifiers. Just as 'good' has its comparative 'better' and superlative 'best', both derived from a different stem, so too does 'bad', namely 'worse' and 'worst'. Close cousins of 'bad' are 'weak' and 'poor'. Something is said to be bad if it is not of adequate or expected quality; depending upon what the adjective qualifies, it may be poor, weak, harmful, worthless, defective,

The Moral Powers: A Study of Human Nature, First Edition. P. M. S. Hacker.
© 2021 John Wiley & Sons Ltd. Published 2021 by John Wiley & Sons Ltd.

or deficient in one way or another. This, however, is superficial inasmuch as it fails to elucidate the varieties of badness and the conceptual relations between them, and between them and the varieties of goodness. It is important to note, before turning to analysis, that being bad or a bad so-and-so is sometimes privative – that is, some things are bad if they are *lacking* in good-making qualities. A knife is a bad one if it does not cut well; a joke is bad (feeble) if it is not funny; one's memory is bad (poor) if one cannot remember things others typically can remember. But there are also numerous non-privative kinds of badness: food has gone bad if it is rotting or infected;[1] a pain is bad if it is severe; an examination script may be bad if it contains numerous mistakes. Just as there are varieties of goodness so too there are varieties of its contrary, badness. The domain is rich in conceptual relations.

Following the spirit, if not always the letter, of von Wright's analysis, we may distinguish between medical, technical, artefactual, instrumental, detrimental, and hedonic badness (see list 3.1).

- Someone's health is bad, weak, or poor if they are ill, suffer from a malfunctioning or diseased organ, or are infected or injured.
- An organ is bad (e.g. bad eyes) or weak (e.g. weak heart), and a sense faculty is poor (e.g. poor eyesight, poor sense of smell), when they fail to fulfil their characteristic function due to malformation, disease, injury, or illness, and hence are detrimental to the well-being of the agent to whom they belong, deleteriously affecting the agent's ability to lead a normal life.
- One may follow Aristotle and add that a *thing* is medically bad if it is bad for the health and fitness of a being; hence smoking is bad for one, as are many kinds of foods, both those that have gone bad and those that deleteriously affect one's health. The notion of health, Aristotle taught us, is a focal one.
- A professional person or craftsman is a technically bad one if they are not merely poor in the exercise of their skill, but

---

[1] Of course, it is then not fit for consumption. But it is not fit for consumption not because it lacks good-making qualities, but because it is rotting and decaying. By contrast a meal or dish may be bad because it lacks good flavour, is inadequately cooked, or is tasteless.

incompetent. A poor craftsman or member of a profession may be better than none, but a bad one should be avoided if at all possible, since they are likely to do more harm than good.

- Non-instrumental artefacts (houses, cars, bridges, roads) are bad if they are incompetently designed, inadequately constructed, or made from inappropriate materials, and so fail to fulfil their function adequately.
- Instruments are bad if they are not merely poor of their kind but positively damaging or dangerous to use.
- Things are detrimentally bad if they detrimentally affect the functioning of instruments and artefacts, or the welfare or well-being of a living creature.
- Something is hedonically bad if it is unpleasant to watch or listen to, disagreeable to play or engage in, unpalatable to eat or drink, and so forth.

List 3.1 *Some varieties of badness*

Of course, many other things that do not readily fit into these varieties can be said to be bad. A bad currency is one that is debased, weak, or counterfeit. A bad report, account, or academic paper is one that contains errors, overlooks relevant evidence, or argues poorly. A bad law is one that is unjust, badly drafted, or unwarrantedly vague. A bad debt is one that cannot be discharged. A bad child is one that is naughty, misbehaves, is unruly, or disobedient. The list could readily be prolonged and one might then investigate how to accommodate these kinds of case within the varieties of badness and how to expand the list of varieties if necessary. But our purpose is to investigate the nature of moral badness, which was deliberately omitted from the list as it is premature to decide whether it is a primary or derivative form of badness.

Badness, wickedness, and evil One may also say of someone that he is bad or a bad person – but it would be more common to be specific and characterize the person as cruel, malicious, dishonest, arrogant, or vindictive, or to say without more ado that he is wicked or evil. Cruelty, dishonesty, arrogance, and vindictiveness are morally bad human character traits. Morally bad character traits are vices. Deadly vices, such as cruelty, dishonesty, and arrogance are manifest in wicked or evil deeds.

To say that someone is a bad person, or to say of something done that it was a morally bad thing to do, is the most general form of moral disapprobation. It is also the crudest and least informative. In many contexts, it would be inappropriate or jejune if not risible: 'Hitler was a bad man', 'to commit genocide is a bad thing to do', 'the intention to inflict pain for fun is bad' are ridiculously inept observations. However, 'morally bad' has many relatives. In the previous three sentences the expressions 'wicked' and 'evil' would be more apt. They signify deep forms of moral offence and wrong-doing. 'Wicked' admits comparative and superlative ('wickeder' and 'wickedest'), but 'evil' lacks any comparative or superlative form (other than by addition of intensifiers such as 'more' and 'most'). As we shall see, this is no coincidence. To say of someone or some deed that it is evil is the most severe moral condemnation in our vocabulary.

It is customary to distinguish three different kinds of evil: natural, supernatural, and human. *Natural evils* are simply natural catastrophes that destroy human life, property, crops, and means of livelihood such as earthquakes, tsunamis, volcanic eruptions, avalanches, floods, and droughts. Biological natural evils are plagues and epidemics. The evil lies in their results, not in any conscious agency. The evil done by one such disaster may be greater than that done by another. They are conceived to be evils because such natural phenomena cause the agonized deaths of countless human beings and destroy the means for satisfying the minimal needs necessary for the maintenance of community and individual life.

Supernatural evil has been believed to exist by the large majority of human beings throughout human history and still is. Gods may be evil (e.g. Ahriman, according to Zoroastrian belief), fallen angels are evil (Lucifer); the Devil or Satan is evil; demons, the supernatural servants of the Devil, too are evil; and human beings who have sold their souls to the Devil are evil.[2]

The phenomenon of natural evil has, since mankind began to struggle to understand the world in which we live, called out for an explanation. The dominant explanations in the West and the Middle East have been by appeal to the gods or to God. Polytheism provided a ready explanation in terms of strife between the gods or the wrath of one god interfering in human affairs, often to spite another. In the absence of a scientific understanding of natural disasters, it was the

---

[2] For a brief survey of the sources of diabology, see Appendix 2.

best that could be offered by way of explanation. Monotheism, in particular Christianity with its emphasis on *faith*, characteristically offered a theodicy to justify the ways of God to man, including the occurrence of natural evils. Judaism, by contrast, saw little need for a theodicy but instead advocated complete and uncomprehending *trust* in God and resignation to catastrophe.[3] (Job offers no theodicy and repudiates the attempts of his comforters to provide one. He says to his wife 'Thou speakest as one of the foolish women speaketh. What? Shall we receive good at the hand of God, and shall we not receive evil?' (Job 2: 10).) Dualist religions (such as Zoroastrianism, Zurvanism, and Manichaeism) offered an explanation in terms of a cosmic drama between the forces of good and the forces of evil in which mankind is caught up. According to their beliefs, God's omnipotence is limited and is counterpoised by an evil Godhead (e.g. Ahriman).

Human evil, which is the concern of this and the next two chapters, is characterized as morally depraved, wicked, vicious, corrupt, and above all cruel. The primary subjects of the attribute 'evil' are:

(i)   deeds, namely evil-doing;
(ii)   the upshot or results of deeds, namely the evil done;
(iii)   agents, namely evil-doers.

We should also distinguish within (ii) between self-directed evils done and other-directed evils done. The former include the evils of alcohol, drugs, gambling, and so forth, when taken to excess. These are commonly damaging to others too (e.g. members of one's family).

Evil: both common and ubiquitous

Before outlining the numerous philosophical problems that cluster around the concept of evil, it is appropriate to bring to mind what kinds of things we are going to reflect on. This is not pleasant, for evil is horrific – but it is necessary.[4] It is necessary because we are all too

---

[3] Rabbi Nahum Ish Gamzu (first century AD), the teacher of Rabbi Akiva, is alleged to have advanced a rudimentary form of the argument that all is for the best in the best of all possible worlds that God has created. But this is too thin to characterize as a theodicy.

[4] As will be evident throughout this chapter and the next, I shall give more examples from the Holocaust perpetrated by the Nazis than from other instances of genocide. There are three reasons for this. It provides an unlimited range of evils that no reasonable person can deny. More than any other such catastrophe, it has informed postwar Western discussions of evil. Finally, as a result of Nazi bureaucracy, it is by far the best documented holocaust, so far more is now known about it than about any other genocide.

prone to distance ourselves from evil, thinking that 'it could never happen here' or that 'I would never do that'. But this is an illusion. Monstrous evil can happen in any society, given the appropriate conditions – it *could* happen here. The willing, responsible agents are human beings *like us*, given the right socio-historical, political, and economic conditions, and given the right leaders. Cruelty, injustice, and slaughter are characteristic features of our species, and to think otherwise is, as Edmund put it in *King Lear*, 'an admirable evasion of whore-master man' (I. ii). The *potentiality* for evil is part of our nature. It is humbling to reflect that, statistically speaking, when one is faced, for example, with a genocidal regime (like the Nazi empire) or a polity bent on crimes against humanity (like Stalin's Soviet Union) the odds are very high that one would at the very least have 'gone along, to get along' – as the large majority did. Moreover, put to the test, as Winston Smith is in Room 101 in Orwell's *1984*, almost everyone will break at some point and succumb unwillingly.

Literary and historical examples of evil I know no better description of evil, its horror and its immunity to a theodicy, than Dostoyevsky's in *The Brothers Karamazov*, in a chapter entitled 'Pro and Contra'.[5] Ivan Karamazov confronts his saintly and devout younger brother Alyosha with the problem of evil. All his examples, Dostoyevsky explained, were taken from Russian newspapers of the day.

> By the way, a Bulgarian I met lately in Moscow ... told me about the crimes committed by Turks and Circassians in all parts of Bulgaria through fear of a general rising of the Slavs. They burn villages, murder, outrage women and children, they nail their prisoners by the ears to the fences, and in the morning they hang them – all sorts of things you can't imagine ... These Turks took a pleasure in torturing children too: cutting the unborn child from the mother's womb, and tossing babies up in the air and catching them on the points of their bayonets before their mothers' eyes. Doing it before the mothers' eyes was what gave zest to the amusement. Here is another scene that I thought very interesting. Imagine a trembling mother with her baby in her arms, a circle of invading Turks around her. They've planned a diversion: they pet the baby, laugh to make it laugh. They succeed, the baby laughs. At that moment a Turk points a pistol four inches from the baby's face. The baby laughs with glee, holds out its little hands to the pistol, and he pulls the trigger in the baby's face and blows out its brains. Artistic wasn't it?

------

[5] Dostoyevsky, *The Brothers Karamazov*, II. 5. 4.

Ivan continues remorselessly:

> But I've still better things about children. I've collected a great deal about Russian children, Alyosha. There was a girl of five who was hated by her father and mother, 'most worthy and respectable people, of good education and breeding'. You see, I must repeat again, it is a peculiar characteristic of many people, this love of torturing children, and children only. To all other types of humanity these torturers behave mildly and benevolently, like cultivated and humane Europeans; but they are very fond of tormenting children ... It's just their defencelessness that tempts the tormentor, just the angelic confidence of the child who has no refuge and no appeal, that sets his vile blood on fire. In every man, of course, a demon lies hidden – the demon of rage, the demon of lustful heat at the scream of the tortured victim, the demon of lawlessness let off the chain....
>
> This poor child of five was subjected to every possible torture by those cultivated parents. They beat her, they thrashed her, kicked her for no reason until her body was bruised all over, then they went to even greater refinements of cruelty – shut her up all night in the cold and frost in the privy, and because she didn't ask to be taken up at night (as though a child of five sleeping its angelic, sound sleep could be trained to wake and ask), they smeared her face and filled her mouth with excrement, and it was her mother, her mother did this. And that mother could sleep, hearing the poor child's groans! Can you understand why a little creature, who can't even understand what's done to her, should beat her little aching heart with her tiny fist in the dark and the cold, and weep her meek unresentful tears to dear kind God to protect her? ... Do you understand why this infamy must be and is permitted? Without it, I am told, man could not have existed on earth, for he could not have known good and evil. Why should he know that diabolical good and evil when it costs so much? Why, the whole world of knowledge is not worth that child's prayer to 'dear, kind God!'.

Should anyone think that these tales describe the horrors of a backward nineteenth-century empire in which some people commit the most appalling evil, but that things are different here and now, let them but open the pages of the daily press for news about the abuse, and in particular sexual abuse, and murder of children and young women.[6]

---

[6] To give but a few examples: Josef Fritzl built a cellar in his house in Amstetten, Austria. On 24 August 1984 he imprisoned his daughter Elisabeth in it and kept her there for twenty-four years, raping her daily from the age of 11 until the age of 35. She had seven children by him. All this was carefully concealed from his wife. Between 1963 and 1965, near Manchester in England, Ian Brady and Myra Hindley brutally

Gibbon gloomily pointed out that 'history is little more than the register of the crimes, follies and misfortunes of mankind', and that 'the history of empires is the history of human misery'. Let us remind ourselves, for current purposes, of some of these crimes that historians record. Crassus lined the Appian Way from Brindisi to Rome with more than six thousand crucified slaves from the Spartacus revolt. Caesar boasted in his *Gallic Wars* that he had slaughtered more than a million Germans and extirpated whole tribes. The crusaders of the First Crusade, on breaking into Jerusalem, ran amok for three days slaughtering men, women, and children, until the streets ran with blood up to their ankles, before repairing to the Church of the Holy Sepulchre to give praise to the God of love. During the Albigensian Crusade, Arnaud of Amaury, head of the Cistercian order and papal legate, wrote to Pope Innocent III of the sacking of Beziers (1209): 'Nearly 20,000 citizens were put to the sword, regardless of age or sex. The workings of divine justice have been wondrous.' Prior to the assault, the barons under his command objected to slaughtering all the inhabitants, since about half of the population of Beziers was Catholic. Amaury replied, 'Kill them all, God will know his own!' The Albigensian Crusade is widely held to be a prototype of religious genocide.[7]

Genocide   Genocide,[8] though by no means unique to the twentieth century, has certainly been a twentieth-century speciality, beginning with the German genocide of the Herero and Nama tribes in German South West Africa between 1904 and 1908. Up to 100,000 Hereros and 10,000 Namaquas – men, women, and children – are thought to have been slaughtered. The official publication of the German General Staff in Berlin, *Der Kampf*, glorified the genocide:

> This bold enterprise shows up in the most brilliant light the ruthless energy of the German command in pursuing their beaten enemy. No

---

raped and murdered five children between the ages of 11 and 17. They were described by the trial judge in his closing remarks as 'two sadistic killers of the utmost depravity'. Between 1967 and 1987 Frederick West and Rosemary West in Gloucestershire in England raped, tortured, mutilated, and murdered at least twelve young women, dismembered their bodies, and buried them under concrete in the cellar of their house.

[7] The crusade (1209–29), initiated by Innocent III, was against the Cathars of Languedoc, centred on Albi in the County of Toulouse. Catharism was a version of the Dalmatian and Bulgarian Bogomil form of Gnostic Christianity, which was itself a development out of Manichaeism.

[8] The word was introduced into our language in 1943 by Rafael Lemkin (1900–59) and adopted by the United Nations in 1948. It is now part of the vocabulary of international law.

pains, no sacrifices were spared in eliminating the last remnants of enemy resistance. Like a wounded beast the enemy was tracked down from one waterhole to the next until finally he became a victim of his own environment. The arid Omaheke [desert] was to complete what the German army had begun: the extermination of the Herero nation.[9]

Survivors, mainly women and children, were sent to death camps such as Shark Island, where they were starved, worked to death, and used for medical experiments by doctors who subsequently held chairs at prestigious German universities.[10] Between 1915 and 1917 Ismail Enver Pasha and Mehmed Talaat Bey, leaders of the Ittihadists (the Young Turks) during the First World War, instigated the dispossession and merciless slaughter of a million and a half Armenian Christians – men, women, and children.[11] The Soviet Union under Stalin committed both genocide (the Holodomor (death by starvation) in the Ukraine in which some seven million Ukrainians are said to have perished) and crimes against humanity[12] (the gulags established by Lenin and Stalin, and the reign of terror that led to the deaths of tens of millions). The Nazis' anti-Semitic genocide involved the murder of 6 million Jews by mass shooting and gassing, as well as by starvation, disease, brutal beatings, and hangings in the ghettoes.[13]

---

[9] Quoted in Helmut Bley, *South-West Africa under German Rule* (Heinemann, London, 1971), 162.

[10] Prisoners on Shark Island who suffered from scurvy were injected with opium or arsenic, and the results analysed on autopsy, among others, by Dr Eugen Fischer, who later became vice-chancellor of the University of Berlin. He taught subsequent Nazi doctors who made deadly and horrific medical experiments on inmates of concentration camps. Three hundred Herero and Nama skulls are said to have been sent to Germany for the medical analysis of inferior races.

[11] Twenty-eight countries have currently (2019) recognized this holocaust as genocide. Britain has not, probably for fear of damaging trade and its military alliance with Turkey, where mention of this episode as a genocide incurs three years' imprisonment.

[12] A category of crime in international law that was successfully advocated by Sir Hersch Lauterpacht (1897–1960) in connection with the Nuremberg trials of Nazi war criminals. The history of the struggle to get *crimes against humanity* recognized in international law, and Lemkin's parallel struggle to get *genocide* so recognized is recounted in Philippe Sands's *East West Street: On the Origins of Genocide and Crimes against Humanity* (Weidenfeld & Nicolson, London, 2017). Sands favours the former category. The latter is, I think, no less useful and important.

[13] To be sure, over and above the anti-Semitic Holocaust, the Nazis also slaughtered Roma, homosexuals, mentally impaired children and adults, Jehovah's witnesses, Russian prisoners of war, and political dissidents of any kind with equal ferocity and unmitigated evil.

When, in the death camp of Treblinka, the murderers ran short of poisoned gas, babies were thrown alive into the furnaces to save Zyklon B. Horrific medical experiments were carried out on men, women, and children.

The uniqueness of the holocaust The Holocaust was unique, and not merely because of its scale and extreme cruelty. Eight reasons for holding it to be unique are given in list 3.2. Although individual reasons characterize other horrors, the combination is unparalleled.

(i)   The meticulous and systematic bureaucratic organization of the Holocaust.

(ii)   The transportation of millions of victims over huge distances for the sole purpose of murdering them.

(iii) The diversion of transport, funding, and manpower from the war effort for the sake of transporting victims from all corners of Europe to the concentration camps and death camps in Poland.

(iv)   The technology and mechanization of slaughter in the murder camps.

(v)   The industrialization and commercialization of death (e.g. hair shorn from women before gassing for use in upholstery, gold teeth extracted from corpses and melted down for bullion, bones used for the manufacture of soap).

(vi)   Its universality (the aim was to slaughter not only all Jews within the Reich, but all Jews within any captured territories).

(vii) The fact that it occurred in one of the most civilized nations in the world.

(viii) That it aimed to shatter the moral order that is a mark of humanity and to destroy the restraints and limits imposed upon human beings by morality.[14]

List 3.2   *The uniqueness of the Jewish genocide committed by the Nazis*

---

[14] In relation to (viii), it might be observed that: Hermann Rausching, Nazi Gauleiter for Danzig, recorded the following remark made by Hitler in the early 1930s: 'We terminate a wrong path of mankind. *The tables of Mount Sinai have*

Since the Second World War, crimes against humanity have been committed in Mao's China, in Pol Pot's Cambodia, in Khomeini's Iran, and in Assad's Syria, and genocidal slaughter was committed in East Timor, Nigeria, and Rwanda. And so the tale continues. We have the mark of Cain upon our brow.

The horror of it   The ending of Conrad's novella *The Heart of Darkness* is altogether apt for concentrating our minds on the evil in human nature. Kurtz, the subject of the tale, had gone out to the Belgian Congo as an ivory trader, full of high hopes of bringing civilization and education to the natives. He had, however, succumbed to the fierce joys of unlimited power. The narrator Marlow sardonically describes the jungle station far up-river where Kurtz dwelt. The human heads on poles lining the path to his house on the hill

> only showed that Mr Kurtz lacked restraint in the gratification of his various lusts, that there was something wanting in him – some small matter which, when pressing need arose, could not be found ... Whether he knew of this deficiency himself, I can't say. I think the knowledge came to him at last – only at the very last. ...

> ... Believe me or not, his intelligence was perfectly clear – concentrated, it is true, upon himself with horrible intensity, yet clear; ... But his soul was mad. Being alone in the wilderness, it had looked within itself, and by Heaven's, I tell you, it had gone mad. I had, for my sins I suppose, to go through the ordeal of looking into it myself. No eloquence could have been so withering to one's belief in mankind as his final burst of sincerity. He struggled with himself too. I saw it – I heard it. I saw the inconceivable mystery of a soul that knows no restraint, no faith, and no fear, yet struggling blindly with itself.

> ... One evening coming in with a candle I was startled to hear him say a little tremulously, 'I am lying here in the dark waiting for death.' The light was within a foot of his eyes. I forced myself to murmur, 'Oh, nonsense!' And stood over him as if transfixed. Anything approaching the change that came over his features I have never seen before, and

---

lost their validity. Conscience is a Jewish invention ... It is our duty to depopulate, just as it is our duty to provide appropriate care to the German population. ... Natural instinct commands every living being not only to defeat the enemy but to destroy him. In earlier ages there existed the good right of the victor to exterminate whole tribes, whole nations' (quoted by Gunnar Heinsohn in 'What Makes the Holocaust a Uniquely Unique Genocide', *Journal of Genocide Research* 2:3 (2000), 411–30; I am grateful to Bernard Harrison for drawing my attention to this).

hope never to see again. Oh, I wasn't touched, I was fascinated. It was as though a veil had been rent, I saw on that ivory face the expression of sombre pride, of ruthless power, of craven terror – of an intense and hopeless despair. Did he live his life again in every detail of desire, temptation, and surrender during that supreme moment of complete knowledge? He cried in a whisper at some image, at some vision – he cried out twice, a cry that was no more than a breath:

'The horror! The horror!'

2. The grammar of evil: preliminary clarification

Characterization of evil Evil is the deliberate infliction of death, severe bodily or mental harm, or extreme suffering on another human being or sentient creature without adequate warrant or in excess of what is required for the attainment of a justifying or excusing goal.[15] Adequate warrant may be medical requirements to save the life of, or to cure, a patient; self-defence; the necessities of war; the preservation of the state. To be sure, there is widespread disagreement, both within and between cultures, about what counts as a warrant, or what counts as excess. There is a broad grey area, rather than merely a fuzzy boundary, between what is wicked and what is evil, between what is a necessary evil (an evil warranted in order to avoid an even greater one) and what is evil *simpliciter*, and between the licit and the evil. I shall refer to this domain of dispute as 'the borderlands of evil'. It provides material for important individual, historical, and legal reflection upon actual cases. Such contestable cases will be avoided here, since our concern is to shed light on the conceptual and moral character of central and indisputable cases. That is a prerequisite for casuistical investigations.

Harm caused by evil deeds The suffering, infliction of which is evil, may be physical or non-physical (mental or psychological, in one meaning of these polysemic terms) and commonly both. The harm caused by evil deeds may be fatal to the victim; it may deprive the victim of potentialities that would have enabled him or her to lead a full life; and it may leave lasting mental and physical scars. Jean Améry attested from his experience at the hands of the Gestapo:

---

[15] I am much indebted to John Kekes's thoughtful book *The Roots of Evil* (Cornell University Press, Ithaca, NY, 2005).

Anyone who has been tortured remains tortured ... Anyone who has suffered torture never again will be able to be at ease in the world, the abomination of the annihilation is never extinguished. Faith in humanity, already cracked by the first slap in the face, then demolished by torture, is never acquired again. ... slight pressure by the tool-wielding hand is enough to turn the other [the victim], along with his head, in which are perhaps stored Kant and Hegel, and all nine symphonies, and *The World as Will and Representation* – into a shrill squealing piglet at slaughter.[16]

One who is being tortured immediately knows that all moral limits have been removed, that one is wholly within the unconstrained power of violence of one's torturer. One's moral universe is inverted, since instead of being able to appeal to others to help and to ameliorate one's suffering, one's agony is an invitation for one's torturers to inflict more. It is this that makes evil deeper than badness and wickedness, both of which remain within the pale of morality. All torture involves humiliation that can burn deep into one's soul whether or not it is accompanied by physical torture. Indeed, humiliation may destroy, and be intended to destroy, the very soul of the victim by forcing him to betray or besmirch what is most sacred to him. This is patent in a horrific example of evil cited by Rai Gaita from Chaim Kaplan's Warsaw Diary:

A rabbi in Lodz was forced to spit upon a Torah scroll that was in the Holy Ark. In fear of his life he complied and desecrated that which is holy to him and his people. After a short while he had no more saliva. To the Nazi's question, why did he stop spitting, the rabbi replied that his mouth was dry. Then the son of the 'superior race' began to spit into the rabbi's mouth and the rabbi continued to spit on the Torah.[17]

---

[16] Jean Améry was a pseudonym of the Viennese Jewish author Hans Mayer (1912–78). He worked for the Belgium Resistance and was caught and tortured by the Gestapo. He was deported to Auschwitz, and thence first to Buchenwald and subsequently to Bergen-Belsen. He survived the war and published a volume of five essays *Jenseits von Schuld und Sünne* [*Beyond Guilt and Atonement*] about his experiences in 1964 (translated as *At the Mind's Limits: Contemplations by a Survivor on Auschwitz and its Realities* (Indiana University Press, Bloomington, 1966). He committed suicide in 1978.

[17] Raimond Gaita, *Good and Evil*, 1. Our horror here goes beyond the infinite humiliation of the victim, for it is also horror at the deliberate, gleeful shattering of the moral order of humanity (see item (viii) in list 3.2).

Some people may never recover from such evils and be incapable of leading a normal human life. Others may partly recover, but be sorely and irreparably scarred (like Améry himself, or Primo Levi). And yet others may overcome the horrors of their past and lead a fulfilled life. What makes it possible to triumph over such adversity I do not know. It is a token of the potential greatness of the human spirit.

Evil and cruelty   The primary evil vice is cruelty.[18] Its primacy consists in its involvement in deliberate but unwarranted infliction of suffering. But it is not necessary, since the unwarranted infliction of death is evil, even if it is painless. It is not sufficient, since there are forms of cruelty that are wicked rather than evil. But it is a criterion for evil-doing. Evil-doing may be accompanied by indifference towards the suffering inflicted or may be sadistic. Taking pleasure in inflicting suffering on other living beings, finding joy in destroying their souls, and delighting in shattering the moral order are the most extreme form of depravity.

No degrees of evil-doers   As noted, 'good' and 'bad' have comparative and superlative forms, 'evil' does not, but requires the addition of intensifiers. An evil done may be *greater* or *lesser* than another, depending upon the amount of harm done or suffering caused. But one person cannot be 'eviller' than another or the 'evillest' among a group of evil people – but only more vicious, depraved, or sadistic. This is suggestive. Of course, one may say that one person is more evil than another – but it is moot whether this is ever correct. If a person has done evil, then he is evil. If he does further evil, that does not make *him* 'more evil' – rather further confirms that he is evil. If we say of one person (say, Lavrentiy Beria) that he was even more evil than another (say, Nikolai Yezhov) that simply means that he *did* more evil. One cannot *be* more evil than someone who tortures a child for fun. Being responsible for two million deaths does not make one more evil than if one is responsible for only one million. All that is true is that the former *génocidaire* did more evil than the latter.

---

[18] Keith Thomas points out a wide consensus in the seventeenth century in viewing cruelty as 'the extremest of all vices' (Montaigne) and as a breach of 'the law of nature' (Hobbes). It was held to be 'inhuman' in being incompatible with humanity, 'unnatural' in being contrary to natural law, 'unchristian' in that it conflicts with the teachings of the Gospels, 'savage', 'barbarous', and 'uncivil' because inconsistent with the moral standards of civilized society. It is striking how the multiplicity of reasons all converge (Keith Thomas, *In Pursuit of Civility* (Yale University Press, New Haven, CT, 2018), 149).

Types of evil agency   Evil may be done by a solitary person; by a person in, and persons acting as, a group – a local gang or a well-organized group, such as the Mafia or Ku Klux Klan; by tribes, especially in intertribal warfare, for example among the pre-Columbian Meso-American tribes, who practised hideous human sacrifice on a monstrous scale; by militias, such as the 'Brownshirts' (Sturmabteilung, or SA, formed in 1921) in Weimar Germany, military groups sanctioned by the state dedicated to slaughter such as the Teskilât-i Mahsusa in the Ottoman Empire during the Armenian Genocide (1915–17) and the Hamidiye irregular cavalry during the reign of the Sultan Hamid for the purpose of the Armenian slaughter in 1894–6 and again for the massacres of the Assyrians in 1915,[19] or the SS Einsatzgruppen starting in 1939 in German-occupied Poland and later in the Soviet Union.[20] Evil may be perpetrated by an artificial person, such as a company that knowingly sells dangerous goods, or a pharmaceutical company that carries out medical experiments on unwitting victims or knowingly sells contaminated medicines. Evil may be a characteristic of an institution, such as the Spanish Inquisition, the evil-doing of which may not be under the control of any one individual and may persist against the will of the majority of members of the institution. It may be done by the state by means of legislation (e.g. the anti-Semitic and racial Nuremberg Laws) or orders to state organs and organizations.

Non-human patients of evil and wickedness   Evil may be done to animals by cruel killing and trapping, destruction of habitat, substandard caging, starvation of animals in captivity, brutality towards working animals or pets. Nature may be harmed – by the destruction of species and habitat, by wilful damage to landscape,

---

[19] The Teskilât-i Mahsusa ('Special Organization') were formed to expedite the Armenian Genocide. It was responsible for the slaughter of hundreds of thousands of Armenians on forced marches without food or water into the Syrian desert: 80,000 women and children were burned alive in stables and haylofts, 50,000 were loaded onto boats that were sunk at sea by these 'butchers of humanity', as they were called by Vehib Pasha, commander of the Ottoman Third Army. The Hamadiye were Kurdish light cavalry, originally formed for the Armenian massacres under Sultan Hamid in 1894–6, but also deeply involved in the Armenian Genocide in the First World War, and responsible for the slaughter of 10,000 Assyrians in 1915 (see Taner Akçam, *A Shameful Act* (Constable, London, 2007)).

[20] The SS Einsatzgruppen ('Task Forces') were paramilitary death squads responsible for mass shootings between 1939 and 1945 in Nazi-occupied Poland and the Soviet Union. They are estimated to have killed more than 2.2 million people. Many of the leaders of the Einsatzgruppen were highly educated men, with doctoral degrees.

by pollution of the atmosphere, by the poisoning and pollution of the oceans, and by causing global warming. Such deeds, in so far as they do not involve cruelty, are wicked rather than evil. Similarly wicked rather than evil is doing deliberate harm to beautiful artefacts – slashing great paintings for the sake of publicity (e.g. the Rokeby Venus in 1914 by a suffragette), destroying great works of architecture (Monte Cassino, the Imperial Summer Palace in Beijing, the Winter Palace in St Petersburg (Leningrad as it then was), destroying great works of arts and crafts (as was done to much of the Chinese national heritage during Mao's Cultural Revolution, and in the destruction of the colossal Buddhist monuments in Afghanistan by the Taliban).

So much for the subjects and objects of evil. Evil is characteristically done deliberately, with malice aforethought. But evil may be done spontaneously, intentionally, and without deliberation, as in the case of some pogroms (chillingly described by Isaac Babel in 'The Tale of My Dovecot'). Cold-blooded evil involves planning. So it harnesses rational deliberation to evil intent. Can evil be done accidentally or inadvertently? Accident or negligence may produce evil, but unless the negligence is culpable neither the deed nor the agent are evil.

Etymology and translations of 'evil' The etymology of 'evil' is revealing, since Germanic languages provide distinctions absent from Romance languages, and from Latin, ancient Greek, and ancient Hebrew. The English *evil* is derived from Old English *yfel* and Middle English *uvel*, and from Old Frisian *evel*. It is related to Old High German *ubil*, which is the source for the German *übel*. The root meaning is 'exceeding due measure' or 'overstepping proper limits' – which, as we shall see, gives us a clue how to construe the important differences between bad, wicked, and evil.

German is very fluid, inasmuch as the use of *übel* has multiple branches: *Übeltäter* means evil-doer, but *Übelkeit* means nausea, and *übel nehmen* means to take offence. Indeed *übel* can also do service for 'bad', 'ill', 'vile', 'nasty', 'wicked', and 'foul'. Other expressions available in the German toolbox are *schlecht* which, in one context or another may mean bad, poor, poorly, ill, or sick; *böse*, which may signify evil, bad, wicked, angry, cross, naughty, sinister, malign, villainous, or ferocious; *gemein* (*Gemeinheit*), which may indicate mean, vile, villainous, or wicked; *gottlos*, which may mean godless, wicked; *teuflisch*, which may mean devilish or fiendish; and *scheusslich*, which may mean dreadful, terrible, or nasty.

Latin is well equipped too: *malus* signifies bad, wicked, harmful, or noxious. Other terms that can be roped in are *nequam* (worthless, bad, good-for-nothing, vile); *nefas* (contrary to divine law, impious deed, wrong); *improbus* (wicked, bad, cruel, shameless, dishonest, presumptuous); *pravus* (crooked, depraved, perverse, wicked, bad); *perversus* (perverse, bad, distorted, askew, awry); *turpe* (foul, base, nauseous(; *vitiosus* (vicious, faulty, defective, corrupt, bad); and *corruptus* (corrupt, depraved, ruined, wicked, bad, immoral). However, there is no expression in Latin that corresponds precisely to 'evil' as opposed to 'bad' or 'wicked'.

The absence of an expression corresponding precisely to 'evil' is characteristic of ancient Greek too. *Kakós* means bad, wicked, or evil indifferently, spiteful or mischievous, depending on context; *kakia* signifies wickedness, malice, spite, or being rotten. *Aischros* means ugly, dishonourable, shameful, or filthy. *Diephtharmenos* means corrupt, depraved, or decadent; *fáulos* signifies vicious, nefarious, wicked, or unrighteous; and *parephtharmenos* means corrupt. *Ponēros* means hurtful, worthless, bad, wicked; *mochtheros* signifies sinister, vicious, or malicious.

Ancient Hebrew, despite the rendering of the Old Testament in the King James Version, has no expression that distinguishes evil from wickedness and badness. *Ra* signifies bad, wicked, or evil indifferently: the Tree of the Knowledge of Good and Evil is *Etz ha-da'at tov va-ra*; *pesha* means 'crime', 'transgression', or 'breaking a relation'; *khet* means 'sin', 'breaking a covenant with God', 'missing the target', or 'failure to fulfil an obligation'; *avon* indicates 'wrong', 'twisted', 'trespass', or 'crooked'; and *zadon* means 'malice'.

Germanic languages in general, and English in particular, are fortunate to have a direct means to distinguish the most extreme form of moral iniquity from badness and wickedness. *Wicked* and *wickedness* are more specialized than *bad* or *badness*, in that they explicitly signify moral opprobrium. But *evil* signifies what is beyond the pale. We shall explore this clue below. The grammatical network within which *evil* is located is rich and ramified (see table 3.1) and displays the multiple foci of attention and association that can be highlighted in English.

These grammatical surveys give us an idea of the domain of thought we are about to investigate. With these in mind, we may now turn to investigate philosophical problems associated with evil.

bad	wicked	evil	depraved	foul	ignoble	vicious	demonic
wrong	iniquitous	malevolent	degenerate	vile	dishonourable	malicious	diabolical
	sinful		corrupt	repulsive	base	villainous	devilish
	nefarious		outrageous	repugnant	degrading	atrocious	black-hearted
	immoral			odious	despicable	heinous	fiendish

Table 3.1   *The conceptual network of evil*

3. Philosophical problems: does evil exist?

Denials that evil exists With the above examples in mind, it seems obvious that evil exists: that evil deeds are done, that human beings produce evil as the result of their evil deeds, and that there are evil human beings. Evil is ubiquitous in human cultures, manifest in slaughter of the innocent, wanton cruelty, abuse and maltreatment of women and children, torture, mutilation in war and peace, degradation, and humiliation. This is as evident in the daily press and television news as it is in recorded history. So it is surprising, indeed extraordinary, that it is quite common to deny that evil exists. Phillip Coles, for example, avers that 'we ought to dispose of the concept of evil. The argument is not that we *can* do without it, but that there are good moral and political reasons why we *should* do without it.'[21] Alan Macfarlane agrees that the concept is obsolete.[22] Richard Rorty holds that 'there is no such thing as an intrinsically evil desire' but only desires that get in the way of our project of maximizing the overall satisfaction of desire.[23] Such people cannot *see* evil, even though they are surrounded by it, if not in their personal experience, then on radio and television and in the newspapers.

Blindness to evil and its existence An analogy might be drawn here. In the Middle Ages, people did not merely believe in miracles: they *perceived* them. With the rise of science, educated people in the West, on the whole, ceased to see miracles and advanced empirical explanations for what would once have seemed the intervention of God. Today, it seems, many educated people cannot see evil when it is before their eyes. But there is this difference between not seeing miracles and not seeing evil: one may say 'It seemed miraculous, but it was merely a cleverly devised illusion, as are all such things', but someone who witnessed Treblinka cannot say, 'It looked as if it was evil, but that was merely an illusion, as are all such things.' Moreover, whereas a God seems necessary for miracles, all that is necessary for evil is the black heart of man and its manifestations in deed. That evil does not exist is surely false. But what is interesting is

---

[21] Philip Cole, *The Myth of Evil* (Edinburgh University Press, Edinburgh, 2006), 235.

[22] Alan Macfarlane, 'The Root of All Evil', in David J. Parkin (ed.), *The Anthropology of Evil* (Blackwell, Oxford, 1985), 57–76.

[23] Richard Rorty, *An Ethics for Today: Finding Common Ground between Philosophy and Religion* (Columbia University Press, New York, 2010), 15.

to discern why so many people think or have thought this true to the point of not even seeing evil when confronted by it at second hand in the news or in books.

There seem to be five main reasons adduced to support the claim:

   (i)   that the very expression is obsolete;
  (ii)   that avowing the existence of evil commits one to a weird ontology;
 (iii)   that evil is not an explanatory concept;
  (iv)   that evil can be explained away;
   (v)   that evil is privative: it signifies nothing other than the absence of good.

Examining these suggestions and suppositions will shed some light on the nature of evil and on reflections on it.

Is the concept of evil obsolete?   (i) *The expression 'evil' is becoming rare as applied to actions and obsolete as applied to human beings*, according to the *Oxford English Dictionary*. I doubt whether 'evil' is any more obsolete than 'virtue' or 'virtuous', and it is surely more commonly used than 'nefarious' and 'iniquitous', which are in no sense obsolete. ('Zounds', 'methinks', or 'Alack' are obsolete.) It is only as rare as discussions of human nature and of the cruelties and viciousness of mankind. Even if it is infrequently used, when it *is* used it is very useful indeed for drawing distinctions that are morally important. In so far as it *is* becoming obsolete as applied to persons, that is lamentable. It is *not* rare but common to find the genocides of the twentieth century being characterized as evil; there is nothing outmoded in describing the Mafia, the Blackshirts, or the Ku Klux Klan as evil organizations; it is not obsolete, but wholly appropriate to aver that Enver, Mussolini, Franco, Salazar, Hitler, Stalin, Mao, Pol Pot, Assad, Khomeini, and their like were evil men.

It is striking that in English law a judge will sometimes declare, 'You have been found guilty of murder. I say in addition that you are a truly evil man.' Such an addendum to a formal declaration of guilt is apt in cases of particularly savage and horrendous murders. This, I suggest, confirms the analysis that is under consideration. One who does evil, as opposed to wickedness, has placed himself beyond the moral pale and beyond redemption. He has set himself to destroy the very foundations of human existence in society. Far from the concept of evil being obsolete, it fulfils an important condemnatory role in the law. It invites a harsher punishment and puts in question any possibility of shortening of the maximal sentence.[24]

Ontological queerness of evil
(ii) *To aver that evil exists is to commit oneself to a strange mythological ontology.* This claim may take slightly different forms. One is the idea, similar to John Mackie's 'argument from queerness' against the existence of objective values, that if evil did exist it would be a very strange 'entity' (thing) – a supernatural power.[25] But to insist that evil exists is not to assert that it is an objective force in the universe, a force that one might encounter or experience (as in Tolkien's *Lord of the Rings*) – a kind of moral equivalent of magnetism or gravitation. It is to hold that there are evil people who kill or inflict unbearable suffering on other human beings or animals without adequate warrant. One might encounter such people, experience the evil they do, or observe the evil done. Nothing mysterious or 'metaphysical' is involved. Anyone who holds that evil does not exist for such reasons must also hold that goodness does not exist either – that there are no morally good people or morally good deeds done. But this does not deserve to be taken seriously.

Is the concept of evil primitive?
Another suggestion is that evil is an idea rooted in primitive agrarian societies and associated with primitive beliefs in devils and demons.[26] Evil, thus conceived, is shadowy, mysterious, threatening, and associated with darkness and night (when crimes are commonly committed) and with secrecy. It was also linked in popular imagination with rituals of sorcery, sacrifice, and exorcism. This may all be perfectly true. But it does not follow that the concept of evil is essentially tied to such mythological beliefs. It was not so linked in the Hebrew Bible (by contrast with the New Testament) or in classical rabbinical teachings on the nature of *ra* and *yetser ha'ra* (the inclination to do wrong or evil). It neither is nor should be so linked in current secular debates about the most extreme forms of wrong-doing and cruelty.

Is the concept of evil explanatory?
(iii) *The characteristic of being evil is not explanatory.* A common reason for rejecting evil as a characteristic of certain human beings is that the concept of evil is not explanatory. One may ask why someone performed a terrible deed, but the answer 'He did it because he is evil' is alleged to

---

[24] I am grateful to Mark Freedland for drawing my attention to this usage.
[25] Cole, *The Myth of Evil*, ch. 1.
[26] Alan Macfarlane, 'The Root of All Evil'.

be empty – it explains nothing. At best it is a mythological explanation that carries any force only in conjunction with unacceptable beliefs in evil 'forces' in the world.[27] However, the fact, if fact it be, that something is non-explanatory does not mean that it is a fiction. For the primary role of the utterance 'He is evil' may be verdictive rather than explanatory. Secondly, 'He did it because he is evil' is neither more nor less explanatory than 'He did it because he is forgiving', that is, than any explanation by reference to pronenesses or tendencies. Such explanations are weak but not empty. They tell us that the event or action that occasions puzzlement is of the kind that one might reasonably expect of an agent with such a nature. Thirdly, 'He killed the innocent child because he is evil' seems neither more nor less explanatory than 'He gave away his cloak because he is good'. But does one really wish to claim that 'because he is good' can have no explanatory force whatsoever?

(iv) *Evil cannot be explained away.* An idea, common today, militating against the existence of evil is manifest in a form of scientism that has two roots, one in determinism, the other in the medicalization of moral responsibility. The determinism may be Laplacean causal determinism or it may be historical determinism (e.g. Marxism). The science may be psychology (Diderot or La Mettrie in the eighteenth century, phrenology in the nineteenth, psychoanalysis and behaviourism in the twentieth, or cognitive psychology in the twenty-first) or cognitive neuroscience. These lend themselves to various combinations and permutations, the upshot of which is to purport to show 'scientifically' that any appearance that certain human beings are evil is merely a surface phenomenon, since in truth ('scientific truth') we are all victims of forces beyond our control. The evil of evil-doers can therefore be explained away by showing that they themselves are victims.[28] (The notions of freedom and responsibility will be examined in Chapter 7.)

In vulgar form, similar ideas have become widespread among educated elites in the West since the 1970s, much encouraged by neuro-

---

[27] See, for example, Cole, *The Myth of Evil.*

[28] Such views informed some of the famous pleas of the great American attorney Clarence Darrow (e.g. the Leopold–Loeb trial, discussed below) and some of the controversial recommendations of Baroness Barbara Wootton, to which Herbert Hart replied in his 'Punishment and the Elimination of Responsibility', reprinted in *Punishment and Responsibility: Essays in the Philosophy of Law* (Clarendon Press, Oxford, 1968).

mania. On such a view, all motives and actions are due to circumstances beyond an individual's control, such as the economic system and poverty, the social system and class-ridden society, the political system and concentration of power, colonialism, globalization, the unconscious mind, a painful childhood and childhood trauma, and so forth. It is these that are to blame for human wrong-doing. So human beings are not responsible for the evil they cause, and they cannot themselves be evil. This view will be shown to be misguided in Chapter 4.1, where the forms of explanation of evil are examined.

Christianity on the illusoriness of evil (v) *Evil does not exist because it is privative; it is merely the absence of goodness, and in itself it is nothing at all.* This was a standard Christian view advanced by the church fathers in their endeavour to formulate a theodicy. The problem of evil for Christian monotheism is indeed sore. For the existence of evil seems inconsistent with the omnipotence, omniscience, and benevolence of God. It appears that to account for the existence of evil, one must either limit God's benevolence, curtail his omnipotence, or constrain his omniscience. But these are not options for monotheists. A variety of solutions to the predicament were posed. One prominent solution, advanced by Pseudo-Dionysius the Areopagite (ca. 500) is that God, being benevolent, cannot be the source of evil. But since God created everything, evil cannot be independent of God. Therefore evil must be nothing. Its existence must be illusory. It is merely the absence of good. It is privative, like darkness – merely a deficiency, without any substantial being. The view was repeated by John Damascene – 'Evil is nothing else than a lack of the good' – and by Western theologians too. To be sure, this doctrine had to be reconciled with the existence of the Devil, who plays so prominent a role in the New Testament.

Is evil privative? The idea that evil is privative, that is, it consists in the absence of good, stripped of its theological trappings, is unconvincing. There is nothing privative about taking pleasure in the agony of others, or feeling joy at the sight of their torment. Even if evil is privative like darkness, it is unclear why that relieves God of the responsibility of allowing it – after all, he could presumably have created a universe of light or not have created the universe at all. If sadism, hatred, cruelty, and vengefulness are corollaries of giving human beings free will, God must have known that free will would lead to unlimited wickedness and evil. If this is the price of free will, and hence of moral agency, one may question, with Ivan Karamazov, how such a luxury can be paid for with the suffering of innocent children. If it is said that a world with free will is better than a world without it, one may ask: better for whom? Certainly not for the innocent children. Nor does this sophistry explain how natural evils (such as the great Lisbon Earthquake in 1755, which aroused Voltaire's wrath) can be permitted by a benevolent God, and how the suffering of animals is compatible with the goodness of God. To be sure, Christian theologians have attempted to meet all these challenges.

The appeal of the denial of evil
The various contemporary claims that there is no such thing as evil, that evil does not really exist, that the expression 'evil' is obsolete, or that it is essentially a religious concept with no place in secular thought and secular morality, or that it has a place only among primitive societies with supernatural beliefs, are decidedly odd. Among other things, the supporting arguments are demonstrably weak and easily undermined. Nevertheless their proponents advance them in good faith and often with much fervour, and the ideas have considerable support among intellectuals. The moot question is why? What makes them so appealing?

If the concept of evil is wholly detachable from religious contexts and if it can be completely removed from any links with primitive anthropological beliefs, if there really are evil people and if evil deeds are indeed done, then the power of reason is much more limited than the liberal Enlightenment and their heirs envisaged. For evil people appear to be beyond the reach of reasoned argument with respect to the most fundamental of moral considerations: the infliction of death without warrant and cruelty without constraint. Many contemporary thinkers, psychologists, and social scientists find this thought deeply disturbing. One reaction is to insist that to apply the concept of evil to a human being is itself atavistic, vindictive, inhuman, and unforgiving. It is to suggest that some human beings, who are sane, in control of their deeds, and answerable for what they do are nevertheless, on some specific topic or topics, beyond the reach of reason and reasonable debate. This challenges the ideals of the liberal Enlightenment concerning the brotherhood of mankind under the aegis of reason. It is therefore not surprising that the medicalization of evil is commonly favoured to show that those who do such deeds are themselves victims of forces beyond their control and are not responsible, or not wholly responsible, for their crimes. For then they cannot be viewed as being 'beyond the pale' or as being irredeemable. No human being, it may be held, can be evicted from the brotherhood of man. This subject will be discussed in Chapter 5.4. Maybe no one can be evicted, but, as we shall see, perhaps those who do evil cast themselves out.

4. Philosophical problems: can evil be explained?

Religion's need for a theodicy   Hardly less surprising than the denial that evil really exists is the view that evil or the existence of evil cannot be explained. It is perhaps intelligible if advanced by theologians who hold that evil is privative. But that claim is unconvincing, and attempts at a theodicy are no less so. One alternative, an alternative taken by classical rabbinical Judaism, which had no truck with devils,[29] is to declare that the existence of natural evil (earthquakes, plagues, etc.), although sometimes sent as a punishment, is often beyond our power of understanding. However, the existence of human evil is explained by reference to the wickedness of those who choose of their own will to do evil. But even this does not explain why God does not intervene in the monstrous evils inflicted by human beings on other human beings. It is not surprising that some said that God died at Auschwitz, just as it had been said in the seventeenth century that God was killed at Magdeburg.[30] Job's answer, as we have seen, is trust in God and acceptance of destiny.

---

[29] As is shown in Appendix 2, there is no Devil in the Hebrew Bible, and neither Lucifer (who does not even occur there) nor satans are devils. Any suggestion to the contrary is due to mistranslation or misinterpretation. Diabological beliefs, however, were rife among the many religious movements in the Middle East during and after the Second Temple period (as is evident from Christianity itself), as is patent in the pseudepigrapha, especially the apocalyptic writings of the books of Enoch, Jubilees, and Wisdom. Though they are rejected by rabbinical Judaism and excluded from the compilation of the sacred writings of the Tanach, there is a presumption that they affected popular Judaism and some Judaic sects.

[30] The Protestant city of Magdeburg was sacked in 1631 by the Catholic imperial forces under the field marshal Graf zu Pappenheim. Of 30,000 citizens only 5,000 survived. Pappenheim wrote: 'I believe over twenty thousand souls were lost. It is certain that no more terrible work and divine punishment has been seen since the Destruction of Jerusalem. All of our soldiers became rich. God with us' (quoted in Hans Medick and Pamela Selwyn, 'Historical Event and Contemporary Experience: The Capture and Destruction of Magdeburg in 1631', *History Workshop Journal* 52 (2001), 23–48). By 1648, when the Thirty Years War ceased, the population had dropped to 450.

Unintelligibility of evil is only apparent

Putting aside the peculiar worries of monotheism, it is noteworthy that many have felt that there is something ultimately unintelligible about evil. It has been said that

> Radical evil seems ... to surpass the boundaries of moral discourse; it embodies a form of life and a conceptual scheme that is alien to us. We seem unable to evaluate such acts from a moral vantage point because they are as incomprehensible to us as would be the behaviour of people who did not share our concepts of time and space.[31]

Focusing upon the Holocaust in particular, Jean Améry (whose voice we have already encountered) averred:

> There really is nothing that provides enlightenment on the eruption of radical evil in Germany ...

> This evil really is singular and irreducible in its total inner logic and its accursed rationality. For this reason all of us are faced with a dark riddle.[32]

It is indeed true that, confronted by histories of gross evil, and even when confronted by deeds of evil, we are prone to assert, 'I can't understand how anyone could do such things!' Scholars of the Armenian Genocide or Jewish Holocaust may well, after extensive reading of records and memoirs, say 'I still can't understand it'. But to take this as indicating the inexplicability of evil seems to me to involve a misconception of understanding, and a failure to realize the import of this phrase in this context. Someone studying the Holocaust may read the words of Franz Stangl, the Nazi commandant of the death camps of Sobibór and Treblinka, who was responsible for the murder of 900,000 innocent men, women, and children: 'My conscience is clear. I was simply doing my duty.' In an interview with Gita Sereny he remarked on his organization of industrialized death: 'That was my profession. I enjoyed it. It fulfilled me. And yes, I was ambitious about that, I won't deny that.'[33] Reading these horrifying words, one may well exclaim 'I just can't understand this!' But having studied the

---

[31] C. S. Nino, *Radical Evil on Trial* (Yale University Press, New Haven, CT, 1996), ix.

[32] Jean Améry, *At the Mind's Limits* (Indiana University Press, Bloomington, 1980), xviii.

[33] Gita Sereny, *Into that Darkness: From Mercy Killing to Mass Murder*, 2nd edn (Pimlico, London, 1974), 200.

facts and read the interviews, confessions, and analyses what more is needed in order to understand? What additional piece of evidence or theory would finally make things clear? There is no more revelatory evidence than what has already been amassed in tens of thousands of documents and detailed psychological and sociological studies. So it is the inclination to make this utterance that needs explaining.

The exclamation 'I still can't understand how anyone could do such things' or 'I just can't understand how it was possible' is an utterance, but not an utterance of failure of understanding. Note that we are inclined towards the modal form here: not 'I don't understand', but rather 'I can't understand'. The utterance is analogous to 'I can't believe it!' said in the face of firm and irrefutable knowledge. What that amounts to is: 'It runs contrary to everything I thought possible.' Hence, in the case of reflecting on terrible evil, what the utterance amounts to is: 'I am horrified to know that human beings can sink to such levels of depravity.' It is an exclamation of horror, not an utterance of failure of understanding.

There is a further possible root of one's reaction of incomprehension. Confronted by a tale of monstrous evil, we are disposed to think 'How could anyone do such a thing?' – unreflectively presupposing that evil-doers go in one step from normal human life, with its complex weave of good and bad, to doing monstrous evil. We forget that the path to evil is often long and involves innumerable small steps, and slow but steady corruption as each step is taken. For the most part, especially in organized evil, one sells one's soul to the devil slice by slice.

What needs explaining   We must ask whether there are explanations for:

 (i)   how there can be evil in a universe created by an omniscient, omnipotent, and benevolent God;
 (ii)   the propensity of *Homo sapiens* to cruelty, destructiveness, sadism, and so on;
 (iii) collective deeds of gross evil (e.g. genocide, crimes against humanity);
 (iv)   the evil nature of evil people; and
 (v)   individual evil acts.

Man's potentiality   Before trying to answer these questions, I shall lay
for good and evil   out the general framework within which I shall
                        approach them. Contrary to Christian doctrine and
interpretation of the tale of Adam and Eve and the Fall, human beings are not born in sin (although psychopaths appear to be born without

the capacity for sympathy and fail to evolve the power to comprehend or internalize standards of moral value). Contrary to Rousseau and the Romantics, man is not born good and only later corrupted by society. Humans naturally evolve a potentiality to do what is right and morally admirable, and also a potentiality to do what is wrong and morally reprehensible, evil being the extreme case. They also evolve a second-order ability to learn the difference between right and wrong in the course of character formation through habituation and emulation (see Chapter 4). The potentiality to do good is implicit in our natural propensity to sympathy (see Chapter 2), but it needs refining and transforming by moral education, experience of family life, the hurly-burly of social life, and reflection. Furthermore, the propensity to sympathy admits of degrees – some people having a greater propensity than others to respond sympathetically to the sorrows, grief, and suffering of mankind. Moreover, it is readily crushed by brutalization and circumstance. It is much reduced, on particular occasions, by alcohol and fatigue. And it is curtailed by beliefs in the subhuman character of victims.

The potentiality for wickedness and evil is latent in our natural motive of selfishness; in our natural competitiveness and single-mindeness in pursuit of chosen goals to the neglect of the interests and well-being of others; in greed, envy, and jealousy; in acquisitiveness and lust for power; in indolence (regarding global warming, for example); and in indifference to the suffering of others. Which of these potentialities prevails in the life of any particular person depends upon their natural endowment of sympathy; moral education; parental love, instruction and example; schooling; socio-economic circumstances; prevailing political conditions; and moral luck (see fig. 3.1).

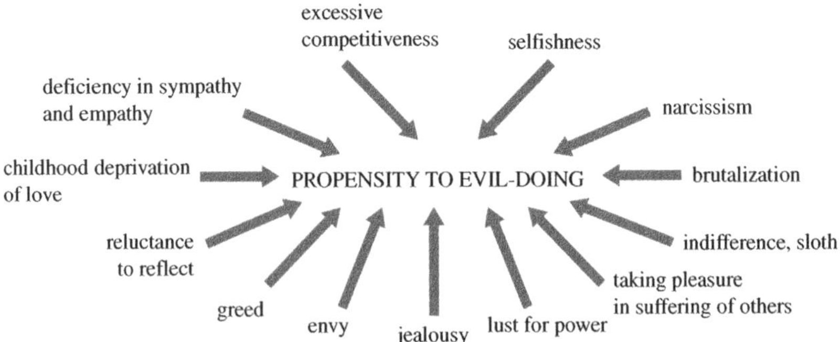

Figure 3.1 *Factors conducive to our propensity for evil-doing*

Let me now turn to the questions. With regard to (i) *Is there an explanation of how there can be evil in a universe created by an omniscient, omnipotent, and benevolent God?* I have nothing further to add to my expression of scepticism about the possibility of a cogent theodicy by reference to the privative character of evil. A different theological explanation advanced by Augustine explained the existence of evil suffered by mankind by reference to inherited human evil. *All* human beings, *from birth*, are tainted by original guilt (*originalis reatus*) that is inherited from Adam's original sin and transmitted by sexual conception. So *all merit punishment*, and no blame attaches to God for the suffering of the innocent, since there are no innocent. The doctrine is repulsive. The grounds upon which Augustine rests his case are five biblical passages, three of which are mistranslations and two of which are misinterpretations.[34] In addition, the matter is of no concern for a philosophico-anthropological account of good and evil.

(ii) *Is there an explanation for the propensity of Homo sapiens to cruelty, destructiveness, sadism, and so on?* Can the theory of evolution shed light upon the general human propensity to do evil? Animals cannot do evil in the sense in which humans can, although they may be the cause of natural evils. For the doing of evil requires reason and reasoning, deliberation and intention formation, reflection on deeds about to be done as well as on deeds already done, remorse and expiation, and knowledge of good and evil (see Appendix 1). There are no evil, non-language-using animals, only more or less aggressive ones and more or less violent ones (chimpanzees are notoriously much more violent that bonobos). So evolution cannot explain the presence of evil among animals, since there is none. It can, however, explain the animal roots of human evil. It has evidently favoured the lion who is disposed to kill all the cubs of the females in a pride he has just taken over from an older, less fit lion, as it has favoured the cuckoo, which is an innate free-rider and killer.[35] Alpha males among chimpanzees maintain their dominance and sexual privileges through aggression, sometimes savagely beating younger males in the tribe. So aggression may pay in the currency of sexual success. It may also pay in terms of group success in dominating territory, as is patent in the savagery with which

Animal roots of human evil

---

[34] See Christopher Kirwan, *Augustine* (Routledge, London, 1989), ch. 7.

[35] Whether the so-called Cinderella effect of the relative probability of a stepfather killing his stepchild by contrast with a father killing his child (see Chapter 7.1) is comparable to lion behaviour and has a comparable evolutionary explanation is still moot.

chimpanzees attack members of another tribe that have strayed into their territory. But there is no *evil* here, only behaviour patterns and pronenesses that humans have inherited and can curb.

Evolution displaced by history and culture Evolution has obviously favoured language-using hominids. However, once a moderately well-developed language is in place, human beings, while retaining the genetic behavioural propensities of the species, develop new ways of social change. To be sure, a whole host of behavioural tendencies are inherited. These can be studied by psychology. In addition, a host of behavioural dispositions are the products of culture. History displaces evolutionary *social development* of the kind found among other social and tribal animals, including hominids other than man. Historical explanation is logically unlike evolutionary explanation. Different human societies evolve according to diverse social pressures: individual leadership, economic circumstances and development, social structures and class differentiation, belief systems and religions, the existence of propertied priestly classes, knowledge and its development (both technical and theoretical), and environmental conditions. There is no reason for thinking that history favours either brutal or less brutal societies, or that it penalizes societies prone to moderation in their relationships with other societies and in their penal conventions, or favours societies prone to cruelty in warfare and punishment. It is not evident that the Roman Empire (during the Nerva–Antonine age, the most successful society that the bloodstained continent of Europe has ever seen) would have lasted less long or longer had it abolished gladiatorial games and crucifixion earlier than it did.

Explanations of collective evil (iii) *Are there explanations of collective deeds of gross evil, for example genocide and crimes against humanity?* The difficulty is not that there is no explanation, but that there are so many kinds of explanations, and often multiple explanations for the same event. We do know, with reasonable accuracy, why Caesar slaughtered more than a million members of Germanic tribes. It is a very complex explanation, intelligible only by reference to Caesar's personality, the class structure of Rome in the first century BC, the antecedent history of the corrupt republic, the character of the Roman army and the changing relations of the troops to their generals, and so on and so forth. We know why Suetonius Paulinus sought to exterminate the Iceni tribe of East Anglia. Here too a complex historical explanation as well as a personal one is needed. We know why the crusaders slaughtered innocent women and children in Jerusalem in a three-day orgy of killing. We know why Pope Innocent III induced the horrendous Albigensian Crusade. So too, there is no mystery about why, in the late nineteenth century, the United States government committed genocide upon Native American tribes.[36] Nor is there any puzzle about the Herero, Armenian, or Jewish genocides, the crimes against humanity committed by Stalin and Mao, the Rwandan genocide, the Pol Pot slaughters, and Khomeini's murder of more than 30,000 Iranian citizens in 1988. The point is not lack of explanation, but rather that very different socio-historical conditions can have much the same genocidal upshot. The social and political conditions leading to mass murder in the Ottoman Empire, the Third Reich, Stalin's Russia, Assad's Syria, Rwanda, and Khomeini's Iran were vastly different. The explanations we can give are *historical, political, economic,* and *sociological.* These may invoke nomological elements (in both economics and sociology), although they are predominantly ideographic. The ideographic explanations are primarily teleological, involving human goals pursued with reasons, done deliberately under conditions of uncertainty, with antecedent intention formation, exemplifying complex motives. But psychological explanations of the behaviour of pivotal figures may often include nomological elements concerning personality types, behavioural dispositions, emotional propensities, and so forth.

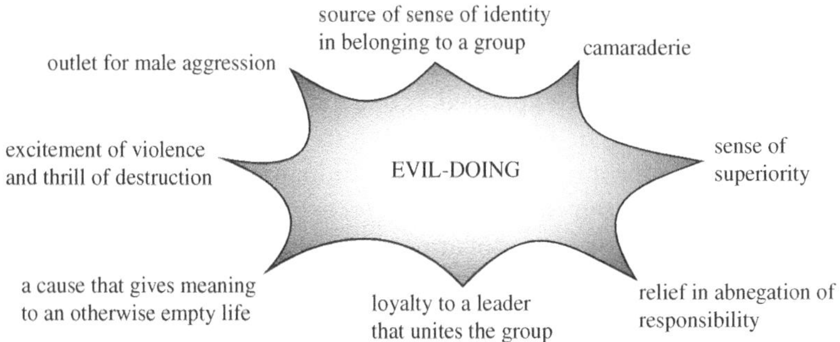

Figure 3.2 *The attractions of group evil*

---

[36] The slaughter of the Northwestern Shoshone at Bear River in 1863, of the Cheyenne at Sand Creek in 1864, of the Blackfoot at Marias River in 1870, and of the Lakotas, culminating in the massacre at Wounded Knee in 1890. See Dee Brown, *Bury My Heart at Wounded Knee*(Holt, Rinehart, & Winston, New York, 1970).

Given appropriate social, economic, religious, and ideological conditions that vary from one epoch to another, from one society to another, from one historical development to another, and from one people to another, we can offer various *socio-psychological explanations* of the attraction, in those circumstances, of evil (see fig. 3.2). Organized violence, as manifested by the SA in Germany in the 1920s and later the SS, by Teskilât-i Mahsusa and the Hamidiye cavalry in the Armenian Genocide, by Mussolini's *squadristi*, and by the Ku Klux Klan in the southern United States, gives point and purpose to male aggression. These organizations create a collective identity, giving members a sense of belonging. They mobilize and unify their troopers, separating them from the rest of society and marking them out as superior, dedicated to a higher cause and to a supreme leader. Such groups provide for their members constant excitement, the opportunity of unrestrained violence against others (aliens, inferiors, enemies of the people, subhumans) in the name of the *Volk*, the faith, or a millennial future. This provides for their members the intoxication of power, the freedom and thrill of destroying and slaughtering with impunity, and the promise of receiving accolades and recognition. It also commonly involves much booty and extensive rape.

Baron-Cohen's science of evil (iv) Bearing this in mind, we can now turn to the question of the possibility of *explaining the evil nature of the people who engage in such horrific deeds*. To be sure, given the widespread interest in cognitive neuroscience, it is not surprising that neuroscientists should aim to initiate a science of evil. Simon Baron-Cohen at Cambridge University sought to do so. He asserted that the key to understanding human evil is deficiency of empathy.[37] Empathy, he explained, is our ability to identify what someone else is thinking or feeling and to respond to their thoughts and feelings with appropriate emotions.[38] Empathy occurs, he averred, when we suspend our single-minded focus of attention and instead adopt a double-minded focus of attention. The former involves thinking only 'about our *own* mind, our current thoughts and perceptions', the

---

[37] Like most contemporary psychologists and cognitive neuroscientists, he employs the term 'empathy' to mean a confused amalgam of *Einfühlung* and 'sympathy' as the latter was used prior to Titchener's introduction of the expression 'empathy' into English in 1909. The idea of sympathy is pivotal to the naturalist explanations of morality given by Hume and Adam Smith. It is quite distinct from empathy, which is not an emotion but a cognitive power. For detailed analysis, see *TP*, ch. 12.

[38] Simon Baron-Cohen, *The Science of Evil: On Empathy and the Origins of Cruelty* (Basic Books, New York, 2012), 17.

latter when 'we keep in mind someone *else's* mind at the very same time'. So when empathy is 'switched off' we think only about our own interests, but when it is 'switched on' we focus on other interests too.

Misconceptions about empathy and sympathy This is not an adequate basis from which to start experimenting. These are not neuroscientific insights, but fumbling philosophical forays into terrain already well explored in the eighteenth century by Hume and Adam Smith. The fumblings are clumsy. We rarely think about our own *mind* (save in old age when one often fears one is losing one's intellectual powers); we do, tautologically, think our own thoughts, but that does not imply thinking *about* one's own thoughts; we rarely think *about* our perceptions, but often reflect on *what* we perceive; however, what we perceive is commonly no different from what others with us perceive. Single-minded attention does not imply selfishness – one may be doing something for one's spouse, children, or friend. Focusing on the interests of others does not imply feeling any empathy for them – one may be an accountant doing the tax returns of a client. One's dentist, one hopes, is empathetic in attending to, and understanding, the plight of his patient, but one hopes that his attention will be wholly undivided. One may have a superb ability to know the mind of another and yet use that knowledge for one's own evil ends, as does Cathy Ames in Steinbeck's *East of Eden* (that is why we need both the concept of sympathy *and* that of empathy (*TP*, ch. 13)). She is a woman innately totally bereft of sympathy, lacking any compassion or kindness, without gratitude or remorse, concerned only with her own self-interest and using others exclusively as means to her ends. She is a monster. She murders her own parents, shoots the man who saves her life when he stands in her way with his love for her, and poisons a woman who took her in and showed her every kindness, in order to take over her property. She is, no doubt, a psychopath, but perfectly sane (psychopathy is quite distinct from psychosis).

Empathy (in either sense) is not the ability to *identify* what someone else is thinking or feeling. When someone whips out a knife at me, I don't *identify* that as a threat – I react to it. When someone laughs delightedly at something funny, I don't *identify* their laugh as amusement – I apprehend it, and probably laugh sympathetically. When someone says 'The behaviour of so-and-so is shameful' I don't have to *identify* his thought – he has just *told* me what he thinks. The idea that we are in a constant process of *identifying* what others think and feel is a manifestation of deep-rooted confusion about the nature of our knowledge of others. The minds of others are not hidden behind a facade of their behaviour; they are expressed in their behaviour. The fact that autistic children have great difficulty in apprehending the thoughts and feelings expressed by others does not mean that they cannot *read the minds* of others, but rather that they *cannot read the behaviour* of others – although, to be sure, thinking and feeling are not behaviour.[39] There is indeed such a thing as reading the mind of another, but it is akin to reading between the lines.

Empathy is not the ability to respond to the thoughts and feelings of other people with appropriate feelings. That one responds to the thoughts and feelings of a bigot, fanatic, and prejudiced person with outrage and indignation does not imply that one feels any empathy for him or her, even though one's feelings are wholly appropriate to the repulsive reflections and emotions that person has expressed.

Are there cortical empathy pathways? It is on the basis of such inadequate reflections on empathy that Baron-Cohen decided that the roots of wickedness and evil lie in lack of empathy. It is, he averred, deficiency in empathy that explains why human beings do evil. Having, to his satisfaction, clarified this, he proceeded to experiments, endeavouring to identify a cortical 'empathy pathway'. This, he claimed, can be shown by fMRI (functional magnetic resonance imaging) to involve ten distinct interconnected areas of the cortex.[40]

However, it is not evident that, psycho-pathological cases apart, evil people may not display what Baron-Cohen calls 'empathy' towards their comrades and pets – Hitler was fond of dogs and enjoyed the company of members of the inner circle of his followers. While evil doers may well be incapable of companionate love (see Chapter 5), but that does not imply that they cannot take pleasure in

---

[39] It is noteworthy that Baron-Cohen's previous research was on autism, and important that he studied under Uta Frith, one of the main proponents of the mistaken doctrine that every normal human being has a theory of mind ('theory-theory' as opposed to 'simulation theory'), acquired in infancy and early childhood. But, as argued in *TP*, these putative theories are chimerical.

[40] Much doubt has been cast during recent years on the reliability of fMRI, not least by its inventor Seiji Ogawa. See, for example, Maxwell R. Bennett, Sean N. Hatton, Daniel F. Hermens, and Jim Lagopoulos, 'Behaviour, Neuropsychology and fMRI', *Progress in Neurobiology*, 145–6 (2016), 1–25; M. R. Bennett, L. Farnell, and W. G. Gibson, 'Quantitative Relations between BOLD Responses, Cortical Energetics, and Impulse Firing', *Journal of Physiology* 119 (2018), 972–89. Results based on fMRI should be treated with great caution.

company or display some forms of care for others. So the idea of 'empathy circuits' is, I hope, starting to lose whatever remote plausibility it had.

It is not the business of philosophy to interfere in empirical matters, and what activities in which parts of the brain are necessary for a person to feel empathy (or to sympathize) with another is an empirical question. But if the investigation begins with such profound misconceptions of what empathy is, what it is to feel empathy, and what it is to empathize with another, it is doubtful whether significant neural results are likely to be obtained. Given that the scans are carried out on immobile patients, who are given photographs of smiling, angry, or tearful faces, or are instructed to imagine someone they love in the arms of another, it is implausible to suppose that what is being tested is cortical responses to anything other than photographs and casual fantasies. In the stream of life, one's *sympathy* for another may involve joyous response to their joy, quiet relief that they have succeeded and are pleased, or pride in their achievement; it may involve sorrow that accompanies their sorrow, compassion for their suffering, pity for their predicament, an impulse to console, a desire to be with them, anguish and despair at their suffering, an overwhelming desire to ameliorate their lot, and so forth. 'Empathy' (or sympathy) may be triggered by face-to-face encounter, by a telephone call, by letter, or by hearsay. It may be manifest in indefinitely many forms, depending upon the circumstances, and, of course, its manifestation may well be deferred until one encounters the object of the empathy. The endlessly complex interweave of history and circumstance, of knowledge, thought, emotion, and volition, not to mention perception, consciousness, awareness, and realization, make it prima facie implausible to suppose that there is anything that can be deemed to be an 'empathy circuit' that is active whenever 'empathy' is felt.

The very idea of a *science* of evil seems as preposterous as a science of virtue. Whether or not it is true that 'Just because it is evil' is an explanation or an answer to the question 'Why did he do that evil deed?', we do not lack explanations of why people do evil deeds. This is a subject of the next chapter, where question (v) is investigated. But even without reflection it is surely evident that people may do evil deeds out of a multitude of motives (out of jealousy, rage, envy, hatred, ambition, etc.) and for manifold reasons (because one's victim insulted, humiliated, and ruined one, or because he is an infidel, an apostate, a Jew, a kulak, a bourgeois, an aristocrat).

None of these doubts about the prospects of neuroscience to discover the neural roots of evil imply comparable scepticism about psychological, social-scientific investigations into childhood development, into parental background and home environment, into education, into general and practical intelligence, and into the social adaptability of those who have been found to display a proneness, and with time a criminal proneness, to wickedness and worse. There is a respectable social science of criminology. Much can be learnt from such investigations. This is not a matter to which philosophy can contribute. But *the forms* of humdrum psychological explanations may be grist for philosophical mills. These forms will be examined in the next chapter.

4

Explanations of Evil

1. The variety of explanations

Evil is perennial   Human evil is perennial. Some of it is a function of the historical stage of society. There was a time when bear-baiting, cock-fighting, dog-fighting, and other forms of vicious cruelty to animals were amusing entertainment. Bull-fighting still is in Spain. There was a time when the public took pleasure in gladiatorial games; today such spectatorial enjoyments have been reduced to watching boxing, wrestling, and the violence of cinematic entertainments. There were times when slavery was an accepted public institution, child labour was the norm, and sexual exploitation of children and young teenagers was to be expected. Some of these evils still exist, and the struggle against them persists. Genocide and crimes against humanity have always existed – but at least they are now legally classified and internationally recognized as crimes. The evils of warfare change only in their weaponry. The evils and wickednesses of bureaucracy are as old as well-developed bureaucratic hierarchies. Individual evil or the wickedness of one adult against another, of an adult against a child, and of a son or daughter against a parent change but little over the ages.

Evil is the form of deadly vices   Just as being virtuous is not a character trait, being evil is not either. Just as being virtuous is not a virtue, so too being evil is not a vice. Rather, one might say,

The Moral Powers: A Study of Human Nature, First Edition. P. M. S. Hacker.
© 2021 John Wiley & Sons Ltd. Published 2021 by John Wiley & Sons Ltd.

it is the common form of all deadly vices – vices that are soul destroying (see Chapter 5.2). If a person's dominant character traits are deadly vices, then that person is a wicked or evil human being. It is evil deeds motivated by evil intentions that make a person evil. Mere wicked thoughts that never reach action may be ugly or even hideous, but they are not evil making. Mr Pugh in *Under Milkwood*, while taking morning tea up to Mrs Pugh, whispers on the stairs:

> Here's your arsenic, dear.
> And your weedkiller biscuit.
> I've throttled your parakeet.
> I've spat in the vases.
> I've put cheese in the mouseholes.
> Here's your ... [door creeks open]
>   ... Nice tea, dear.

But he is not an evil man, merely an unhappy one who entertains malicious fantasies.

Explanations of evil-doing
Evil-doers, like anyone else, have character traits that may form recognizable patterns with explanatory weight. They have beliefs, often religiously or ideologically driven, or imbued with racial bigotry or xenophobia. These guide their actions. They have hopes: to attain wealth, power, and recognition. These drive their actions. They may harbour resentments for actual or imaginary insults, humiliations, and failures. These provide powerful motives for their evil deeds. Evil-doers produce reasons for their evil-doing and offer justifications for their evil deeds. These reasons may be very varied – ranging from specifications of means to personal selfish ends, through the reasons advanced by religious fanatics and bigots, to the reasons underlying ideological fervour. They pursue goals, ranging from personal ambitions, to the greater glory of God, the triumph of the proletariat, the extermination of native peoples, or the purification of the *Volk*. Their evil-doing may be fuelled by powerful emotions of hatred, pride, and rage. Each of these elements may vary in different ways, may combine with many others in complex patterns, and may depend upon background conditions, circumstances, and opportunity for their actualization in evil-doing. It is by reference to such familiar forms of explanation that instances of evil and particular evil-doers can be understood.

These forms of explanations may, of course, be reinforced by social-scientific explanations. Social psychology investigates correlations between proneness to gross immorality and parental over-indulgence or indifference, neglect, and brutality, a domineering or submissive father or mother, childhood molestation or rape, and single parenthood or step-parenthood. Similar correlations are made with success or failure at school and university, sexual proclivities and frustrations, and various forms of alienation and abuse. Socio-economic factors provide further complexity: gang warfare is a phenomenon of run-down neighbourhoods and poverty; terrorism is, and violent anarchism was, largely a lower middle-class phenomenon. So too do political conditions: periods of political disintegration, as in Russia prior to the First World War, Weimar Germany between the world wars, and central America over the last few decades, involve the state losing its effective monopoly in the use of force and facilitate the formation of violent extra-legal organizations. These may range from anarchist cells to party organizations, drug cartels, and independent militias such as the Freikorps or the Brownshirts (SA) (see fig. 4.1).

2. Reasons and motives for doing evil

Understanding evil   I have suggested that there can be no science of evil. There can be no science of good either. Nor can there be a science of morality, despite the fact that experimental psychologists, psychiatrists, and now cognitive neuroscientists too study the moral character of human beings and their moral and immoral behaviour. Psychological experiments may indeed establish important correlations and statistical probabilities that may be crucial for the formation of intelligent social policy. But one cannot understand *a good or evil human being* by subjecting a person to psychological tests or brain scans alone, for to understand a human being is to understand a life lived in the hurly-burly of a society in the white waters of history. And to understand *a wicked or evil deed* is to understand it as performed in the stream of life, with an antecedent history and a specific configuration of circumstances, in a particular socio-historical context, performed by a person with a 'biography' and 'autobiography', and inherited as well as acquired characteristics. Understanding good and evil deeds and evil-doers is largely ideographic rather than nomothetic, and involves practical reasoning rather than causal determination – largely, but not exclusively. Psychology and neuroscience contribute to knowledge of tendencies, liabilities, and susceptibilities within the framework of which human

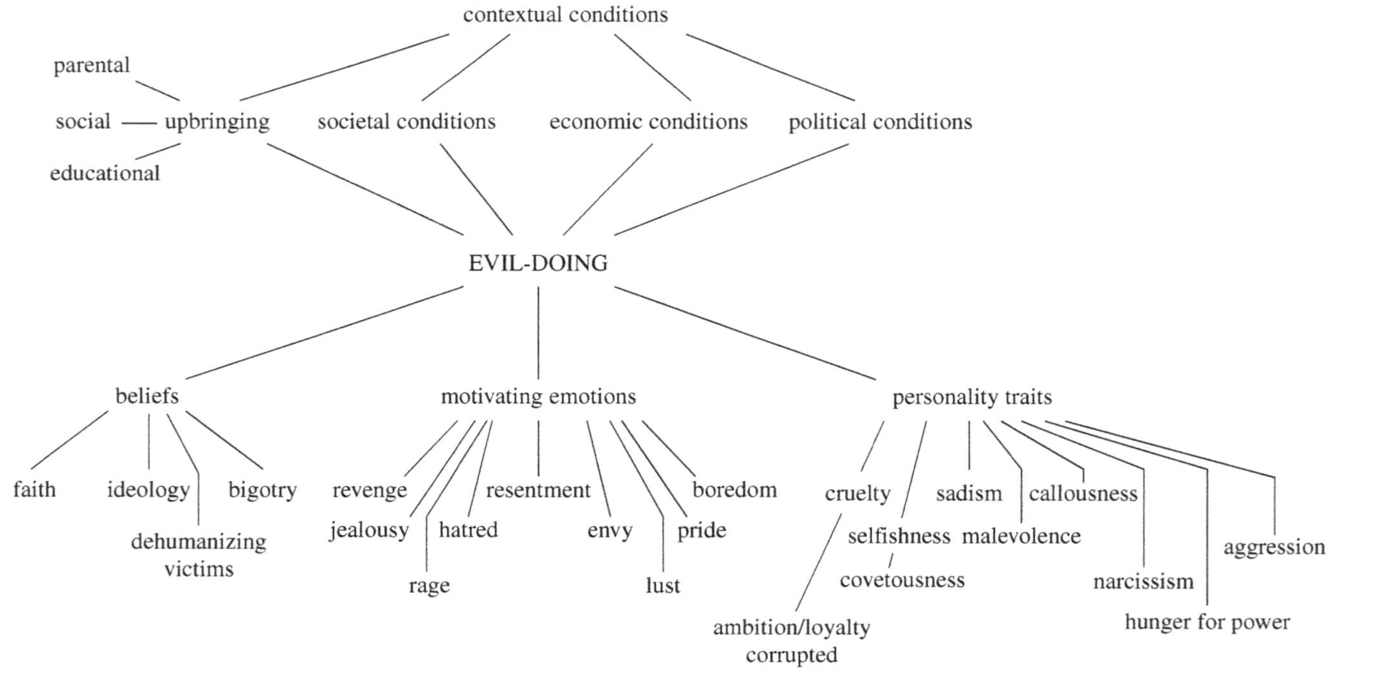

Figure 4.1 *Explanatory factors in evil-doing*

beings make the choices they make, adopt the goals they choose, form their intentions, and act for reasons. They may study cortical lesions that explain cognitive and volitional deficiencies. All these forms of explanation are complementary – all contribute to the understanding of evil deeds and evil-doers.

Literature and ideographic understanding   The greatest students of the place of evil in human life and of the subjective psychology of evil in a complex social setting are great novelists, dramatists, and poets. Consequently, here, as in *The Passions: a Study of Human Nature*, I shall draw copiously on illustrative examples from literature. The fact that they are fictional is irrelevant, since they depict human possibilities. Great works of literature are able to present characters in a fine-grained manner that is rarely, if ever, possible in life or in biography, for we know too little about each other. We are hardly ever transparent to each other, but at best only translucent. Imaginative writers labour under no such constraints, and their greatness is proportional to the depth of their insights into human possibilities and human nature (in addition to their mastery of their craft).

Motives for evil-doing   There are indefinitely many reasons and motives for wickedness and evil-doing, and usually more than one for any given wrong. These may occur conjunctively or in blends. Pure malevolence – ill will and the desire to cause suffering – is in itself a motive. Claggart, the master-at-arms in Herman Melville's novella *Billy Budd*, was moved simply by hatred of the benevolence, innocence, popularity, and good looks of Billy, and his own delight in inflicting suffering on the innocent. He found joy in corrupting others to spy upon their shipmates for him. Claggart could not look upon any human kindness, generosity, or good will with equanimity.[1]

> With no power to annul the elemental evil in him, though readily enough he could hide it; apprehending the good but powerless to be it; a nature like Claggart's, surcharged with energy as such natures almost

---

[1] Melville is inclined to characterize Claggart's evil as madness, rooted in envy, which takes the form of unlimited hatred of goodness as such. Not so in E. M. Forster's libretto for Benjamin Britten's opera, in which there is nothing mad about Claggart. Peter Ustinov's film *Billy Budd* similarly presents Claggart as evil through and through, taking sadistic pleasure in a flogging and determined to destroy Billy simply because he cannot abide the sight of goodness and delights in its destruction or corruption. Claggart was brilliantly portrayed by Robert Ryan.

invariably are, what recourse is left to it but to recoil upon itself and like the scorpion for which the Creator alone is responsible, act out to the end the part allotted it.[2]

This is a case of what we may called 'absolute evil' or 'pure evil'. For there is no *further* goal to the evil that Claggart did – he merely desired the cruel destruction of goodness. Like pure vandalism, pure evil may be enjoyable. Doing evil doubtless gave Claggart the pleasure of power and reinforced his sadism. But there was no ulterior motive.

Pure malevolence may be mixed with other motives such as selfishness, greed, lust, envy, jealousy, and so forth. So, for example, the faithful family steward Mackellar raises the question of the motives of James Durie, the charming but malevolent heir to the Master of Ballantrae:

> Was the man moved by a particular sentiment against Mr Henry [his brother]? or by what he thought to be his interest? or by a mere delight in cruelty such as cats display and theologians tell us of the devil? Or by what he would have called love? My common opinion halts among the three first; but perhaps there lay at the spring of his behaviour an element of all. As thus: Animosity to Mr Henry would explain his hateful usage of him when they were alone; the interests he came to serve would explain his very different attitude before my lord [their father]; that and some spice of a design of gallantry, his care to stand well with Mrs Henry; and the pleasure of malice for itself.[3]

In Graham Greene's *Brighton Rock*, Pinky's natural malevolence is coupled with a desire to dominate his gang and to protect his position by murder.

Ill will towards others is rarely a natural disposition but usually the *consequence* of various motives. Cassius, both in history and in Shakespeare's play, was moved to murder Caesar out of envy and resentment:

> I was born as free as Caesar; so were you,
> We both have fed as well, and we can both
> Endure the winter's cold as well as he:
> ... And this man
> Is now become a god, and Cassius is

---

[2] Herman Melville, *Billy Budd*, section 12.
[3] R. L. Stevenson, *The Master of Ballantrae* (1889), ch. 4.

> A wretched creature, and must bend his body,
> If Caesar carelessly but nod on him
> …
> Why, man, he doth bestride the narrow world
> Like a colossus, and we petty men
> Walk under his huge legs, and peep about
> To find ourselves dishonourable graves.
> (Julius Caesar, I. ii)

Iago was motivated by resentment and hatred at having been passed over for promotion in favour of Michael Cassio, a less experienced, upper-class officer:

> Despise me if I do not [hate Othello]: three great ones of the city,
> In personal suit to make me his lieutenant,
> Oft capp'd to him: and by the faith of man,
> I know my price, I am worth no worse a place.
> But he, as loving his own pride and purposes,
> Evades them, with bombast circumstance
> Horribly stuff'd with epithets of war;
> And, in conclusion,
> Nonsuits my mediators; for, 'Certes', says he,
> 'I have already chosen my officer.'
> (Othello, I. i)

Later he adds jealousy to his motives:

> I hate the Moor
> And it is thought abroad that 'twixt my sheets
> He's done my office; I know not if't be true …
> Yet I, for mere suspicion in that kind,
> Will do as if for surety.
> (Othello, I. iii)

Unrestrained hedonism coupled with a corrupt ethos is likewise a spring of gross immorality. In *Les Liaisons dangereuses* (1782), Vicomte de Valmont, a libertine, seduces Madame de Tourvel (a faithfully married woman) and Cecile de Volanges (a teenager just out of a convent) for amusement and to win a bet with the equally depraved Marquise de Merteuil. He and the marquise have created a private world of their own based on a degenerate scale of values and cynical view of humanity that is threatened by the virtue of Madame de Tourvel. Both women's lives are destroyed for the sake of

Valmont's pleasure in the seduction and manipulation of others, and the enjoyment of swapping tales of depravity with the marquise. Dorian Gray, in Oscar Wilde's eponymous novel, luxuriating in his good looks and charm, and corrupted by Lord Henry's hedonistic cynicism, moves through life indifferent to others and their welfare in a single-minded search for ever more decadent forms of sensual pleasure, while Basil Hallward's painting of him, which has become a portrait of his soul, slowly grows more and more hideous as his wickednesses multiply.

Boredom too, in the post-Byronic era, appears to have been an appealing motive for wickedness among the upper classes, if the novels of the day are anything to judge by. Pechorin in Lermontov's *A Hero of our Time* (1840–1) and Onegin in Pushkin's *Yevgeny Onegin* (1833–7) are both young aristocrats who ruin or destroy the lives of others out of no more than Byronic ennui, the one seducing a girl for casual fun, the other killing a good friend, Lensky, in a duel that he himself has frivolously provoked by flirting with his fiancée. In a later time and different (bourgeois) vein, Ibsen's Hedda Gabler (1891) embarks on her destruction of Eilert Lovberg and his masterwork partly out of intolerable boredom, partly out of sexual frustration, and partly out of rage at herself for her inability to break the chains of convention that bound her into a loveless marriage by contrast with Thea, with whom she had been to school together more than a decade previously. Thea has had the courage to leave her repulsive husband, and has spent two years assisting Eilert Lovberg write a masterpiece. Hedda both envies Thea and is jealous of her. Having dispatched Lovberg to commit suicide in despair over the loss of the manuscript of his book, she burns the precious manuscript, which unbeknown to him had come into her possession, realizing what impact this would make on Thea:

> [Throws one of the quires into the fire and whispers to herself.]: Now I am burning your child, Thea! Burning it, curly locks! [Throwing one or two more quires into the stove.] Your child and Eilert Lovberg's. [Throws the rest in.] I am burning – I am burning your child.

In yet another social context, this time a lower-class one in an imaginary future, boredom may be a motive for horrendous violence by uncouth, uneducated, and savage teenagers, as is made vivid by the rampage of violence by Alex and his gang in Anthony Burgess's ghastly dystopian novel *A Clockwork Orange*.

Resentment at one's natural misfortunes may lead one to seek revenge against the world, as is patent in the case of Shakespeare's Richard III ('But I, that am not shap'd for sportive tricks ... I that am rudely stamp'd, and want love's majesty ... I, that am curtail'd of this fair proportion, Cheated of feature by dissembling nature'). Edmund, the illegitimate son of Gloucester in *King Lear*, is moved to do evil by bitter resentment at his bastardy coupled with boundless ambition ('Why bastard? wherefore base?'). Hatred, pure unadulterated hatred, can be a powerful motive for evil-doing, as is evident in Hugh Walpole's *Judith Paris*, in which Uhland Herries murders his cousin John for no other reason than personal hatred rooted in festering childhood dislike and (like King Richard) rage at the world for his own misshapen body.

Character traits and evil-doing Character traits associated with evil-doing are similarly manifold and likewise occur in blends. Some afford motives for action, given appropriate circumstances; others merely provide a fertile seedbed for wickedness and evil, given the right conditions. (A high proportion of accusers in the East Anglian witch-hunts in 1646–8 owned property adjacent to that of the old women they accused. Languedoc was the wealthiest region of France, which fact played no small part in motivating the northern barons to join the holy crusade against the Cathars.) Callousness and either indifference or sadism are characteristic of evil-doers. The former pair are non-motivating negative vices, inasmuch as the suffering of others is held to be *no reason* for helping them, for protesting against their maltreatment, or for intervening to stop it, and *no reason* for feeling sympathy or compassion for them. Cruelty likewise may be a negative vice, inasmuch as the prospect of others' suffering is no reason for *not* inflicting suffering on them. It may, however, be a positive vice inasmuch as the prospect of others' suffering *is* a reason for inflicting suffering upon them, and the infliction of such suffering is enjoyable for the sadist.[4]

The road to Auschwitz, Ian Kershaw remarked, was built with hatred and paved with indifference. Cruelty and sadism are common characteristics of those who readily do evil (brilliantly depicted in Isaac Babel's *Red Cavalry*, which describes the behaviour of the Cossack irregular cavalry during the Russian civil war in the Ukraine).

---

[4] I am indebted to Anselm Mueller for the distinctions between positive and negative vices and between motivating and non-motivating vices.

It is no coincidence that pride (arrogance), lust (concupiscence), and anger (irascibility) are among the seven deadly sins. It was pride that led Browning's Duke (Alfonso d'Este) to murder his duchess (Lucrecia di Cosimo de' Medici).[5] It was King David's lust for Bathsheba that led him to bring about the death of her husband Uriah the Hittite, and anger that led to Alexander of Macedon's murdering his friend and adviser Black Cleitos.

Character traits that may be admirable may also, if corrupted, incline a person to wickedness and evil. Ambition, which, properly controlled, can be a striving for excellence and achievement, can also be a powerful motive for immorality, wickedness, and evil, given callousness, selfishness, and ruthlessness.[6] This was evidently the case among many of the leading Nazi murderers, such as Stangl, Heidrich, Eichmann, and Rudolf Höss (commandant of Auschwitz), who succumbed to evil partly out of ambition. (It is noteworthy that one cannot *succumb* to good. One succumbs to evil inasmuch as one *allows oneself* to do evil, just as one succumbs to temptation when one allows oneself to give in to it.) Likewise, loyalty corrupted can lead men to evil. It may be corrupt in one of two ways: it may be attached to wicked causes or evil people, and it may become corrupted through failure to realize its proper limits. Both were patent among the Junker officer class in the Wehrmacht in Nazi Germany, who swore an oath of allegiance to the Führer, and held the bond of loyalty to be indefeasible. Even more disturbingly, love corrupted

---

[5] Robert Browning, 'My Last Duchess'.

[6] See Richard of York's soliloquy:

> Why, I can smile, and murder whiles I smile,
> And cry 'Content!' to that which grieves my heart,
> And wet my cheeks with artificial tears,
> And frame my face to all occasions.
> I'll drown more sailors than the mermaid shall,
> I'll slay more gazers than the basilisk,
> I'll play the orator as well as Nestor
> Deceive more slyly than Ulysses could,
> And, like a Sinon, take another Troy.
> I can add colours to the chameleon,
> Change shapes with Proteus, for advantages,
> And set the murderous Machiavel to school.
> Can I do this, and cannot get a crown?
> Tut, were it further off, I'll pluck it down.
>   (*Henry VI, Part III*, III. ii)

can also lead to evil, often in response to fear of losing the love, to jealousy and bitter resentment, and to uncontrolled possessiveness (*TP*, ch. 10).

Beliefs and evil-doing Blends of emotions and concatenations of character traits may be further reinforced by systems of apparently warranting beliefs. One such set of beliefs, common in some instances of genocide and of crimes against humanity, as well as in the enslavement of native peoples and the institution of slave trading and slavery is that the victims of one's inhumanity are not truly human, being infidels, *kafirs* as the Arab slavers called them (including two million Europeans taken to the Barbary Coast for sale), legitimate booty of war, as the African tribes who sold their prisoners as slaves to Europeans thought, or sub-human savages, as Europeans viewed the African slaves they shipped to the Americas. There is an important difference between considering one's victims to be (i) inferior, (ii) sub-human, or (iii) non-human. Warfare encourages the first, and some forms of slavery (where slaves are deprived of all rights) the second. The third is common among *génocidaires*, but is also to be found in those who commit crimes against humanity (e.g. Lenin's and Stalin's eradication of the kulaks) who describe their victims as insects, cockroaches, parasites, vermin, or animals or as diseases in the body politic – a cancer that has to be cut out of the organism of the nation (a favoured anti-Semitic trope).[7] In Franz Werfel's great novel *Forty Days of Musa Dagh* Dr Johannes Lepsius, an actual Protestant pastor who strove valiantly, if futilely, to halt the genocide, confronts Enver Pasha and pleads for peace between the Turks and the Armenians (the words are derived from Dr Lepsius's report):

Here for the first time Enver Pasha laid bare his deepest truth. His smile no longer had any reserve in it, a cold stare had come into his eyes, his lips retreated from a strong and dangerous set of teeth.

'There can be no peace,' he said, 'between human beings and plague germs.'

---

[7] For example: 'Do not think that you can fight a disease without killing the causative agent, without destroying the bacillus, and do not think that you can fight racial tuberculosis without seeing to it that the nation is freed from the causative agent of racial tuberculosis. The influence of Judaism will never fade as long as its agent, the Jew, has not been removed from our midst' (from Hitler's table talk around 1920, quoted by Heinsohn, 'What Makes the Holocaust a Uniquely Unique Genocide').

A German platoon leader in Russia, after his platoon had carried out a particularly horrific murder of 150 Jews – men, women, and children – by locking them into a barn and then burning it and shooting those who jumped in flames from the windows, told his remorse-ridden troops:

> 'You and your sensitive feelings! Men, you cannot go on like this. This is war! One must be hard! They are not our people. The Jew is not a human being! The Jews are the cause of all our misfortunes! And when you shoot one of them it is not the same thing as shooting one of us – it doesn't matter whether it is a man, woman, or child, they are different from us. Without question one must get rid of them. If we had been soft we should still have been other people's slaves.'[8]

Hence too the Nazi attempt to dehumanize Jews in concentration camps by shaving their heads, forcing them to don demeaning clothes, requiring them to wear yellow stars of David, tattooing numbers on their arms, starving and beating them, and humiliating and torturing them. Humiliation of their victims was thought to make it easier for Nazi troops to slaughter them in a detached manner without feeling defiled.[9] It is noteworthy that there is an existential contradiction in humiliating one's victims, while at the same time insisting that they are not human but germs, cancers, insects, or animals. For one cannot humiliate pathogens, carcinomas, or cockroaches, and, although one may maltreat animals, torture and torment them, one cannot humiliate them. Only humans can be humiliated.[10]

The need to hate   There is in many human beings a deep-rooted need to hate, which commonly counter-balances and suppresses anxieties, feelings of inadequacy, and resentment. It is

---

[8] Simon Wiesenthal, *The Sunflower*, 2nd edn (Schocken Books, New York, 1997), 49.

[9] Franz Stangl, commandant of the death camps Trobibor and Treblinka, confessed: 'To tell the truth, … one did become used to it … They were cargo. … I think it started the day I first saw the *Totenlager* in Treblinka. I remember Wirth standing there, next to pits full of blue-black corpses. It had nothing to do with humanity – it could not have. It was a mass – a mass of rotting flesh. Wirth said "What shall we do with this garbage?" I think unconsciously that started me thinking of them as cargo. … I rarely saw them as individuals. It was always as a huge mass. I sometimes stood on the wall and saw them in the tube – they were naked, packed together, running, being driven with whips' (Sereny, *Into that Darkness*, 200–1).

[10] See Avishai Margalit and Gabriel Motzkin, 'The Uniqueness of the Holocaust', *Philosophy and Public Affairs* 25 (1996), 65–83.

complementary to an equally profound need to blame others for the ills that beset society. Hatred strengthens a weak sense of identity, patent in the violent behaviour of spectators of sports (the Blues and the Greens in Byzantium, football hooligans today). Blaming a subordinate group in one's society for its ills enables the ruling order or the majority to evade responsibility for the failings of society by finding a scapegoat (e.g. the Jews in Christian Europe; the Catholics in post-Elizabethan England, the Armenians in the late Ottoman Empire). When cultivated by unscrupulous leaders, especially in times of political disorder and economic stagnation or decline, these twin roots (the need to hate and the need for a scapegoat) can grow into the most poisonous of plants. Vasily Grossman's Anna Sergeyevna describes the phenomena during the period of the Soviet Union's extermination of the kulaks (known in Russian newspeak as 'dekulakization'):

> I was only a girl – and during meetings and special briefings, from films, books, articles and radio broadcasts, from Stalin himself, I was always hearing the same thing: that kulaks are parasites, that kulaks burn bread and murder children. The fury of the masses had to be ignited against them – yes, those were the words; it was proclaimed that the kulaks as a class must be destroyed, every accursed one of them … I too began to fall under this spell. It seemed that every misfortune was because of the kulaks; if we were to annihilate them immediately, then happy days would dawn for us all. No mercy was to be shown to the kulaks. They were not even human beings; goodness knows what they were – some kind of beasts, I suppose.[11]

The upshot of this was the slaughter of some five million kulaks (out of eleven million peasant deaths between 1929 and 1932).

Religious belief and evil-doing
Another class of beliefs conducive to great evil are religious. We have already noted the evils of crusades, both against Islam and against deviant branches of Christianity such as the Cathars. Major splits within Islam and Christianity have generated rivers of blood and savagery, for example between Sunni and Shia to this day, in the French wars of religion which culminated in the ferocity of the St Bartholomew's Day massacre of Huguenots in August 1572, and in the bloodbath of the Thirty Years War (1618–48) between Catholics and Protestants in

---

[11] Vasily Grossman, *Everything Flows*, published posthumously in the USSR in 1989 and in an English translation in 2011 (Vintage Classic, London, 2011), 127.

which a third of the population of German lands died. Spanish Catholicism contributed its share of atrocities not only in Latin America, but also in the tortures and auto-da-fés carried out by the Inquisition in Spain (1478–1834). The Islamic State (or self-declared Caliphate) has been a recent manifestation of this form of evil.

Ideology and evil-doing
From the late eighteenth century onwards ideological beliefs played a major role in stimulating evil, not for the greater glory of God but for the future glory of man. Some of the ideologies that sprang up in the nineteenth century were benevolent in intent and peaceful in method (e.g. various early forms of socialism and anarchism). Others were not (e.g. Jacobinism, communism, fascism). The latter characteristically justified evil means to achieve their desired social and political ends to the point at which whatever beneficial goals (if any) they may originally have embraced became completely corrupted. They were typically murderously intolerant and crushed freedom of thought and speech. Over the past two and a half centuries opposition to the implementation of ideology and failures of policies pursued by ideological dictators has characteristically been met with fury and fanatical rhetoric in support of the persecution, expropriation, and murder of scapegoats.

The first major social movement that bent an ideology to wholly evil policies were the Jacobins during the French Revolution. They initiated the reign of terror in the name of the Rights of Man. This peaked in 1793–4 at the hands of Robespierre and Saint-Just, who lauded reason and brought forth monsters while it slept (brilliantly presented in Georg Büchner's play *Danton's Death*). Robespierre proclaimed, 'To punish the oppressors of humanity is clemency; to forgive them is cruelty', to which Saint-Just added, 'The republic consists in the extermination of everything that opposes it.' In a speech to the Convention on 5 February 1794, Robespierre declaimed:

> If the attribute of popular government in peace is virtue, the attribute of popular government in revolution is at one and the same time *virtue and terror*, virtue without which terror is fatal, terror without which virtue is impotent. Terror is nothing but justice: prompt, severe, inflexible; it is then an emanation of virtue.

Such double-speak was to become the hallmark of twentieth-century ideological dictatorships, in which the ideological ends justify the means, no matter how evil; morality is held to be bourgeois deviationism; and the means by which the end is to be achieved are held to be 'revolutionary justice'. The dialectic of communist revolutionary justice in the Soviet Union is brilliantly depicted by Arthur Koestler in the dialogues between Rubashov, an old Bolshevik who has strayed from the party line and his interrogator, Ivanov.

> 'My point is this', he [Ivanov] said, 'one may not regard the world as a sort of metaphysical brothel for emotions. That is the first commandment for us. Sympathy, conscience, disgust, despair, repentance, and atonement are for us repellent debauchery. ... That is an easy solution. The greatest temptation for the likes of us is: to renounce violence, to repent, to make peace with oneself. Most great revolutionaries fell before this temptation, from Spartacus to Danton and Dostoievsky; they are the classical form of betrayal of the cause. The temptations of God were always more dangerous for mankind than those of Satan. As long as chaos dominates the world, God is an anachronism; and every compromise with one's own conscience is perfidy. When the accursed inner voice speaks to you, hold your hands over your ears.'[12]

A further set of characteristic beliefs of evil-doers in evil regimes is associated with the adulation of a leader (e.g. Mussolini, Stalin, Hitler, Franco, Mao, Pol Pot, Kim Jong-il, Mugabe) to whom superhuman powers are often assigned – of virtue, intelligence, and dedication to the cause and to the people. Belief in such things unifies the followers, empowers them to carry out monstrous deeds with impunity, releases them from the burden of conscience, relieves them of all responsibility, and encourages the further belief that what they are doing is in the service of the leader, the party, the state and the people and is therefore meritorious. It also elevates loyalty and fidelity to one's oath to the ultimate overriding value.

3. Can evil be a motive?

So, there can be many motives for doing evil. But can evil itself be a motive? Plato argued that no man can do evil knowingly. Aristotle held that, whatever we aim at, we necessarily conceive it to be a good.

---

[12] Arthur Koestler, *Darkness at Noon* (Jonathan Cape, London, 1940), §7, 'The second hearing'.

Some, like Kant, have held that to do evil is contrary to reason. These views are partly right and partly wrong.

Those who do evil know perfectly well what they are doing in one sense: they are killing, severely harming, or inflicting terrible suffering on men, women, and children (or on animals or an animal) who have done no wrong or a wrong that cannot warrant the ensuing suffering. In the case of absolute evil, malevolence pure and simple provides a motive out of which an evil-doer may act. Claggart, for example, does evil for its own sake, and derives from it satisfaction or even sadistic pleasure. A child may take pleasure in pulling the wings off a fly; young teenagers may pour petrol on a cat and ignite it, finding its agonized shrieks, leaps, and writhings amusing.

Justifications and excuses
More commonly, those who do evil entertain, or are given, a justification or excuse that purports to warrant such murder or slaughter, such torture and torment, such hideous humiliation and degradation – so they think that they aren't *really* doing evil at all, or that the evil they are doing is excusable. Some may really believe the putative justification: that the 'enemy' (the kaffirs, the Armenians, the Jews) really are sub-human or non-human. Such believers are men of evil and corrupt conscience, for they are responsible for harbouring such wicked unsupportable beliefs. Others view their rationale as an excuse, that the ultimate ends justify the means: they are *really* only getting legitimate booty; they may be killing human beings,[13] but they are actually only obeying orders like everyone else and obedience to orders overrides everything else; they are really just teaching the other Mafia 'family' a lesson; they are merely disciplining their child; they are merely engaged in competition for the sake of profit, and what suffering is caused is irrelevant; and so forth. How deep these beliefs really are is, of course, debatable. For the most part, they are con-fabulations, and those who cite them are guilty of self-deception in addition to their crimes.

Human beings do not generally choose to do evil or wickedness *out of ignorance*. But it is correct that *by and large* they would not *characterize* what they do as evil or wicked. Some may embrace evil knowingly and reluctantly, as does an akratic, who cannot muster the

---

[13] One member of a Nazi *Einsatzgruppe* on the Eastern Front confessed that he found murdering adult Jews in the thousands terrible, and refused to do so. Instead, he limited himself to shooting little children since if they were let free they would die of starvation in the forest or be eaten by wolves, so it was merciful to kill them.

strength of will to resist his evil impulses. Yet others may embrace evil both knowingly and willingly, as do Shakespeare's Richard III, Iago, or Macbeth, taking their evil deeds to be warranted by their ambition, by resentment, or by despair. Others may view ideas of good and evil as obsolete, part of the ethos of the feeble-minded and lower orders (e.g. Plato's Thrasymachus or Callicles, Nietzsche's Zarathustra). And others may view morality as a luxury they cannot afford (e.g. Alfred Doolittle, who tried selling his daughter to Professor Higgins in Shaw's *Pygmalion*: asked 'Have you no morals, man?' he replied, 'Can't afford 'em, Governor. Neither could you if you was as poor as me' (act II)).

It was a salient feature of Aristotle's account of practical reason that every action and pursuit is thought to aim at some good, upon which Aquinas embroidered: what is willed is always willed *sub ratione boni*,[14] which is sometimes rendered 'under the guise of the good'. In so far as 'the good' signifies something that is morally meritorious or thought to be so, then this is surely not *generally* true (and not what Aristotle had in mind). It *can* be said to be true in the case of ideologically committed *génocidaires*, who present their victims as the source of all evil in society. So killing them is alleged to be right and good. However, the Vicomte de Valmont did not seduce Madame de Tourvel believing that it was morally good to do so – he did it for fun, for the excitement and pleasure, and to show that stimulated desire in his female victims can override any moral restraint they have. Slavers did not enslave their victims in the belief that it was moral, but that it was profitable, coupled with the belief that the slaves were sub-human. But it is true that whatever one does with intent, and not merely because one 'felt like it', must be conceived to have some feature or features by reference to which one can justify or purport to justify what one does. For there must be some purported reason for forming the intention to do it in advance of acting. It must be appealing in some respect or other, otherwise what is the point of doing it? The appeal may lie in the act itself, or it may lie in the further goal for which the act is a means.

The forms of appeal are clear enough. Doing wicked things may be hugely profitable – as it was for Joe Keller (in Arthur Miller's *All My Sons*) who sold defective aircraft engines to the US Air Force during the Second World War. Similarly, the illegal surreptitious testing of a

---

[14] Aquinas, *Summa Theologiae* 1ª2ᵃᵉ, 8, 1.

pharmaceutical drug in Africa and the subsequent concealment of its serious side effects for the sake of the company's financial interests is depicted in John le Carré's *The Constant Gardener*. Doing evil, especially in groups, may be exciting, engendering a sense of belonging and of comradeship, and alleviating boredom, restlessness, and alienation (as described in Burgess's *The Clockwork Orange*). The joint planning may be thrilling, and achieving the goal may afford satisfaction in this life or the promise of it in the next (the Crusaders or jihadists). Doing evil involves pleasurable self-assertion; inflicting suffering may be enjoyable – it may give one a sense of power over weaker and defenceless victims and consequently a satisfying sense of superiority. In depraved cases it may even involve sexual excitement. It may give meaning and direction to otherwise meaningless lives, inasmuch as it is advancing the cause of Christ or Allah,[15] or of the party and the perfect society of the future.

Evil and rationality   Is evil irrational? If being irrational means an absence of reasons – arbitrariness – then clearly evil is not irrational save in some cases of psychotic or psychopathic savagery and mob rampage. If a deed is irrational when it is done spontaneously without reflection or premeditation, then many evil deeds are not irrational, since crimes are usually carefully planned. (In fact, spontaneity does not imply irrationality but only absence of deliberation.) If one form of rationality is calculating and adopting efficient means to a given end, then evil-doing is often rational. The Mafia is notorious for the meticulous planning of its murders. The Holocaust was done with exemplary efficiency by the Nazis. The Mafiosi do evil to protect their financial interests and to maintain their power over the districts they control and terrorize. These are their reasons. In cases of genocidal evil, the evil-doers accept what is patently false as a justifying reason, for example anti-Semitic blood libels (canards originating in medieval times in Lincoln in 1155) or the Protocols of the Elders of Zion (a Russian police forgery in 1903). The religious or ideological believer unreflectively adopts a religious

---

[15] It is striking that many of the youthful suicide bombers in militant Islam today are also motivated by the promise of going to paradise immediately on death, where they will be met by seventy-two virgins to satisfy their sexual fantasies, and the assurance that the *shahid* in addition earns intercessionary rights for seventy members of his family. This is well presented in the documentary film *Path of Blood* and in the eponymous book by Thomas Small and Jonathan Hacker (Simon & Schuster, New York, 2014).

or ideological mantra with implied consequences: *Deus vult* – they are heretics and enemies of the one true God (the Crusades); *Allahu Akbar* – kill the infidels (jihadis); 'They are enemies of the republic' – and deserve to be guillotined (the French Reign of Terror); 'They are Armenians' – and must be exterminated (genocidal Ittihadists); 'They stand in the way of history' – and must be swept aside by the onward march of the proletariat (communist ideologists in the Soviet Union). In the individual case of evil-doing, the belief advanced, while it may be true, does not license the deed: the pawn-broker Alyona Ivanovna in *Crime and Punishment* was indeed a repulsive old hag, but that did not give Raskolnikov a warrant to murder her and steal her wealth, even for the good and selfless cause of giving financial assistance to his impoverished mother and sister. An evil-doer may simply lack any moral sense or sensibility, viewing other human beings as no more than means for achieving their ends (e.g. Jonathan Wild, the Great, in Fielding's eponymous novel). Evil people may murder without compunction to attain their ends, as is illustrated by Shakespeare's Richard III and by Cathy Ames in *East of Eden*. Doing evil may, therefore, in a narrow sense, be rational, that is it may involve reflection and planning, calculating effective means to an evil end. At the same time, and in another sense, doing evil can-not be rational, since there can be no *legitimate* justification or excuse for *voluntarily* doing evil (as opposed to doing evil under duress), and those justifications or excuses that are offered cannot have good grounds with which to back them up. Moreover, doing evil can never be *reasonable* (*HNCF*, ch. 7). For being reasonable involves taking the legitimate concerns of others into account and is linked to the appreciation of values and their multiplicity. Above all, to be evil is to be in bondage to all the baser and destructive instincts and motives in mankind.

Can evil be a motive? It is true that 'because it is evil' is not an intelligible answer to the question 'For what reason did you do that?'.[16] To characterize a deed as evil is to give it the most extreme form of condemnation, and so the best possible reason *not* to do it. So it cannot also be advanced as a reason *for* doing it. Unsurprisingly then, it cannot, by itself, be cited as a motive either. So it is not surprising that we are reluctant to say that someone did

---

[16] What of Satanists? Presumably this is an answer they may give for the evil they do. But they do not do evil for its own sake – they do it to please or appease the satanic powers they worship.

something 'out of evil' or 'for the sake of evil' (as opposed to 'Out of pure malice' or 'For the pleasure of watching them suffer'[17]). That is why Milton's Lucifer is not, I think, altogether intelligible:

> So farewell hope, and with hope farewell fear
> Farewell remorse: all good to me is lost;
> Evil be thou my good; by thee at least
> Divided empire with heaven's king I hold
> By thee, and more than half perhaps will reign.'
> (*Paradise Lost*, IV. 108–12)

It is striking that Milton subsequently elaborates and eliminates the conceptual stress:

> To do aught good never will be our task,
> But ever to do ill our sole delight,
> As being contrary to his high will
> Whom we resist. If then his Providence
> Out of our evil seek to bring forth good,
> Our labour must be to pervert that end,
> And out of good still to find means of evil.
> (*Paradise Lost*, IV. 159–65)

Accordingly, evil is a means for Lucifer to continue his struggle against God and to thwart his will. But there are also cases, both in fiction and in fact, in which, influenced perhaps by Nietzsche, someone comes to think of himself as so superior as to have passed beyond the vulgar constraints of good and evil. Thus Raskolnikov thought of himself as such a superior being, akin, he thought, to Napoleon, unconstrained by the morality of the herd. Similarly, Leopold and Loeb thought of themselves as *Übermenschen* (see p. 148).

Nevertheless, 'Because he is evil', like 'Because he is good', *is* an acceptable answer to the question 'Why did he (does he) do that?' – although, as already remarked, it is not an explanation in

---

[17] Simon Baron-Cohen tells a chilling tale: A Nazi officer condemned an inmate of Auschwitz to hang, and ordered the victim's best friend to kick away the chair in order to enjoy the friend's anguish. The condemned man kissed his friend's hand, and kicked away the chair himself. This is an awesome example of maintaining dignity and defiance even in hell on earth. Here too is a patent case in which the perpetrator of evil not only wished to hang an innocent man in a most painful way, but also to destroy the soul of his friend. Instead, the victim in effect committed heroic suicide.

terms of the agent's reasons, but rather a dispositional explanation: that's the sort of person he is. People of that bent tend to do such terrible things.

4. Knowledge of good and evil

Adam and Eve ate of the Tree of the Knowledge of Good and Evil, and their eyes were opened. The ancient legend of Genesis characterizes humanity in terms of the knowledge of good and evil, which they attained contrary to the will of God. No other creature on earth possesses such knowledge. Yet it is conceptually peculiar in comparison to other forms of knowledge. Aristotle pointed out that one does not forget the nature of excellences or virtues (*Nicomachean Ethics*, 1100[b]17) and again that practical wisdom, unlike the skills of crafts, cannot be forgotten (*Nicomachean Ethics*, 1140[b]29). Gilbert Ryle picked up this sapient hint and raised the question of why one cannot forget the difference between right and wrong. It would be nonsense to answer the question of whether one does not know the difference between right and wrong with the words 'Well, I did learn it once, but I have forgotten it'.[18] And what goes for right and wrong goes equally for good and evil.

Forgetting the difference between right and wrong

The impossibility of forgetting the difference between right and wrong is not because we are prenatally imprinted with such knowledge, as Locke supposed. We do have to learn the difference between right and wrong as it is understood in the society in which we grow up. But, just as one cannot forget the difference, so too one cannot remember it. One cannot explain one's having desisted from a wicked deed by saying, 'Then I recollected the difference between right and wrong, so I didn't do it.' Similarly, Ryle elaborated, one cannot be reminded of the difference between right and wrong. So one can't explain why there is no such thing as forgetting by saying that everyone remembers such a thing. For then it would make sense for some exceedingly forgetful person to forget the difference and to be reminded of it; but it does not. The impossibility of forgetting is not an empirical one. One may object that it may be possible to expostulate with someone planning to do something evil. One may exclaim,

---

[18] The following remarks are indebted to Ryle.

'But that would be murder!' or 'You can't do that, it would be treachery!' Are these not reminders? No, they are remonstrations and expostulations. They could not be replaced by 'Have you forgotten that this would be murder?', 'Don't you remember that torture is wrong?', or 'I must remind you that this is treachery'. Nor would the perpetrator respond to such expostulations by saying 'Thank you so much for reminding me' or 'Yes, of course – it slipped my mind'.

Remembering the difference between good and evil
Remembering the difference between good and evil is not akin to remembering a practical skill, for, as Aristotle pointed out, one can forget a skill if one has not practised it for long enough. But no one can say, 'I have forgotten the difference between good and evil, for I have not had to draw the distinction for so many years.' One cannot excuse one's misdemeanours, as one can excuse one's incompetences, by lack of practice. One cannot engage in practising fairness and justice to keep one's hand in, as one may practise one's German or tennis, nor can one's honesty become rusty through desuetude, as one's knowledge of mathematics may. To be sure, moral corruption may set in as a result of professional practices in the cut-throat competition of business or banking, and moral deterioration through the decline of youthful public-spiritedness. One may become mean through prolonged financial hardship, and one's sympathies for others' suffering may dry up. But this is not akin to loss of a skill.

Are ethics and aesthetics one?
It is not for nothing that it has been said, by Kant (*Critique of Judgement*, §59) and the young Wittgenstein (*Tractatus*, 6. 421) among others, that ethics and aesthetics are one. Aesthetic judgement is educated taste and cultivated preference. Some people may innately have a greater aesthetic sensibility than others, just as some people may have a greater proneness to feel sympathy than others. Those who possess a natural aesthetic sensibility will take to the appreciation of literature, music, or painting with great ease, and find it easier to learn the standards of excellence appropriate to different genres and to apprehend the differences between what is good and what is bad. Moreover, they will not only acquire powers of discrimination; they will also learn to appreciate and judge excellence. Commonly, they will also admire it, and take pleasure in what is aesthetically superior – but not inevitably: one may grant the excellence of Rubens, but dislike his flamboyance or his ideal of female beauty. The judgement of excellence in aesthetics may involve much subtlety, sensitivity, and imagination, as does moral judgement in complex cases. Indeed, do not the two converge in discussions of great literature?

Is there not a kinship between learning to make aesthetic judgments and learning the difference between right and wrong? For, although one's refined taste may deteriorate and one may lose one's taste for poetry, one cannot be said to have *forgotten* the difference between good and bad poetry, just as one cannot be said to have forgotten the difference between right and wrong. Although one cannot forget, one may change one's mind about aesthetic value (e.g. one may cease to admire a given artist and regret one's past (mis)judgements) just as one may change one's mind about what is right or wrong and what is good or evil (e.g. death penalty, abortion, euthanasia). One may come to regret one's earlier, perhaps immature, judgements, just as one may come to feel remorse about what one previously did, mistakenly having thought it to be right or licit. Someone who has acquired an educated taste has also acquired a set of preferences. One comes to love or at least to appreciate what is superior. Similarly, someone who has learnt the difference between right and wrong has come to care about the differences, to feel ashamed or guilty at doing wrong, and to esteem others less for their having done wrong. One will have scruples about doing something one knows to be wrong. One may be blind to beauty and ugliness, as one may be blind to good and evil. Moreover, one's eyes are not opened to beauty or goodness, ugliness or evil, by reasoning. One cannot reason people into caring.

Nevertheless, ethics and aesthetics are not one, and knowing the difference between right and wrong is unlike knowledge of aesthetics. One can study music, take a course on baroque painting, become an expert on Donatello, Verrocchio, or Giambologna. But there are no moral experts – priests, rabbis, and mullahs are not experts in moral judgement, although they may be great preachers. But preachers are neither moral scientists nor moral connoisseurs. Practical wisdom, not moral fervour, bears a kinship to connoisseurship, but even it is not a form of moral expertise. There are no lessons in morals at school, although there should be moral training. Nor are there courses in morals, but only in moral philosophy, at university. Some people may be able to reason methodically and are skilled in mustering arguments. They may engage in 'practical ethics' and casuistry and their judgements may well be worth listening to. But that does not make them moral experts in the sense in which a skilled cardiologist is an expert, let alone in the manner of a highly proficient astronomer. Some people can learn practical wisdom – in the hard school of experience, and from those who have achieved it. One does not learn techniques and methods (which connoisseurs do, e.g. Giovanni Morelli's methods for identification of paintings), but rather good judgements. There are loose rules, but they do not form a system, and only experienced people can apply them correctly.[19] One may love and know much music but be indifferent to the visual arts; one may be concerned only with Renaissance art to the total neglect of Byzantine art or Impressionism. But one can hardly know the difference between good and evil and be indifferent to its manifestation in business or politics or in warfare. It is true that one may change one's aesthetic judgements as one may change one's moral judgements (e.g. concerning the death penalty, abortion, euthanasia). But such changes in one's moral judgements are typically reversals of judgement, whereas changes in one's aesthetic judgement are more typically refinements and rankings.

To have a sense of duty or a conscience is not akin to having good taste. A lapse of taste is a reason for feeling embarrassed, but a moral misdemeanour is a reason for guilt. To learn the difference between right and wrong is to learn to do what is right and to eschew what is wrong. It is to come to feel remorse and guilt at one's own transgressions, and indignation and resentment at the transgressions of others. It is to come to appreciate the benevolence of others and to condemn and feel shocked at their malevolence. It is also to learn when and how to apologize and make amends for one's own wrongs and misdemeanours, and how and when to reproach others for theirs. If one is not adequately brought up, one's moral sensibility, having grown on poor soil, may remain stunted. We are not like specimen trees, which will naturally grow to their full glory if isolated from other trees. We are more like rosebushes, which need to be trained and cultivated. The person who knows the difference between right and wrong does not possess any specialized knowledge.[20] He knows that murder is wrong, that telling lies is reprehensible, that injustice and unfairness are to be condemned. He knows not to murder, lie, be unjust or unfair – as we all do, save psychotics and psychopaths. He does not possess any special skills or habits (he is not *in the habit* of telling the truth or of keeping his word – he just does so, having been habituated to do so). He knows to do what is right, even if he occasionally lapses in greater or lesser ways.

---

[19] Cf. Wittgenstein's remarks on learning to judge genuineness of feelings (*Philosophy of Psychology*, §355).

[20] But if he ventures into an alien society, he will have to learn the morals and mores of that society. That will indeed be specialized anthropological knowledge.

5. Experimental psychology: Milgram's and Browning's explanations of evil-doing

Hannah Arendt's Hannah Arendt's reports of the Eichmann trial in
banality 1961 and her subsequent book on it introduced the
phrase 'the banality of evil'. Her central thesis was
that evil is 'banal'. By this she meant that evil-doers in totalitarian
regimes are no more than cogs in a bureaucratic machine. The evil
that people do in such a system of government, or at any rate the evil
done by Adolf Eichmann, who was responsible for organizing the
slaughter of millions of Jews in death camps in Poland, was allegedly
done unthinkingly. He was, she argued, a mere cog in a machine – a
functionary with no convictions (not even an anti-Semite), concerned
only to obey orders and to advance his career. The fundamental
awfulness of totalitarianism is that it reduces human beings to
machines, crushes thought and reflection, and induces mindless obedience. Examination of Arendt's arguments and evidence for her thesis
would involve too long a digression. It has therefore been allocated to
Appendix 3, in which it is argued that her thesis is itself either banal
or false. It is banal if all it says is that ordinary people can do great
evil. It is false if it claims that Eichmann was a mere cog in the military
and bureaucratic machine that executed the Holocaust. Eichmann
was no cog – he designed a large part of the machine. The primary
bearing of Arendt's book on our concerns is that it gave rise to a further kind of explanation of evil in general, and of the evil of the *génocidaires* in particular. Arendt's misunderstandings and misconstruals
had non-trivial consequences in the field of psychology and social psychology. In the wake of her book, experimental psychologists who
were interested in explaining evil-doing and in understanding the personal psychology of evil-doers turned their attention to social psychology to discover the extent to which doing evil might be explained, as
Arendt had explained it, by mere submission to authority.[21]

Milgram experiment The best-known investigation is the Stanley
Milgram experiments (the results of which were
published in 1963 and 1974) in which subjects were instructed to give

---

[21] There had been an earlier experiment in this vein prior to Arendt's book. This was
Solomon Asch's conformity experiment conducted at Swarthmore College in 1951,
which showed a disturbing proportion of student subjects lying in order to conform
to the judgements of others – 'going along in order to get along'.

electric shocks to a bogus patient strapped to a chair, whenever the patient gave the wrong answer to a question. The shocks apparently increased up to 450 volts, which the subject was told meant increasing the agony for the patient (who screamed with apparent pain), until he fainted. The subjects were assured that the patient would suffer no physical damage, and that the experiment was crucial for scientific purposes. Sixty-five per cent of the subjects went along with the experiment to the point of apparently inflicting excruciating pain on the patient. The experiments were repeated by many other experimenters in different countries with similar results. These results were astonishing and depressing. Human beings do have a frightening disposition to obey someone in authority and are willing to inflict great pain on helpless victims when assured that it is crucially important for the advancement of science and that the victim will suffer no physical harm.[22]

Whether the Milgram experiments showed that the primary explanation of the Jewish Holocaust is the human propensity to obey orders is debatable. It screened out ideology, replacing it by the indeterminate goal of the advancement of science. But Nazi ideology certainly played a role in relieving the Nazi murderers of any sense of human solidarity or responsibility. Whereas Milgram's subjects were assured that no permanent physical harm would come to the patient, the Nazi murderers knew that they were slaughtering men, women, and children. Whereas Milgram's subjects knew nothing about the patient to whom they were ostensibly giving electric shocks, the Nazi murderers knew that their victims were Jews, whom they had been taught to view both as non-human and as responsible for all the ills of the world. Milgram's subjects were not acting as members of a group of killers to whom they felt allegiance; they were not in the army or police, let alone in a war; and they were not following a Führer. Finally, the experiment lasted an hour, not years, and the subjects did

---

[22] An even more appalling experiment was the Stanford prison experiment (which would today be illegal) run by Professor Philip Zimbardo in 1971. The subjects were divided into two groups, prison guards (in uniforms with batons) and prisoners (in humiliating prisoner garb), and placed in a mock prison. The results were horrifying, as the subjects assumed their assigned roles. It had to be called off after six days, during which time some traumatized subjects had nervous breakdowns. Some of the 'guards' manifested extreme sadism in their treatment of the 'prisoners'. The supposition was that role assumption under authority has a fundamental effect on personality and moral responsibility.

not see thousands upon thousands of corpses that were the result of their work. So, although the Milgram experiments are important, they show only that the human propensity to do evil when ordered to do so by an authority is one possible standing background condition for evil-doing.

Browning on Battalion 101   Christopher Browning's *Ordinary Men: Reserve Police Battalion 101 and the Final Solution in Poland*[23] was much influenced by the results of Milgram's experiments and by the Arendt thesis of the banality of evil. Browning made use of extensive interviews conducted in 1962 with 210 survivors of a battalion of Hamburg police reservists who had been responsible for the murder of 83,000 Jews in Eastern Poland between 1942 and 1943.[24] They were ordinary working-class men who were not fit for front-line duty. None had criminal records. Hamburg had, prior to 1933, been a socialist city with a majority of communist voters. They were, therefore, not likely to be ardent Nazis or ideological anti-Semites. They were too young to have fought in the First World War, and too old to have been educated as Nazis at school. None were punished or sent to the front for refusing to participate in mass executions, nor did they think they would be. Nevertheless only a small handful refused. Browning came to the conclusion that what explains the murderous behaviour of the men of Battalion 101 was not indoctrination, fanatical leadership, or violent anti-Semitism but simply peer pressure, fear of stepping out of line, and reluctance to let their comrades down.

Doubtless this is largely true. It does not, however, confirm Arendt's thesis of the banality of evil. It does not show that the *commandants* of the murder battalions, many of them educated men, were motivated by the desire to keep in line. It does not demonstrate that the atrocities of the SS or SA were motivated by group conformity as opposed to ideology and fanatical loyalty to the Führer. It ignores the extent to which ferocious anti-Semitism was a powerful motivating consideration among the Nazi *génocidaires*.

There is not *one* explanation of individual evil-doing, nor is there *one* explanation of genocide and crimes against humanity. There are multiple explanations depending upon context, circumstances, the national character and traditions of the perpetrators and of their

---

[23] Christopher R. Browning, *Ordinary Men: Research Police Battalion 101 and the Final Solution in Poland* (HarperCollins, New York, 1992).

[24] There were seventeen such battalions.

victims, the beliefs and ideology of the murderers, and the individual personalities of the *génocidaires*. What the Holocaust taught us is that we humans are collectively capable of far greater evil than had ever been dreamt of, that the fond nineteenth-century faith in the steady moral and political improvement of mankind was chimerical, that horrors lurk beneath the surface of every society and can be released by a variety of socio-economic and political circumstances.

5

Evil and the Death of the Soul

1. Body, mind, and soul

Destroying one's soul In his book *East/West Street*, Philippe Sands relates that when the Nazis occupied Lemberg (Lvov) in Galicia, the savage persecution of Jews began immediately. Professor Maurycy Allerhand, a distinguished professor of law at the University of Lemberg, who had taught both Hersch Lauterpacht and Rafael Lemkin, was immediately interned at Janowska, the main concentration camp in Lvov. Seeing a German officer mercilessly beating another inmate, Allerhand went up to the German and asked '*Have you no soul?*', whereupon the German took out his revolver and shot Allerhand dead. Martha Gellhorn, in her report on the Eichmann trial, which was considerably more thoughtful than Arendt's, remarked that Eichmann is a warning to us all that we must *guard our own souls*.[1] The first section of Chapter 3 concluded with a long quotation from Joseph Conrad's *Heart of Darkness*, in which Kurtz peers into the

---

[1] 'We consider this man, and everything he stands for, with justified fear. We belong to the same species. Is the human race able – at any time, anywhere – to spew up others like him? Why not? Adolf Eichmann is the most dire warning to us all. He is a warning *to guard our own souls*; to refuse utterly and forever to give allegiance without question, to obey orders silently, to scream slogans' (Martha Gellhorn, 'Eichmann and the Private Conscience', *Atlantic Monthly* 209:2 (1962), 52–9 (emphasis added)).

darkness of his own soul and can find nothing there but horror. In Vasily Grossman's *Life and Fate*, Anna Semyonovna writes in her last letter to her son Viktor Shtrum before she is murdered by the Nazis:

> I've seen that the people who shout most loudly about delivering Russia from the Jews are the very ones who cringe like lackeys before the Germans, ready to betray their country for thirty pieces of German silver. And strange people from the outskirts of the town seize our rooms, our blankets, our clothes. It must have been people like them who killed doctors at the time of the cholera riots. And then there are people *whose souls have just withered*, people who are ready to go along with anything evil – anything so as not to be suspected of disagreeing with whoever's in power.[2]

None of these writers is religious. That they all invoke the soul of man is striking and important. To do evil is to destroy one's own soul.[3] This is not a causal statement. Evil-doing is intrinsically related to the death of the soul.

In *Human Nature: the Categorial Framework* I suggested in passing that it is a constitutive truth that human beings have souls, that the soul is distinct from the mind, and that the contrast between the soul and the flesh is quite different from that between one's mind and one's body (*HNCF*, 234–5, 275 n.). It is now time to clarify this secular concept of the soul, to elucidate what legitimate role it has in our conceptual scheme, to contrast it with our concept of the mind, and to relate it to our ideas of good and evil.[4]

The mind   The mind, we have seen, is not a something, but it is not a nothing either, because it is not a thing (or, more pretentiously, not an 'entity') of any kind. All our talk of the mind boils down to talk of our distinctive intellectual powers and of our intellectual capacities (second-order powers to acquire intellectual powers through learning and experience) and their exercise (*HNCF*, ch. 8). But it would be misleading to assert that the mind *is* a set of intellectual powers. For,

---

[2] Vasily Grossman, *Life and Fate* (Vintage, London, 2011), ch. 18 (emphasis added).

[3] Interestingly Sir Kenneth Branagh, discussing his role as Reinhard Heydrich in the television drama *Conspiracy* (2001), remarked on the difficulty of playing a man without a soul.

[4] I am much indebted to Ilham Dilman's illuminating book *The Self, the Soul, and the Psychology of Good and Evil* (Routledge, London, 2005). What I have to say about the soul would have to be variously modified for German, which has only *Seele* and *Geist* to do the work that, in English, is divided between *mind*, *soul*, and *spirit*. (It also has *Verstand* which may sometimes fulfil the role of 'intellect'.) Other cultures, in particular Far Eastern ones, may meet the evident need in different ways.

to be sure, to make up one's mind is not to make up one's intellectual powers but to decide, and to change one's mind is not to change one's intellectual powers but to revoke a previous decision. The powers of intellect and will, possession of which is constitutive of having a mind, are not powers *of* the mind, but of the being that *has* a mind. It is the human being, not his mind, that thinks, reasons, and plans effectively, *if* he has a good mind. They are corollaries or consequences of the fact that human beings are essentially language-using animals (*Homo loquens*, as I have often emphasized, rather than *Homo sapiens*). In this sense of 'mind' (which approximates Aristotle's 'rational *psuchē*'), non-human animals cannot be said to have minds, since they are not language users and they lack the intellectual powers distinctive of human beings (see Appendix 1). They also lack souls, since only animals with a mind can have a soul – and that too is a 'grammatical' (conceptual, constitutive) truth. For only creatures capable of knowing the difference between good and evil can be said to have a soul.

The body one is and the body one has   Beings that possess a mind and a soul are bodies, in one sense of that polysemic word (*HNCF*, 9.3–9.4) – that is they are living spatio-temporal continuants consisting of matter. Unlike all other living things, they are not only self-moving and sentient but also self-conscious, in possession of intellect and (rational) will. Human beings are *not* embodied, but rather, as Aristotle said, are *ensouled* (*empsuchos*) – animals endowed with abilities constitutive of possessing a mind. Although they *are* bodies, they also *have* bodies, but – to put things pithily although obscurely – the body a human being *has* is distinct from the body the human being *is* (*HNCF*, ch. 9). To make matters clear: the body a human being has is not a distinct *body* from the body he is. Rather, all talk of the body someone *has*, all talk of someone's beautiful, athletic, aged, feeble, bruised, sunburnt body is no more than talk of *somatic* (bodily) *characteristics* of the human being in question. Everything true of our body is true of ourselves, but not everything true of ourselves is true of our body. I may (but my body cannot) be thinking, since thinking is not a somatic (bodily) characteristic.[5] Talk of our body is therefore complementary to talk of our mind, which is concerned with the *powers of intellect and will* that a human being – the animate material substance that we are – possesses and exercises. Both styles of discourse focus upon

---

[5] To be sure, it would not be *incorrect*, but only insufferably obscure, to say that the body I am (*this* spatio-temporal living material continuant with rational powers) is thinking.

different classes of features of the living human being. How then does talk of the soul fit into this schema?

Manifold meanings of 'soul'   Like 'mind', 'person', 'self', and 'body', 'soul' too is polysemic. 'She is a merry old soul' is synonymous with Scottish English's 'She is a merry old body' – both terms meaning person. In 'There were 276 souls on board when the ship went down', the word 'souls' means human beings. In 'The estate near Moscow was sold together with more than 300 souls' it means serfs. 'He sighed, and breathed no more – his soul had departed this world' may mean no more than 'he died', the 'soul' here signifying the principle of life or life force that makes the difference between being alive and being dead. These uses are of no interest to us in this context. More to the point are the numerous powerful idioms that we use: we speak of a 'soul in turmoil', of 'a storm raging in one's soul', of an evil person's 'black soul', of someone who has lost his way in his moral life as having 'darkness in his soul'. Raskolnikov is a 'lost soul' and Coleridge's ancient mariner, after having shot the albatross, is a 'damned soul'. We may describe a person as a 'noble soul' or a 'gentle soul', or as having a 'beautiful soul'. Some idioms stretch the use of 'soul' beyond the moral domain, as when we speak of 'soul-destroying' jobs – a dulling of sensitive powers by drudgery. Here too one's soul can be said to wither. But it would have been more accurate to speak of the withering of the spirit. For it is one's liveliness, responsiveness to experience, positive engagement with fellow human beings, and capacity for joy that are crushed.

Need for a concept of the soul   Dualist conceptions of the soul as a temporarily embodied substance that is logically independent of the body, that pre-exists human incarnation, or that survives the death of the body, are still rife. This is the domain of religion, theology, and Platonist metaphysics. The notion of the soul as a unitary spiritual substance that can exist in disembodied form is incoherent (*HNCF*, ch. 8). It lacks criteria of identity and individuation. It presupposes the intelligibility of possessing psychological attributes dissociated from the logical possibility of behavioural expression. It takes for granted the intelligibility of possessing concepts of psychological attributes without grasping the behavioural criteria for their ascription to others. However, the Platonic metaphysical conception of the soul is of great interest irrespective of its informing both ancient and Renaissance neo-Platonist ideas about the soul and its immortality, and, via Augustine, ultimately moulding the misconceived Cartesian conception of the soul. One reason why it is of interest is that it contains much insight into human beings as evil-doers. These insights can be stripped of their ontological and metaphysical trappings that Plato added to the Socratic conception. *Correctly formulated*, this notion of a soul fills a gap left by the neo-Aristotelian conception of the mind as constituted by the powers of the intellect and will. It answers to a need in forming our conception of human nature.

Soul and moral powers   In the *Crito* (47d), Socrates remarks that the soul is 'that in us which is improved by right conduct and destroyed by wrong'.[6] So human beings have a soul. This demands clarification since for us the question arises of the relation between the mind, possession of which presupposes the intellectual and volitional powers of man and their exercise, and the soul, possession of which is associated with knowledge of good and evil, our powers to do right and wrong, our sense of justice, our conscience, our susceptibility to remorse, and our feelings of compassion. Socrates presents the soul as something *within* a human being, but we must take the idea of *being within* here with a pinch of salt, since the soul is no more literally within us than our powers are literally within us. Finally, according to Socrates, the soul of a human being can be destroyed by doing wrong. This too needs elucidation, since it follows that human beings may lose their soul (as intimated by Martha Gellhorn, quoted above). Hence too, they may exist without a soul, as Professor Allerhand intimated and died for so doing. Gitta Sereny, author of a profound book about the Nazi architect and minister of munitions Albert Speer, wrote in similar vein that he

> had never killed, stolen, personally benefited from the misery of others or betrayed a friend. And yet, what I felt neither the Nuremberg trial nor his book had really told us was how a man of such quality could become not immoral, not amoral but, somehow infinitely worse, morally extinguished.[7]

---

[6] Of course, the Socratic and Platonic conception of the *psuchē* is quite distinct from the Aristotelian one. The latter is primarily a biological concept, the former largely an ethical (Socrates) or ethical and metaphysical one (Plato).

[7] Gitta Sereny's Introduction to *Albert Speer: His Battle with Truth* (Picador, London, 2017). Sereny's subsequent observations in her conclusion are directly relevant to my claims concerning both sympathy and the death of the soul. She writes: 'Speer himself killed no one and felt no enmity, hatred or even dislike for the millions in Eastern Europe, Christians and Jews, who were systematically slaughtered: he felt nothing. There was a dimension missing in him, a capacity to feel which his childhood had blotted out ... Pity, compassion, sympathy and empathy were not part of his emotional vocabulary. He could feel deeply but only indirectly – through music, through landscapes, eventually through visual hyperbole, often in settings of his own creation: his Cathedral of Light, the flags, the thousands of men at attention motionless like pillars. But then ... at long last he acknowledged Hitler's madness; through the revelations of Nuremberg and the confrontation with the reactions of the civilized world came his realization and horror at what had been done, his feelings of personal guilt, his wish, almost, for death and yet fear of execution, the shame of being spared.'

These Socratic points require scrutiny.

In the sequel Socrates elaborates:

> There is a part of us which is improved by healthy actions and ruined by unhealthy ones. If we spoil it by taking the advice of non-experts, will life be worth living once this part is ruined? The part I mean is the body. ...
>
> What about that within us which is mutilated by wrong actions and benefited by right ones? Is life worth living with this part ruined? Or do we believe that this, whatever it may be, in which right and wrong operate, is of less importance than the body?
>
> (*Crito*, 47d–e)

The soul and the flesh The medical analogy is profound. We care a great deal about our good health and physical integrity. We view the loss of a limb as a great misfortune, for it deprives us of the ability to function as a normal human being. Moreover, with the loss of such abilities, we also lose many of the opportunities for action available to those that have the relevant healthy organs. For opportunities are correlative to abilities. Plato presses the analogy: are there not features of our non-bodily nature that can be damaged, perhaps irremediably damaged, by abuse and misuse – by doing evil? And with the loss of such powers, are human possibilities, opportunities for doing, becoming, and being, not foreclosed? The thought is echoed by the remarks of Vasily Grossman's Anna Semyonovna that human souls may wither. Should we not care for our soul at least as much as we care for our physical constitution and health?

In the *Phaedo*, Plato's Socrates draws another construction line:

> The body fills us with loves and desires and fears and all sorts of fancies and a great deal of nonsense with the result that we literally never get an opportunity to think at all about anything ... we are slaves to its service. (66a)

Subsequently Socrates takes this line of argument too far, imagining the bliss of the soul without the body. We need not go down that dead end: it is, on analysis, unintelligible. But refusing to follow such fanciful asceticism does not prevent one from recognizing that the appetites, which we share with animals, are firmly bound to our physical nature, needs, and cravings (*TP*, 7–12, 26–8). The appetites may hold us in bondage if egotism, self-indulgence, and hedonism constantly triumph over rationality, reasonableness, and self-restraint. We may be enthralled by our acquired appetite for alcohol; we may be victims of our own gluttony; and we may be enslaved by concupiscence. These are not somatic characteristics – one's body (the body one has) is not gluttonous, nor is it an alcoholic or beset with the priapic afflictions of Don Juan or the insatiable lusts of Messalina. It is for this purpose that we have the expression 'the flesh', which we contrast not with the mind, but with the soul.[8] The dividing line between the soul and the flesh is quite different from that between the mind and the body. The appetites, not being somatic features of a human being, are not allocated to the body one has. Not being intellectual powers of a human being, they are not allocated to the mind one has. They are allocated to the flesh.[9]

The soul as form of representation

Human beings have a soul. That is not an empirical statement, but a constitutive (grammatical, conceptual) one. It characterizes the nature of mankind, as does the statement that human beings have a mind. These are not informative propositions, but explicative ones. They constitute a form of representation – our way of thinking and reasoning about ourselves – and belong to our methods of representation. They may serve to remind one that if a creature can be said to be a human being, then it follows that it can *intelligibly* be said to have a mind and a soul. This is, in effect, a grammatical proposition, a rule for the use of the expressions 'human being', 'mind', and 'soul'. Does it follow that *every* human being has a mind or soul? No – only that it *makes sense* to speak of the mind or soul of any being that is human, and that it is part of our *species nature* to have a mind and to have a soul. But one may lose one's mind (go mad) and one's mind may be destroyed by severe injury to the brain or during the final stages of senile dementia. So too one may lose one's soul (while keeping one's mental faculties intact), one may destroy one's soul, and one's soul may be irretrievably damaged, twisted, or scarred either by

---

[8] It may well be true that this idiom, and indeed this distinction, is rooted in Christian asceticism and fear of sexuality. But it would be a genetic fallacy to suppose that it cannot readily be detached from whatever religious connotations it might originally have had. Thus detached it draws an important distinction.

[9] Animals, which lack the rational powers requisite for the ascription of minds (see Appendix 1), are nevertheless subject to appetites, addictions, and cravings (see *TP*, 7–12). But not having minds or souls, they are not subject to conflicts between the soul and the flesh. So their appetites are not allocated to the flesh at all, but *only* to the animal as a whole.

one's own evil actions or by what one has been forced to do or undergo. (But one's soul cannot be damaged, twisted, or scarred by doing good.)

When is a soul acquired?   Is every human being born with a soul? This is a delicate matter, akin to the question of whether neonates are persons. It is clear that newborn children are no more sensitive to moral and aesthetic considerations than they are able to engage in reasoning and are sensitive to reasons. Nor can they be said *simpliciter* to be born with the capacity (second-order ability) to acquire knowledge of good and evil, any more than they can be said *simpliciter* to be born with the capacity to learn a language. In both cases, these second-order abilities naturally evolve in the course of the early phases of life. Just as they naturally evolve the capacity to acquire the powers of reason, so too it is part of the nature of human beings to evolve the capacity to acquire a moral conscience, knowledge of good and evil, and a moral and aesthetic sensibility. For various reasons pertaining to how small children must be treated, one may resolve to characterize them as having minds and souls in virtue of their innate proneness to evolve second-order constitutive powers. Alternatively, one may resolve to describe matters differently. One may then say that children will naturally acquire a mind as they master a language, learn to reason, and begin to raise the question 'Why?' So too, one may resolve to describe the manner in which moral beings come into existence by saying that children only gradually develop a soul as they come to know the difference between right and wrong, acquire a sense of justice and fairness, and learn to assume responsibility for their deeds. Their soul evolves side by side with their personality as they slowly grow out of the egocentricity of childhood.[10] Does this mean that *if* there are children who are born psychopaths, naturally and irremediably indifferent to moral considerations, then they are born without a soul and

---

[10] Nicely expressed by Ludmila Ulitskaya's Victor Yulievich: 'Moral maturation seemed to be as valid a dimension of human development as the biological processes that unfolded in tandem with it. But moral awakening occurs in different ways, and the framework within which it takes place varies according to the individual cast of mind, and other contingencies. Moral awakening or "moral initiation" as he called it, occurs in boys between eleven and fourteen years of age, most often spurred by unfavourable circumstances – an unhappy or difficult family situation, an assault on one's sense of self-worth and dignity (or the dignity of those one holds dear) or the loss of a loved one. In short, an internal upheaval that calls the soul to life' (Ulitskaya, *The Big Green Tent*, trans. Polly Gannon (Farrar, Straus & Giroux, New York, 2015), ch. 5).

without the potentiality for one? Alas, it does. Given that nature occasionally produces children who are physically and mentally profoundly defective, it should hardly be surprising that it produces innately morally defective children too. But one must bear in mind the fact that not all psychopaths become evil. Some may become highly effective front-line soldiers, lions of industry, or tigers of finance.

What one has when one has a soul Like the mind, the soul is not an 'entity' of any kind. Just as the primary philosophical question concerning the mind is not 'What is a mind?' but 'What is it for a creature to possess a mind?', so too the primary philosophical question about the soul is not 'What is the soul?' but 'What is it for a creature to possess a soul?' – that is, what has to be true of a being for it to be said to have a soul? To possess a mind is, to a first approximation, to be sensitive to reasons for thinking, feeling, and acting and to be able to reason. Similarly, very roughly and to a first approximation, to possess a soul is to know the difference between good and evil, to have a conscience, and so to be susceptible to remorse and guilt for wrong-doing. So it is to have and to exercise moral powers. These, as we have seen in Chapter 2, presuppose the emotional dispositions of sympathy. The moral powers are acquired by a human being in a community in the course of maturation, through training, teaching, practice, experience, and reflection. They involve reasoning and recognition of reasons. Hence possession of a soul presupposes possession of a mind. But the powers presupposed for possession of a soul are moral powers and moral sensibility. The soul is also associated with aesthetic sensibility and with the appreciation of beauty and pathos, with which conscience has no connection. This will not be discussed here. Without our knowledge of good and evil we cannot fulfil our nature as human beings, overcome our infantile drives and natural egocentricity, or our mature egotism and selfishness. This practical knowledge of good and evil is exhibited in the way we live, in the forms of our self-awareness, in the attitudes we assume, and in our relations with our fellow human beings.

Conscience With possession of the soul are associated not merely knowledge of good and evil, but also the moral virtues and vices, the former (to continue Socrates' medical analogy) being marks of the health of the soul and the latter signs of ill health, diseases of the soul, and of corruption. It was not for nothing that the worst vices were in the medieval age characterized as 'mortal' or 'deadly' sins, for they spell the death of the soul (see *TP*, 136 n.). Possession of a conscience is constitutive of having a soul. A conscience is acquired in the course of internalizing moral norms and values and *making them one's own*. It *should* be refined by experience and reflection (including critical reflection on current moral norms and attitudes). One's moral sensibility should develop with age. But the path to moral maturity is long and labyrinthine, and many lose their way. Susceptibility to guilt, remorse, and repentance are corollaries of a conscience and are constitutive passive powers of those who have a soul. Such powers are strengthened by transcending the demands of selfishness, self-centredness, and narcissism; by cultivating gratitude, loyalty, trust, respect for others, and self-respect; by a sense of justice and fairness; by compassion; and by the humility that is essential for honesty with oneself, self-knowledge, and self-understanding. To achieve moral maturity is an asymptotic *telos* of the life of a human being (see the quotation from Steinbeck at the beginning of Chapter 3).

Although having a conscience is a corollary of having a soul, the idea of having a soul is the wider notion. The soul, unlike the conscience, is linked to moral sensitivity: to susceptibility to pity and compassion, to sympathy and empathy. 'Only someone with iron in his soul could resist such a plea for mercy,' one may say.

2. The death of the soul

The soul, Socrates averred, can be damaged, mutilated, and destroyed. We have seen this thought advanced by modern secular authors too. It is completely detachable from theological considerations, but it requires elucidation. In the previous two chapters I have drawn a distinction (*explicative* in Carnap's sense of the term) between what is wicked and what is evil by reference to the idea that doing evil crosses borderlands that, once transgressed, allow of no return, whereas doing what is wicked allows for remorse, repentance, and making good the wrong done. Those who do evil destroy their own souls. The wicked damage their souls. The morally indifferent, the bystanders to evil, allow their souls to wither. What does this mean?

Destroying one's soul
To do evil is to destroy one's soul. In destroying one's soul one becomes indifferent to inflicting terrible suffering and humiliation on others and to the slaughter of others. One may become depraved, and positively take pleasure in the suffering one inflicts and the slaughter one effects, and delight in the power one wields and the sense of superiority it gives one. Indeed, as already noted, one may delight in destroying *the souls of others* by forcing them to violate their most fundamental moral

convictions, as detailed in Kaplan's Warsaw diary,[11] or forcing a man to hang his best friend (as described earlier), or, as depicted in the fearful novel *Sophie's Choice* by William Styron, forcing a mother to choose which of her two children is to be shot by a Nazi officer. The evil-doer commonly aims, explicitly or implicitly, to destroy the moral order of humanity and the moral restraints on behaviour that are constitutive of moral beings. In doing the evil he does, he destroys his own moral sensibility, sometimes with glee. This was patently true of many of the Nazi murderers. It is striking that barely any Nazi war criminals expressed any remorse for their deeds – they had sunk too deep in evil to find their way back. Nor, as far as I know, has much been heard of the remorse of the Turkish *génocidaires* for their slaughter of Armenians, or of the Soviet Cheka, OGPU, NKVD, or KGB officers and commissars guilty of crimes against humanity.

Having one's soul destroyed   One may lose one's soul or one's soul may wither through the terrible suffering one undergoes, in which one's moral sensibilities are battered to dust, and from which one cannot recover. One aspect of the evil of evil-doers is the objective of destroying the dignity and self-respect, the moral stature and integrity of their victims. We have seen the upshot of this clearly articulated by Jean Améry, and it was surely true of countless psychologically scarred survivors who lacked his articulacy. It was powerfully and harrowingly depicted by Rod Steiger in Sidney Lumet's film *The Pawnbroker* (1966).

Evil-doer as his own executioner   In deliberately doing evil one destroys one's soul, setting oneself beyond the pale of humanity. Powerful expression of one form of this thought is given by Vasily Grossman's Anna Sergeyevna:

> I asked you how the Germans could send Jewish children to die in the gas chambers. How, I asked, could they live with themselves after that? Was there really no judgement passed on them by man or god? And you said: Only one judgement is passed on the executioner – he ceases to be a human being. Through looking on his victim as less than human, he becomes his own executioner; he executes the human being inside himself.[12]

---

[11] Chaim Aron Kaplan, *Scroll of Agony: The Warsaw Diary of Chaim A. Kaplan*, trans. and ed. Abraham I. Katsh (Hamish Hamilton, London, 1966).

[12] Vasily Grossman, *Everything Flows*, trans. Robert Chandler (Vintage, London, 2011), ch. 14. I emphasize that this is *one* form of the thought. One need not claim that such evil-doers are no longer human beings. Or, to interpret Anna Sergeyevna sympathetically, to say that the moral monster executes the human being *inside himself* is just to say that he destroys his own soul.

It does not follow that one is aware of having destroyed one's soul. (A parallel: someone who has become addicted to pornography is likely, as a matter of fact, to have impaired the possibility of a loving, mature, sexual relationship with another person (indeed brain changes are evident in teenage pornography addicts), but he or she may be wholly unaware of this.) The barriers of evil one has constructed around oneself may exclude any understanding of the good – it has become, one might say, *invisible*. Or it may be distorted, as in a hall of mirrors, appearing to the evil-doer as a form of weakness and stupidity. Or the ramparts of evil that one has constructed around oneself by adopting or acquiescing in an evil ideology or bigoted religion may relieve one's conscience of the evil one does. The results of one's blindness are dreadful. One stultifies one's own life, assuring the impossibility of any kind of maturation and self-fulfilment as a human being. This one may neither know nor understand. For in doing the evil one has done, one will often have lost the ability to recognize that one has destroyed one's moral sensibility – one's very knowledge of good and evil that is the birthright, not the terrible legacy, of the children of Adam.

Suppression of conscience Evil-doers, if they are not born psychopaths, have to suppress the voice of conscience. But sometimes, it seems, that suppressed voice may torment an evil-doer in sleep. This is powerfully depicted in Shakespeare's Richard III, who, during the night before Bosworth Field, has the most fearful nightmares on which he reflects in terror on awakening:

> Soft, I did but dream.
> O coward conscience, how dost though afflict me?
> The lights burn blue. It is now dead midnight.
> Cold fearful drops stand on my trembling flesh.
> What do I fear? Myself? There's nothing else.
> Richard loves Richard; that is I am I.
> Is there a murderer here? No. Yes, I am.
> Then fly? What, from myself? Great reason. Why?
> Lest I revenge. Myself upon myself?
> Alack, I love myself. Wherefore? For any good?
> That I myself have done unto myself?
> No, alas, I rather hate myself
> For hateful deeds committed by myself.
> I am a villain. Yet I lie. I am not.
> Fool, of thyself speak well. – Fool, do not flatter.
> My conscience hath a thousand several tongues,

And every tongue brings in a several tale,
And every tale condemns me for a villain.
Perjury, perjury, in the high'st degree!
Murder, stern murder, in the dir'st degree!
All several sins, all used in each degree,
Throng to the bar, crying all, 'Guilty, guilty!'
I shall despair. There is no creature loves me,
And if I die no soul will pity me.
(*Richard III*, V. v. 132–55)

The net of vices of evil-doers A solitary evil-doer in a normal society condemns himself to fear of being discovered, to suspicion of his neighbours who may disclose his evil deeds. He commits himself to a life of lies and deceit, as he strives to conceal the wrong he has done. A Mafiosi or member of a gang that systematically engages in evil-doing doubtless enjoys the camaraderie and excitement of destruction and domination, as well as the booty. He will lie and cheat to protect himself and his co-criminals from retaliation. But perhaps he will also be paranoid in his fear of rivals and of betrayal. He will very likely not be able to trust even his closest comrades for fear of double-cross in the struggle to maintain his position or to rise in the criminal hierarchy.

A member of an evil-doing militia such as the Brownshirts, or the Cossack bands described by Isaac Babel, will be arrogant – he will look upon others who lack the power he has with contempt. He will be ruthless and merciless, for he can show no 'weakness' before his fellow evil-doers. But can he 'look himself in the face'? Can he look into his own soul (as Kurtz did)? Can he be honest with himself? Can he disclose his hidden fears and nightmares to anyone to gain relief from them? Maybe not. But need he care? He may not be given to introspection or care about understanding himself. He cannot even admit to himself that he is doing evil. He cares not a whit for practical wisdom but only for cunning. But, for sure, he does not lead a life fit for man, precisely because he has destroyed his soul.

An evil ruler will be exhilarated by his successes, self-satisfied, and arrogant. He will be egocentric, for evil is in general the dominion of self-centredness (sometimes denominated 'the self' or 'the ego'). Vindictiveness, vengefulness, greed, hatred, envy, malice, ingratitude, lust, indifference to justice and fairness, and contempt for the powerless are all characteristic of wickedness and evil and are manifestations of egocentricity. Arrogance is a hallmark of dictators and tyrants, who feel, and need to feel, important, powerful, and envied. Others, they think, are there only to serve their purposes. Members of an evil leader's entourage are to be manipulated and controlled, and if they stand in his way they are to be destroyed.[13] He may live in fear of being toppled from the pinnacle of power he has achieved; he may seek for plots everywhere; and, if he finds none, he will confabulate them (as Stalin did). The characteristic features of an evil leader militate against the ability to apprehend the truth about the world in which he lives and about himself – self-knowledge and self-understanding typically lie beyond his grasp. The characteristic passions of an evil leader are infantile and militate against achieving moral and psychological maturity.

Ex post actu scars of evil   The subjective consequences of doing evil doubtless change when the perpetrator puts that life behind him. In this case, exemplified by ex-Nazis who evaded punishment, human psychological defence mechanisms do their work. Memories are suppressed and when revived are distorted. No doubt camaraderie was felt and exhibited in SS reunions, where Nazi songs were bellowed in inebriation, and memories shared, but the reminiscences would have been highly selective – not of burning people alive and bayoneting children, but rather of recollected tomfoolery, of dangers shared and hardships jointly overcome. Such people returned to normal life in a normally functioning society, for the most part concealing their evil deeds by a curtain of silence. They may have returned to a family life with children, to take up civilian jobs, and to have distinguished careers. Can such people nevertheless not be good parents, love their spouse, show compassion for their suffering dog, feel guilty at betraying their spouse, feel ashamed at their misdemeanours in daily life, and in many respects lead a normal life? How then can they be said to have destroyed their soul? Perhaps, in some sense, they destroyed their souls *in the past*, but surely now they can begin again, revive their moral powers, and let the past bury the past?

To this there are two replies. First, our moral life has a history and our past cannot be erased. To have a soul presupposes potentialities for doing what is right and what is good and also for doing what is wrong and what is evil. Evil, we have noted, is not itself a vice but a

---

[13] Vividly described in Simon Sebag Montefiore's *Stalin: the Court of the Red Czar* (Weidenfeld & Nicolson, London, 2003).

general form of deadly vices. One who is evil is someone who possesses a disjunctive set of vices: cruelty, sadism, callousness, and indifference to the suffering of others, vengefulness, deceitfulness, arrogance, contempt for others, envy, lust for power, malevolence, ingratitude, and so on. Deadly vices, no less than virtues, are shown in deeds. These are part of one's autobiography. One may cease doing evil, but that one has done evil does not cease to be true. If one has ceased to do evil, perhaps even strives to do good, can one not, as it were, regain one's soul – become a new person? No. First, it is not given to us to be more than one person in our life. We may change but we cannot cut ourselves adrift from our past, and the evil we have done is an ineradicable feature of who we are. Moreover, the web of vices that characterize evil-doers is not easily discarded. It is not a coincidence that barely any Nazi *génocidaires* showed the slightest remorse when brought to trial.

Secondly, for the most part, those who have done evil can lead a 'normal' life only at the cost of drastic compartmentalization of their experience, suppression of memories, acute duality of personae, dishonesty in their relationships with their children, and living with large areas of taboo in personal relationships. Being evil *ensures* inner disunity of personality. Evil-doing is a psychologically disintegrative force. There is nothing to unify the vices that is comparable to practical wisdom, which unifies the virtues. One may say that, by the evil they have done, evil-doers have foreclosed the possibility of being a member of a moral community, no matter how respectable a member of the social community they may have become. Were they to regain their own moral sensibilities, which they have destroyed by their evil deeds, they would be unable to forgive themselves for what they have done. They would have to damn themselves. However, they may retain their commitment to their evil beliefs, as presumably Lucius Cornelius Sulla did, and continue to think of the evil they have done as warranted, as Eichmann did.

3. Forgiveness and self-forgiveness

Can evil-doers (of whatever stamp) not redeem themselves? Can they not admit their evil to others and to themselves, feel genuine remorse, desire to atone, and strive to live a morally good life? Can they not be forgiven? Should they not be forgiven? After all, to have a forgiving nature is a virtue, and to be unforgiving is a vice. Indeed, it is not merely an other-directed vice manifest in cases where one ought to forgive a wrong-doer; it is also a self-directed, self-harming vice. For those who cannot forgive another person (in particular a parent or child, a friend or lover) lock themselves into past wrongs suffered, and thereby into festering resentments and bitterness that sorely damage themselves. Moreover, Christians believe that God can forgive all sins, given appropriate repentance and atonement. That God can forgive evil-doers, no matter what horrors they may have committed, is surely problematic and disturbing. Would God, if there were a God, have the *right* to forgive moral monsters who have committed atrocities, no matter how much they might come to feel remorse and strive for atonement? Evidently investigation and reflection are needed. The concept of forgiveness has to be elucidated.

Forgiveness   To forgive is to cease to resent a wrong done to oneself or to others who stand in an appropriate relationship to one, to forgo the demand for punishment, vengeance, or recompense, to abandon hatred of the offender. To say 'I forgive you' is a quasi-performative that commits one to abandon holding the wrong-doer to account and to cease resenting his having done wrong. It is akin to 'I believe you' (see *IP*, 212–17). But one may forgive someone without saying 'I forgive you', just as one may believe someone without saying 'I believe you'. Forgiving, like believing, is neither an act nor an event that happens to one. It is neither voluntary nor involuntary. One may forgive a misdemeanour or minor wrong on pronouncement of the quasi-performative. In the case of real wickedness, forgiveness may take time, sometimes a long time. Nevertheless one is answerable for one's forgiving when one should not have forgiven (as Caesar, prior to his assassination, had forgiven many of those who later murdered him), and for not forgiving when one should have done so (e.g. Monsieur Gillenormand in *Les Misérables*, who forgave neither Marius nor his father Colonel George Pontmercy). One may be under an obligation to forgive a wrong-doer (one might forgive him for the sake of the love one bore his mother), or be under an obligation not to forgive (as Hannibal Barca swore to his father Hamilcar never to forgive Rome for the First Punic War), but the obligation cannot be *to* the wrong-doer, since no one has a right to be forgiven. One may beg forgiveness but one cannot demand it. In the case of evil, even begging for forgiveness is overstepping the mark – at most one may beg for mercy. It may be that a victim of wrong-doing *ought* to forgive another in certain circumstances, even though he or she has no obligation to do so. In order not to destroy a valued relationship one ought perhaps to forgive one's spouse for a wrong done. But there are circumstances under which one may be *unable* to forgive someone, just as one may be unable to believe someone. In such cases, the reasons for not forgiving are too overwhelming for one to disregard them, as in the parallel case the reasons for not believing are too weighty for one to overlook. And some people are more forgiving than others by nature.

Forgiving and forgetting Forgiving is compatible with but does not always imply subsequently forgetting. For someone who forgives cannot pretend that the wrong done never took place. When the forgiveness is extended to a close friend who has wronged one and who has expressed remorse in speech and in deed, and made amends, it may well be that with time one will scarcely remember the episode and the friendship may be restored. That one then barely remembers may be a mark of the depth of one's forgiveness. But it may also be the case that, if the friend's wrong-doing was a form of treachery or betrayal, one may forgive him but be unable and unwilling to forget, for one now knows something about him that one did not know before, namely that he is capable of betraying one.

Forgiving, accepting apologies, pardoning Forgiving is not the same as accepting an apology. Catholic children who suffered sexual abuse at the hands of the clergy, and British children who were removed from their parents by social workers and sent to Australia where they were maltreated, may respectively accept an apology from the pope or the prime minister, but that does not imply that they forgive their persecutors. It means that they acknowledge the sincere admission of wrong-doing or wickedness. Whether they also forgive is a further question. Forgiving is different from pardoning, which is a juridical notion, whereas forgiving is a moral one. An authority may pardon criminals, but that does not imply forgiving them. Indeed, while a judge may pardon a criminal for all manner of reasons, he *cannot* forgive the criminal.

Forgiveness and reconciliation What is the relation between forgiveness and reconciliation? Does forgiveness imply reconciliation and restoration of the relationship between victim and wrong-doer? There seems no simple answer here. It depends on the nature of the wrong-doing, on the relation between the victim and the wrong-doer, and on their respective characters. A large variety of permutations is possible, all of which may exemplify forgiving. In the case of antecedent deep friendship or love, in which the wrong-doer manifests genuine remorse, tries to make amends, and begs forgiveness, it may well be that full forgiveness implies reconciliation and attempts to repair the torn fabric of the personal relationship. One may indeed hold that a criterion for full forgiveness is reconciliation and restoration of the relationship. However, in the case of a friend who has betrayed one in some fashion, one may indeed forgive him and forswear revenge and animosity. If one forgives him, the injury he has done one will no longer rankle. One will speak no ill of him, nor shun him, but one may be unable or unwilling to restore the relationship to what it was before. One may still enjoy a lesser form of friendship and some degree of mutual affection, but it may be impossible to restore the trust precisely because one knows him to be capable of betrayal. In the case of a mere acquaintance who has wronged one there is no deep relationship to restore. If appropriate sincere apologies are tendered, one may forgive the wrong done one. The wrong done no longer counts as a reason for harbouring negative attitudes and responses, and civilized relations may be established but perhaps without any desire to restore the relationship to its previous cordiality. *A fortiori* in the case of wrong done to one by a stranger. Here forgiveness is possible given sincere expression of remorse, but there is no relationship to restore and one may indeed never want to see the person again.

Does forgiveness presuppose expressions of remorse and acknowledgement of guilt? Can one, should one, forgive someone who has wronged one but shows no regret, let alone remorse? Surely not. But again, the possibilities are multiple and the rightness of forgiveness is person relative and circumstance dependent as well as culture dependent. Losing one's temper and saying things in anger that should not be said is wrong. It requires a sincere apology. But one may forgive such wrongs without an explicit apology, out of love of one's spouse or child – or, of course, one may just put up with their character flaws. More serious offences that amount to wickednesses do surely require genuine remorse and, where possible, attempts to make amends for the wrong done (as Rochester is forgiven by Jane Eyre or Darcy by Elizabeth Bennett).

Locus standi for forgiveness   However, not just anyone can forgive. In order to be able to forgive one must have an appropriate *locus standi*. The victim, if he survives the wrong-doing, may forgive, but a friend of the victim is not in a position to forgive the miscreant. The parents of a murdered child may forgive the murderer, but, though others may approve of the parents' forgiving nature, they are not in a position to forgive the murderer themselves.

Similarly, the children of a parent who was killed through the fault of another may forgive the other in appropriate circumstances, but not the friends of the children, although they may not hold it against him.

Self-forgiveness When reflecting on forgiveness, we naturally think of one person forgiving another. But there is also such a thing as self-forgiveness. It is by no means uncommon to speak of not being able to forgive oneself or to say that if one did not do such-and-such, one would never be able to forgive oneself. But what is self-forgiveness? After all, it would be odd to think that one has to have any *locus standi* in order to forgive oneself. It would be absurd to suggest that everyone has the appropriate *locus standi* to themselves – this would be akin to holding that in using the first-person pronoun one inevitably successfully refers to oneself (*HNCF*, 9.2). As with many reflexive verbs, there is an asymmetry between the first-person and second- and third-person cases:

(i)   One cannot beg forgiveness from oneself.
(ii) One cannot forgive oneself by sincerely saying to oneself 'I forgive you'.
(iii) In forgiving oneself, one is not foregoing a demand for punishment or vengeance – although one may demand atonement from oneself.
(iv) To forgive oneself is not to manifest the virtue of forgiveness, even if one ought to forgive oneself in a particular case. Hence in forgiving oneself one is not manifesting one's forgiving nature.
(v)   Whereas one may wrong another (e.g. insult them), and on realizing how unjust one was immediately apologize and be forgiven, there is no analogue of this in the case of self-forgiveness. One cannot apologize to oneself for the wrong one has done.

In order to forgive oneself at all, one must have gone through a long period of loss of self-respect, of despising oneself for the wrong one has done, perhaps even of self-hatred. It is not feeling guilt, which is presupposed, but *feeling profoundly ashamed of oneself* that is characteristic of having done wrong and being unable to forgive oneself. There is such a thing as self-forgiveness just as there is such a thing as self-deception which is disanalogous to deceiving another, and there is such a thing as self-love although it is not analogous to loving another. To forgive oneself is to cease hating oneself for what one has done, to stop despising oneself for the wrong one has committed and to cease holding oneself in contempt. It is to regain one's self-respect despite one's awareness of, and guilt for, the wrong one has committed.

4. Evil and the unforgivable

Are there things that are *unforgivable*? It is evident that if evil-doing is understood as leaving the pale of morality, as passing, so to speak, a point of no return, then evil, by contrast with wickedness, *is* unforgivable. What that claim amounts to is that it would be wrong to forgive evil-doers. Can't evil-doers feel genuine remorse and redeem themselves? An example will help our reflections – the Leopold and Loeb case. In 1924 Nathan Leopold (1904–71) and Richard Loeb (1905–36), exceptionally intelligent and gifted students, murdered fourteen-year-old Robert Franks in cold blood simply to prove that they could commit the perfect crime, while also holding that they were Nietzschean supermen who had passed beyond the constraints of good and evil. Both in prison, and after his release on parole in 1958, Leopold dedicated himself wholeheartedly and unremittingly to good works in prison, and subsequently continued to labour for the benefit of prisoners. Did he not thereby redeem himself and return across the dreadful borderlands between wickedness and evil and earn forgiveness? Could others forgive him? Could the parents of the murdered boy forgive him? This seems very unclear. Would they have had a right to do so? Suppose the boy had died in the arms of his parents, begging them with his dying breath to forgive his murderer, as Jesus begged his supposed father to forgive those who had crucified him 'for they know not what they do'? Perhaps. And if Leopold spent the rest of his life trying to redeem himself by good works, might he not succeed? This too is unclear. Perhaps we should ask not whether others might rightly forgive him, but whether he might rightly forgive himself. Arguably not, *if* he had come to understand the difference between good and evil, and to grasp the enormity of his deed. This he did. He evidently strove mightily and *admirably* to redeem himself. But he surely knew that what he had done was unforgivable – that *he could not forgive himself*. To do evil is not to incur a debt that may subsequently be discharged.

Can génocidaires be forgiven? Matters seem clearer in the case of those guilty of crimes against humanity or of genocide. It seems evident in these cases that no human being could possibly have the right to forgive them. Though they are not *more* evil than someone who tortured a single child to death, the magnitude of the evil they have done is monstrous. No one could possibly stand in a special relationship to the vast number of dead that would give them the *locus standi* to forgive such monsters as Enver and Talaat, Hitler and his henchmen, Stalin, Mao, Pol Pot, and so forth. The survivors of such holocausts may be able to forgive their persecutors, but it is *not* evident that they ought to. But be that as it may, they can forgive *only for themselves* – they cannot forgive in the name of the thousands, hundreds of thousands, or millions of dead. Indeed, it lies so far beyond the bounds of intelligibility that it is altogether opaque what would *count* as forgiving the perpetrators, as opposed to asking for mercy for them.

Theists contend that God may forgive the *génocidaires* if they show appropriate remorse and make appropriate atonement. But what would give God the right to do so? Was it the wish of the victims that their torturers be forgiven? How could God forgive Eichmann or Heydrich without the victims rising from their graves and cursing him? The obvious religious answer is that we are all children of God, and he stands to each of us in an appropriate relationship of loving parenthood that *does* give him the *locus standi*. But is this convincing? What sort of a father could stand by and watch the masses of his children being slaughtered? This is a matter of faith (for Christians) and of trust (for Jews). But it is of questionable moral intelligibility (Kierkegaard would presumably agree).

Can evil-doers be happy? Can an evil person be happy? Socrates certainly thought not: in his debate with Polus about Archelaus, the murderous ruler of Macedonia, Socrates confutes Polus: 'But I say that it is impossible [that the evildoer be happy]' (*Gorgias*, 472d) – 'the distemper of evil [will become] engrained and [produce] a festering and incurable ulcer in his soul' (480b). It is striking that when Raskolnikov confesses his crime to Sonya, she exclaims, 'What have you done – what have you done to yourself? ... There is no one – no one in the whole world now so unhappy as you!'[14] So, can those who do evil be happy? Can they be at peace with themselves? Those who are racked with hatred, contempt, rage, fear, vindictiveness, and lust; who derive joy from domination, murder, cruelty, ruthless exercise of power, and malice; who are arrogant,

---

[14] Dostoyevsky, *Crime and Punishment*, V. 4.

narcissistic, and self-centred; for whom others are either sycophantic supporters, adversaries who must be destroyed, or mere means to further their ends, surely cannot be at peace. Perhaps so. But they may be triumphant, exultant, and self-satisfied in attaining their goals, crushing their enemies, and bathing in the admiration of their supporters and sycophants or of the masses they lead. Individual followers of evil leaders may do great evil, as *génocidaires* do, but if they incur no condemnation from their own society they will not necessarily be racked with guilt. Indeed, they may pride themselves on having served their leader and their country, or God and his purposes. They may, in a deep sense, have destroyed their soul – but they are unlikely to miss it. They may have debarred themselves from leading a fully human life, have cut themselves off from the *telos* of man – but that is unlikely to be something they would be capable of understanding.

Can evil-doers love? That too seems moot. To be sure, they may think to do so, mistaking lust for love, and love for a desire to be loved and admired (which seems to be characteristic of great dictators[15]). To love another one must be able to give oneself, to transcend one's egotism and one's selfish concerns. But that possibility exists in inverse proportion to the egocentricity, arrogance, narcissism, and domineering character of the evil-doer. Nevertheless, it seems that familial love may survive evil-doing, as is evident among the Mafiosi, who are noted for their love and loyalty to their mothers and children.

The withering of the soul One may not destroy one's soul, but, as Anna Semyonovna observed, one may merely allow it to wither. It is in this sense that Ian Kershaw said that the road to Auschwitz was paved with indifference. It is this that was manifest, as Victor Klemperer so vividly described,[16] when, after he had been excluded from his university post by the Nazi decree against Jewish academics, his fellow academics in Dresden crossed to the other side of the street when they saw him coming. The withering of one's soul may ultimately lead one to destroy it, as was patent in the Kielce pogrom in Poland in 1946 in which Jewish survivors from death camps, who returned to the Polish village that had been their

---

[15] When Stalin's second wife, Nadya, committed suicide, he is said to have exclaimed 'How could you do this to me?'

[16] Victor Klemperer, *I Shall Bear Witness: the Diaries of Victor Klemperer, 1933–1941*, abridged and trans. Martin Chalmers (Weidenfeld & Nicolson, London, 1998).

home were murdered by the Polish villagers who had expropriated their houses and possessions.

But one's soul may shrivel and darken even when one is a bystander to evil without being indifferent to it. For one may *fail to stand up* to it. In his last novel, *Everything Flows*, Vasily Grossman chillingly describes a successful scientist Nikolay Andreyevitch as he realizes that he has become smug in his refusal to support evil:

> The State had taken on its iron shoulders the entire weight of responsibility; it had liberated people from the chimera of conscience. ...

> Examination of one's own self – how very unpleasant it was. The list of one's despicable acts was unbelievably odious.

> It included general meetings of the Institute; sessions of the scientific council; solemn meetings on important anniversaries; routine meetings in the laboratory; banquets; celebrations in the homes of the important and evil; jokes told during dinners; conversations with directors of personnel departments; letters he had signed; an audience with the minister.

> And the scroll of his life contained all too many letters of another kind: letters unwritten – although it had been his sacred duty to write them. Silence – when it had been his sacred duty to speak; a telephone number it was imperative to ring, and that he had not rung; visits it was sinful not to pay, and that he had not paid; telegrams never sent; money never sent. Many, many things were missing from the scroll of his life.

> And, now that he was naked, it was absurd to take pride in what he had always prided himself on: that he had never denounced anyone; that he had refused, when summoned to the Lubyanka, to provide compromising information about an arrested colleague; that instead of turning away when he happened to meet the wife of an exiled colleague, he had shaken her hand and asked after the health of their children.

> No, he did not have so very much to feel proud about.[17]

The moot question now is what judgement to pass when the political circumstances become so dire that one needs to be a hero, perhaps even a suicidal hero, to stand up to be counted? Out of the blue, Boris Pasternak was telephoned by Stalin and asked for his opinion on his friend the great poet Osip Mandelstam, who had been arrested for

---

[17] Grossman, *Everything Flows*, ch. 3.

sedition. Pasternak was so terrified that he merely stuttered and stammered, too frightened to plead for his friend. Stalin had Mandelstam shot. *We* have no right to criticize Pasternak – but Pasternak could surely never forgive himself. Had Mandelstam been sentenced to decades in the gulag, *he* might have forgiven Pasternak on his release. But would even that have allowed Pasternak to forgive himself?

5. From soul to soul: trisecting an angle with compass and rule

The idea of a secular notion of the soul is fruitful. It enriches the ways of talking about our moral life. It facilitates clarity of thought about evil and evil-doers, enabling us to draw distinctions and to emphasize features that are deep, just as our talk of the mind *properly understood* enables us to emphasize distinctive aspects of the intellectual and volitional powers of mankind. I have emphasized 'properly understood', for the conception of the mind that I have advanced in this tetralogy on human nature is very far removed from the notions of the mind that have dominated European philosophy from Descartes to Wittgenstein (for detailed discussion, see *HNCF*, chs 8–10) and continue to do so. The concept of a soul that has been deployed in this section is equally far removed from the received conception of the soul in philosophy. How are they related? How can one make the transition from the traditional idea of the soul as an immaterial, spiritual, immortal substance to the secular conception of the soul that I have made use of in this chapter to shed light on the moral powers of mankind? Perhaps a Wittgensteinian analogy will help.[18]

Trisecting an angle with compass and rule

Mathematicians in ancient Greece raised a deceptively simple question concerning Euclidean plane geometry. It is very simple to show that there is a way of constructing a bisection of an arbitrary angle with a compass and rule. But is it possible to construct a *trisection* of an

---

[18] It is gratifying to find the following remark in Wittgenstein's *Remarks on the Philosophy of Psychology*, vol. 1, ed. G. E. M. Anscombe and G. H. von Wright, trans. G. E. M. Anscombe (Blackwell, Oxford, 1980), §586: 'Ist es also irreführend, von der Seele des Menschen, oder von seinem Geist zu reden? *So* wenig, dass es ganz verständlich ist wenn ich sage: "Meine Seele ist müde, nicht bloss mein Verstand"' ('Then is it misleading to speak of man's soul or spirit? So little misleading, that it is quite intelligible if I say "My soul is weary, not just my intellect"' (modified translation)).

arbitrary angle? Mathematicians struggled with the problem for centuries. The solution (which is that it is not possible) was discovered only in 1837 by Pierre Laurent Wantzel, who transformed the geometrical problem into a question in algebra and trigonometry. He showed that a certain cubic equation taking trigonometric values cannot be solved by any of the four basic arithmetical operations or by taking square roots (which are the limits of what can be represented by constructions with a compass and rule). This proof would not have been intelligible to ancient Greek mathematicians, ignorant as they were of algebra and coordinate geometry. But suppose we were patiently to explain to a pupil, who knew no more than a competent ancient Greek mathematician, the basic principles of algebra, the principles of coordinate geometry, and the methods of solving cubic equations, and then explained Wantzel's proof to him. He would accept it without ado (he is, after all, perfectly competent). And now we might say that we have changed his idea of what he was trying to do. Indeed, we should have changed his idea of trisection. We should have changed his way of looking at the problem. He might even say 'I see! *That* was what I was really trying to do', although that was not what, in the ordinary sense of the phrase, he was trying to do. But his way of looking at the problem has changed fundamentally.[19]

Usefulness of concepts of mind and soul

One might say something similar about the relation between the ancient Platonic, neo-Platonic, Christian, and Cartesian notions of the soul and the secular, neo-Socratic, concept of the soul that has been utilized above. In craving to understand man's relation to good and evil, Western culture turned to the Christian conception of the soul. We can now show that the traditional Christian notion of an immortal soul makes no sense. We can also show that the Cartesian notion of an immaterial substance that is the locus of good and evil makes no sense. We can go on to demonstrate that the idea that there is a soul that is the locus of good and evil that stands in a causal relation to the brain makes no sense. But we might then explain that, as we know perfectly well, among the powers of a normal human being is knowledge of good and evil, possession of a conscience, the

---

[19] See Wittgenstein, *Philosophical Grammar*, ed. Rush Rhees, trans. Anthony Kenny (Blackwell, Oxford, 1974), 387–92; *Wittgenstein's Lectures on the Foundations of Mathematics, Cambridge, 1939*, ed. Cora Diamond (Harvester Press, Hassocks, 1976), 87–90; *Philosophical Investigations*, §334; *The Blue and Brown Books* (Blackwell, Oxford, 1958), 41.

potentiality to feel compassion and selfless love, the susceptibility to shame, guilt, and remorse. And we may go on to show him that it is fruitful to view such powers and their exercise as constitutive of the soul, just as we view the intellectual and volitional powers of human beings as constitutive of the mind. It must, however, be understood that the concepts of *mind* and *soul* are not concepts of 'things' of any kind, but rather *elements of a fruitful form of representation*. This conception of a soul provides the understanding that was craved in a secularized form readily intelligible to a European cultural tradition and an erstwhile Christian civilization. This is parallel to showing that the problem of trisecting an arbitrary angle with compass and rule cannot be resolved by the methods of Euclidean geometry, but that the understanding that was craved can be achieved by means of a different mathematics. If we have displayed the requisite Socratic skills, our pupil might indeed say, 'Ah, so that was really what I trying to say' or 'That was what I really had in mind'. For we should have led our pupil from one symbolism to another *with his consent*.

Does this show that the concept of a soul is *necessary* or *unavoidable*? No, of course not. It is no more necessary and unavoidable than the concept of a mind. But it is an exceedingly useful concept for the purpose of talking about the lives of human beings and their engagement with each other, and it carries a weight that would be absent if we were to do without it. Were it to become obsolete, that would betoken an impoverishment of our thought and of our humanity unless it were replaced by a different form of representation that served a parallel and comparable purpose.[20]

---

[20] Different cultures may meet this need in different ways. It is of interest to note that Hebrew, both ancient and modern, would not express the destruction of moral sensibility by talking of the death of the soul (*neshamah* or *nephesh*), but rather by speaking of the loss of the image of man (*tselem enosh*), which is itself the image of God. So what an evil person does is, as it were, destroy the divine spark in himself.

PART II

Of Freedom and Responsibility

6

Fatalism and Determinism

1. Of fate and fortune

Abilities, opportunities, and freedom to act We speak and think of ourselves as free agents, facing an open future, free to act or to refrain from acting as and when opportunities arise. Abilities to act are binary and generic, that is they are two-way powers, the power *to perform* and the power *to refrain* from performing *kinds* of act that *satisfy a relevant standard*. One has the ability to hit the bull's-eye, but not to miss it (even though one can miss it on purpose), the ability to swim but not to flounder (even though one is able to pretend to flounder), the ability to speak but not to stammer (even though one may have the ability to act the role of a stammerer). The exercise of an ability to act requires an opportunity. Opportunities are agent relative and are a function of the extent to which an ability has been acquired or a skill has been mastered. What is an opportunity for a master may be none for a beginner. Given an opportunity to act, a free agent can, other things being equal, act or refrain from acting at will. The 'can' here is not the generic 'can' of ability, but the individual 'can' of occasion. When we know what we are able to do, and know that there is an opportunity to exercise our ability, then, other things being equal, we can make free choices. If we decide to act, we are free to act in conformity with our decisions in pursuit of the goals we have in view, unless prevented by a change in circumstances. We do what

The Moral Powers: A Study of Human Nature, First Edition. P. M. S. Hacker.
© 2021 John Wiley & Sons Ltd. Published 2021 by John Wiley & Sons Ltd.

we do freely, save in cases of duress or being obliged, on the one hand, and of uncontrollable impulse or urge, on the other. 'It's your choice', we say, or 'Do make up your mind!' and 'It's up to you'.

Having no choice To be sure, we sometimes say, 'I had no choice', but that means one had no other *eligible* choice, that one's decision was non-causally necessitated by one's commitments that obligate one to act or by the unacceptability of the alternative course or courses of action that oblige one or force one to act. In such cases, the reasons for doing what one did and for avoiding the alternative were overwhelming. 'Here I stand. I can do no other' does not mean that one lacks the ability to do anything other than what one is doing. It means that one is totally committed, as a matter of principle, to the course of action one is taking and will not countenance any other. 'There was no alternative' generally means that there was no *viable* alternative. There *was* an alternative, but it was excluded by overwhelming reasons. The impossibility of acting in any other way is, in such cases, a moral, legal, political, economic, or practical impossibility – not an inability or lack of opportunity (fig. 6.1).

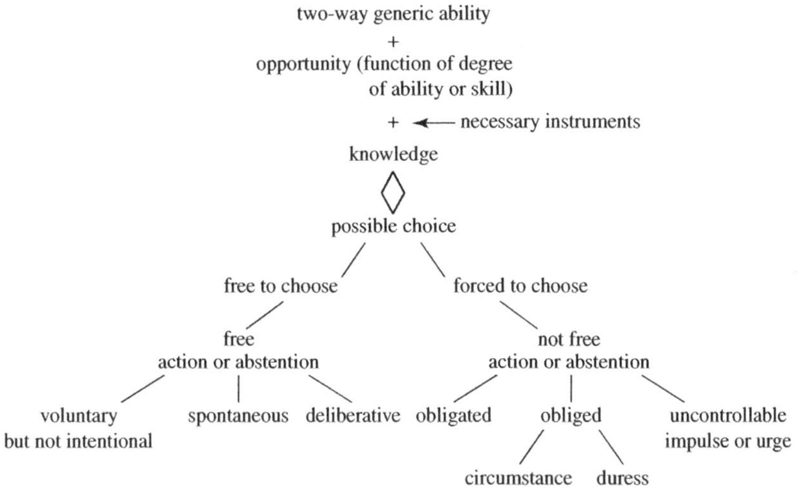

Figure 6.1 *Forms of freedom and lack of freedom*

Could have and would have We have distinct ideas of the modality of decision and action, although it takes uncommon talent to have a clear idea of the grammar of 'can', 'could',

'could have', 'will', 'would', 'would have', and their conditional forms 'could have if' and 'would have if'.[1] We have mastered the use of these expressions, but it is difficult to gain an overview of their use. We think, correctly, that if something happened it does not follow that it was not possible for it not to have happened (if it was beneficial, then we were lucky). Similarly, we think, equally correctly, that if something beneficial did not happen, it does not follow that it could not have happened (we may have been unlucky – you win some, you lose some). 'Could have V-ed' indicates that it *was possible* for someone or something *to have V-ed*. There is a temptation to think that 'could have V-ed' implies that the agent did not V, but actually it can be true that someone could have V-ed whether in fact he did V or did not. That the burglar could have climbed in through the back window is perfectly compatible with his having done so, and equally compatible with his not having done so. But we would not say that he could have if it is known that he did. 'Would have V-ed' implies 'could have V-ed', and indicates which of the possibilities open to the agent would have been taken.

Could have if and would have if   Generally speaking, of all the things that did not happen, some could have happened and others could not have happened. A did not visit me yesterday, but he could have (he is in town so it was possible *for* him to visit me); B did not visit me yesterday either, but he could not have (he is bedridden in hospital); he would have, had he not broken his leg. It was not possible *for* him to visit me, and therefore there is no possibility *that* he visited me yesterday. We distinguish between the conditions necessary or sufficient for the *actuality* (existence) *of a possibility* and the conditions necessary or sufficient for the *possibility of an actuality*. The actuality of a possibility does not imply actualization (that I leave the door unlocked makes it possible *for* you to enter, but that does not imply that you will enter). Hence we distinguish between 'could have if' and 'would have if'. The former indicates what would have made something possible but does not imply its actualization ('I could have entered if I had had the key'). The latter indicates what would have made it actual ('I would have entered if I had had the key'). To be sure, 'I can V, if I want (choose)' and 'I could have V-ed, if I had wanted to (chosen to)' are not implicative. For if I don't want,

---

[1] See A. R. White, *Modal Thinking* (Blackwell, Oxford, 1975), chs 1 and 2. I am indebted to White's analyses.

or didn't want, to V it does not follow that I cannot V or could not have V-ed. Wanting or choosing, in the case of human actions, are not normally conditions on possibilities – but they are in such cases as 'If you *really* want it, you can do it'. Foreseeing a human possibility ('A can ...') is not the same as foreseeing its actualization ('A will ...'), and foreseeing the actualization of a possibility ('A will ...') is not the same as foreseeing its moral, legal, political (and so on) necessity ('A must//has to//V'). An event E_1 or an agent A may cause an event E_2 to occur or make it happen. That means that if E_1 occurs or if A acts appropriately, then E_2 *will* occur. It does not imply that E_2 *must* occur or that it occurs *necessarily*. That there is a causal sequence does not imply that the consequent *has to* occur, but only that it *does* occur. It is an error to project the validity of the inference from cause to effect in accordance with a law onto the necessity of the consequent. But some consequences obviously are necessary: if A takes cyanide, his death is unavoidable: it not only will occur – it cannot but occur. Conversely, it is mistaken to suppose that, because it is foreseen that something will happen, it follows that it *must* happen. To foresee a future actuality is not the same as foreseeing a future inevitability. These conceptual articulations will be examined below.

Reactive attitudes   We hold people responsible for their deeds, unless they did whatever they did by accident or mistake (which was neither negligent nor reckless) or under exculpating duress.[2] We praise people for their attainments and blame them for their misdeeds or culpable omissions. We pride ourselves on our achievements and are ashamed or feel regret, remorse, or guilt for our failures or misdemeanours. This is all commonplace. It is difficult to imagine human life in society or even in solitude without thinking of ourselves thus. For these attitudes towards people are engrained not merely in our thought and talk, but also in our behavioural reactions and responses, both to the deeds of others and to our own

---

[2] There is ambiguity, already pointed out by Bentham in *An Introduction to the Principles of Morals and Legislation* (1789), ch. 8 n. a, in the use of 'voluntary'. Sometimes it is used as the opposite of 'involuntary', sometimes as the opposite of 'constrained'. According to the former usage, every intentional act is also voluntary. According to the latter, acts done under duress are intentional but not voluntary. If so then acts are either voluntary or non-voluntary, and non-voluntary acts are either involuntary or not voluntary. The matter is a locus of extensive disagreement among judges and philosophers of law. For detailed examination of the question, see A. R. White, 'Duress', in *Misleading Cases* (Clarendon Press, Oxford, 1991).

doings. We may feel resentful and angry, and exhibit resentment and anger, at intentional wrongs done to us. We may feel ashamed or guilty for our own intentional wrong-doings. We may be indignant or even outraged at deliberate wrongs or injustices done to others. We view unintentional, non-negligent wrong-doing as excusable. There is all the difference in the world between accidentally knocking a cup of tea all over someone and deliberately so doing. We are free agents tracing a spatio-temporal path through the world in accordance with our inclinations, preferences, choices, intentions, and decisions and in pursuit of our goals, subject to chance and fortune.

Freedom of action This is commonly characterized as 'having free will'. By and large I shall shun this terminology for it is misleading. In so far as the will is a rational power – an ability to act for reasons – it can be neither free nor unfree. As is already evident, we must distinguish between *ability*, *choice*, and *action* (see fig. 6.1). Lack of ability is not lack of freedom, and increasing one's abilities is not increasing one's freedom. That is why it is mistaken to conceive of education as an enhancement of human freedom. But ability is a presupposition of freedom of choice and freedom of action alike. To be able to choose freely and to act accordingly ('of our own free will'), we must possess the requisite generic ability. Lack of an ability may be remedied by experience, training, and education, given the appropriate second-order ability (the capacity) to learn. But one may have the ability needed, only to find it temporarily lost (e.g. through intoxication or injury) or externally constrained (as when one is tied up). The latter is indeed a constraint upon our freedom to act. Our *freedom of choice* may be contrasted with various forms of constraint that necessitate our choice, decision, or resolution. For we may be *caused* to choose or decide, for example by post-hypnotic suggestion, and we may be *forced* to choose or decide, by duress. We may find it practically impossible to choose *not* to perform a given act, being constrained by our commitments or by threats. The former obligate us; the latter force us. *That our acts, actions, and activities are free in one sense is built into the very ideas of a human act, action, or activity.* To act, take action, or engage in an activity is to exercise a two-way power. Absent that and one is dealing merely with movement that is not within the agent's control, as in the case of a reflex movement (misnamed 'reflex action'). Our freedom to act may be limited by lack of opportunity or by lack of the means (including instruments and equipment) to take advantage of an opportunity. Our freedom to act may also be limited by human preventative intervention that may deprive us of an opportunity (e.g. by locking the door), curtailing the exercise of our ability on the occasion of an opportunity (e.g. by tying us up), or by threats of serious harm to ourselves and our interests (duress) or to others whom we cherish (which might be called oblique duress). Acting under duress is done unwillingly and reluctantly. It is not involuntary, since what we do is under our control and done with knowledge of what we are doing. But it may be deemed to be not-voluntary inasmuch as the alternative (being subjected to the execution of the dire threat) is not a serious option for us. In such cases, we take ourselves to have overwhelming reasons not to refuse to do the act demanded.[3] However, acting under unbearable torture is not to act on an overwhelming reason.

So, we conceive of ourselves as active agents limited by the range of our own abilities, often constrained by circumstances over which we have no control and by people who have and exercise the power to frustrate our pursuit of our own goals – power that may be physical, legal, political, or economic.[4] But, within the constraints of chance and fortune, we are free agents. At its most stoical and heroic, this was expressed by the following lines from W. E. Henley's 1875 poem 'Invictus':

> It matters not how strait the gate,
> How charged with punishments the scroll,
> I am the master of my fate,
> I am the captain of my soul.

2. Fatalism

Nevertheless, in many different cultures the nagging thought that we are not masters of our fate is prevalent. Sometimes it is an attitude embedded in a culture, an attitude that may be universal, like *global* fatalism (i.e. that *everything* is preordained), or *individual* fatalism,

---

[3] Duress has been allowed as a defence in English law to criminal damage, arson, theft, handling, perjury, contempt of court, buggery, conspiracy to defraud, accessory to escape from prison, unlawful possession of ammunition, and treason.

[4] There is a philosophical form of fatalism rooted in Aristotle's reflections on tomorrow's sea battle. This was extensively discussed in the Middle Ages, and the issue was revived with the invention of tense logic in the 1950s and 1960s by Arthur Prior. It has more bearing on the logic of tensed propositions and the associated concept of truth than on the ways in which we think of ourselves as free agents. It will not be discussed here.

such as a person's sense of mission or destiny, on the one hand, or sense of something specific, local, being preordained, on the other. The former conceives of a future possibility of greatness or distinction being preordained (and it is up to the agent to seize the opportunity). It is evident in the lives of rare figures in history who have a powerful sense of destiny, a sense that they were born to fulfil a great purpose. This was true of Alexander and Caesar, of Napoleon and Churchill. The sense of something specific being preordained conceives of a specific future event (an actuality) as preordained (such as meeting the person one was ordained to love, or the time and manner of one's death).

Global fatalism   Global fatalism is often expressed by the refrain 'It is written'.[5] According to Sunni Islam, *qadar* (fate, predestination) is one of the six Islamic articles of faith, Allah having written down in the Preserved Tablet all that has happened, is happening, and will happen. According to Sunni belief, this is consistent with free human action; according to Al-Jabiriyah, we have no control over our actions for all is predetermined by God. In the *Rubáiyát*, Omar Khayyam (1048–1131) often appears to express the more extreme view:

> The Moving Finger writes; and, having writ,
> Moves on: nor all your Piety nor wit
> Shall lure it back to cancel half a Line
> Nor all your tears wash out a Word of it.
> (Rubáiyát 71)

Indeed, in *Rubáiyát* 73 it seems that everything has been ordained 'from the dawn of time':

> With Earth's first Clay They did the Last Man knead,
> And there of the Last Harvest sowed the Seed:
> And the first Morning of Creation wrote
> What the Last Dawn of Reckoning shall read.

Global fatalism is an attitude towards life, an attitude of resignation and acceptance of what happens (as is patent in the case of Omar Khayyam). Having such an attitude need not be inconsistent with

---

[5] Perhaps ultimately dating back to the legend of Belshazzar's feast, recounted in the book of Daniel (ca. 160 BC) in which a moving hand writes *Mene, mene, tekel, upharsin* on the wall.

making a maximal effort in pursuit of one's goals. But, if one fails, then one was fated to fail – for 'it was written'. It is also consistent with withdrawing from the hurly-burly of life and becoming a hedonist, as Omar Khayyam did. Global fatalism in the form of predestinarianism is typically, but not exclusively, associated with monotheism rather than with polytheism, and in particular with Christianity and Islam. Secular global fatalism is commonly expressed by the refrain 'What will be, will be', which is not intended as a theorem of tense logic but as an expression of resignation.

Divine intervention: Homer; fate

An individual form of fatalism consists in the belief that specific incidents in a person's life are preordained. Homer held that we are indeed but playthings of the gods, who may interfere in human life wherever and whenever they please.[6] They may determine our thoughts, our dreams, our emotions, our purposes, and our actions. Helen's moral blindness in eloping with Paris was produced by Aphrodite; Teucer would have won in the archery contest, but Apollo begrudged him victory and turned his arrow aside; Menelaus won in a fair fight with Paris, but Aphrodite intervened to save her favourite; and so on. Agamemnon excuses his own gross behaviour in depriving Achilles of Briseis by blaming his behaviour on the gods:

> I am not to blame!
> Zeus and fate and the Fury that walks in darkness,
> *They* are the ones who drove that savage madness in my heart,
> That day in assembly when I seized Achilles' prize –
> On my own authority, true, but what could *I* do?
> For it is a god who impels all things:
> Atë, eldest daughter of Zeus who drives all mad,
> Baneful creature that she is.
>
> (*Iliad*, XIX. 100–7)

The specific destinies of Homeric heroes are determined. Agamemnon was destined to be murdered by Clytemnestra; Hector's death at the

---

[6] A sentiment echoed by Omar Khayyam in characterizing the activities of the 'Master of the game':

> But helpless Pieces of the Game he plays
> Upon his checkerboard of nights and Days
> Hither and thither moves, and checks and slays,
> And one by one back in the closet lays.
>
> (*Rubáiyát* 69)

hands of Achilles was predetermined; it was Odysseus' fate to wander the seas for ten years after the end of the Trojan War. It is the gods who ordain the fates of men.

In Greek mythology, the Moirai (the three Fates) – Clotho (the spinner), Laches (the measurer of the span of life), and Atropos (the inexorable cutter of the thread of life) – represent the inevitable natural order in life and the idea that it is in general wrong to go beyond what has been allotted to one. Such hubris will inevitably end in disaster. Even the gods are subservient to the Moirai. In Roman mythology these dark and mysterious figures are the three Parcae, and much later, in Norse mythology, they are the three Norns. It seems that one could make offerings to them in the hope of appeasing them. The coherence of these beliefs and their consistency with other ancient beliefs in Greek, Roman, or Norse cultures is moot.

Individual fatalism: Sophocles; fate

Some centuries after Homer and Hesiod a different form of individual, local fatalism was expressed by Sophocles in *Oedipus Tyrannus*. Oedipus is described as being destined to kill his father, marry his mother, and beget children by her. But, Sophocles avers, he is not manoeuvred by Apollo, nor does Apollo intervene. On the contrary, everything Oedipus does he does of his own free choice – as he admits:

> Apollo, friends, Apollo
> Has laid this agony upon me,
> Not by his hands; I did it.
> (ll. 1581–3)

For Oedipus freely chose to murder the angry old man who attacked him at the crossroads, not knowing that it was his father Laertes, and he freely chose to marry Jocasta, the beautiful queen of Thebes, not knowing that she was his mother. Nevertheless, despite his free choice, it is still asserted that he was fated (by Apollo, although not through his intervention) to do so. No matter what path he took through life, it would come to this. It is far from clear whether this is coherent.

Local fatalism appears to be common to many different cultures and societies. It is expressed in the Anglo-Saxon poem *The Wanderer* or *The Exile's Lament* (seventh century): 'Wyrd bid ful āraed' ('Fate is inexorable'). The utterance expresses acceptance of the misfortunes of life, but it is compatible with strenuous endeavour. In the poem, it is used to mourn the death of the warrior's lord and of the warriors he loved and the tragedy of life bereft of his comrades and the joys of the mead hall.

Individual fatalism is associated with other important events in human life, such as meeting the person with whom one is destined to fall in love. That meeting, despite its apparent fortuitousness, is often conceived to be ordained by fate. This sentiment is common in the lives of many human beings to this day, especially in cases of love at first sight that flower into a permanent relationship. What was an accidental meeting is thought to be destiny, and the reciprocal love is sometimes expressed by such words as 'You were meant for me'.

In some cases of particular fatalism, the assumption seems to be that, no matter which route one took through life, one was destined to come to *this* point. That is implicit in Sophocles' *Oedipus*. It is nicely articulated by O. Henry in his tale 'Roads of Destiny' in which the protagonist approaches a crossroads and a separate tale is told of his different adventures as he takes each of the three roads, only to end in the same tragic way – 'It was meant to be so!' – although by whom it was meant is not always clear. Fate, in such a context, is merely a personification of its being meant to be so, not an answer to the question of who meant it. This form of individual fatalism is obviously not determinist, since *no matter what the person does*, he is destined to reach the same point. So his reaching that point is not predetermined by a unique set of antecedent conditions, as is required for determinism.

The appeal of individual fatalism

The ubiquity of the various forms of individual fatalism suggests that they have a deep appeal to human sentiment. This is perhaps because such fatalism may relieve one of anxiety in conditions of stress, especially during war. If one is fated to go under, then there is nothing one can do about it and so there is no reason to worry about what is inevitable. If one is not fated to go under, then there is nothing to worry about either. So either way there is nothing to worry about and one might as well just get on with it, shed the burden of anxiety, and face the future with fortitude.

One might say that such fatalism is a peculiar, inverted tribute to, and acknowledgement of, the role of chance and fortune in human life. For, to be sure, there is no destiny, nor is there any predetermining fate that governs or intervenes in our lives. But we are the victims of chance, the recipients of good or bad luck – and at the mercy of Fortuna, whose wheel turns inexorably. This is something to which we must be reconciled, for to rail against fortune is futile, and to resent it is to undermine one's ability to live wisely within its constraints.

Old Testament

As noted above, global fatalism is also associated with monotheism, in particular with Christianity and Islam.

It stemmed at least in part from the apparent difficulties of reconciling God's omniscience with human freedom, and God's freedom to grant us grace with his benevolence. This conception of fatalism played little if any role in the Hebrew Bible. The doctrine of original sin and the consequent doctrine of the necessity of grace are Christian, being prominent in Augustine though originating earlier. Although the Hebrew Bible held that God knows all things, it also insisted upon human freedom to choose between good and evil, and ascribed responsibility to human beings for their deeds. The idea that these two commitments might clash was not taken seriously. In the post-biblical period, according to Josephus, the Sadducees rejected the idea of fate, the Essenes embraced it, and the Pharisees took up an intermediate position. The issue was not extensively debated in the great period of rabbinical theology after the fall of the Second Temple. The most emphatic statement of compatibility of divine foresight and human freedom was given by Rabbi Akiva (ca. AD 50–ca. AD 132): 'Everything is foreseen, and man has the ability to choose freely' (Avot 3: 19).[7] The matter became a focus of argument and controversy in Judaism only in the medieval period (e.g. Saadia Gaon (882–942), Yehuda Halevi (1075–1141), Maimonides (1135–1204), and Gersonides (1288–1344)).

Christianity and divine omniscience   By contrast, in the age of Christianity, the compatibility of free agency and God's omniscience constituted a deep and ramifying problem for theologians from the start. For, if God is omniscient, then he knows in advance what we are going to think, choose, and decide to do, as well as knowing what we shall do. Indeed, he must surely have known, since the dawn of time, precisely what we are going to do. But if so are we really free at all? For, if God knows what we are going to do, we are surely not free to do anything but what we actually do! The paradox of knowledge of the future was vastly exacerbated by the Christian doctrine of the Fall of Man and the consequent dogma of inherited original sin. For, it was argued, given that we are all born in sin, we need the grace of God to resist evil and to be saved from hellfire. What then determines whether we shall be given grace? Surely God must know, at any rate since the Fall (if not since before the dawn of time) to whom he is going to grant grace and from whom he will withhold it. Must he not give grace in accord with desert? That

---

[7] Rabbi Ovadia of Bartenora interpreted Akiva's assertion as meaning that everything is seen and we can choose freely, that is we can choose freely but must remember that God is watching us.

would, according to predestinarians, be an unwarranted limitation on God's infinite power. So grace is granted arbitrarily, quite independently of merit. So our lives (according to global predestinarians), or our salvation (according to one form of local predestinarianism (e.g. Augustine and Luther)), is determined.

The issue of human freedom was intensely debated by Christian theologians in antiquity, throughout the Middle Ages, and well into the early modern era. It was the subject of more than one of Augustine's treatises. In his early writings against Manichaeism he defended a compatibilist view, but as he grew older predestinarianism took hold of him as is evident in his fierce anti-Pelagian polemics towards the end of his life. The matter flared up again during the Reformation, being a central concern of Luther (much influenced by Augustine and his followers) and especially of Calvin. Reform Protestantism found an able predestinarian preacher in Jonathan Edwards in eighteenth-century Massachusetts, whose book on freedom of the will is a high point in the history of American philosophy and theology.

Foreknowledge and freedom
In fact there is no contradiction between complete foreknowledge and freedom, since complete foreknowledge includes knowledge of the existence (actuality) of future possibilities as well as knowledge of their future actualization or non-actualization. If God is omniscient, then it follows that he knows what we are going to do and *also* what we could have done. He knows our abilities and also the opportunities that are available to us. It does not follow from his foresight that everything we are going to do is necessary or inevitable. For although we acted as he knew we would, he also knew that it was possible for us to act otherwise.[8]

To be sure, there are many confusions concerning abilities, possibilities, and when one could, and when one could not, have done otherwise, all of which stand in the way of grasping this resolution of the problem. The matter is of interest to believers, theologians, philosophers, and historians of ideas. It will not be further pursued here inasmuch as it contributes nothing to philosophical anthropology or to a naturalist understanding of human nature and the nature of moral value, and is not the source of modern determinism.

---

[8] Where there *is* a contradiction is in the notion of an atemporal and non-spatial omniscient being who can foresee the future, and the idea of a benevolent and omniscient being who can foresee the evil in the world he creates (especially natural evil), but who bears no responsibility for it.

3. Nomological determinism

Ancient atomism In the ancient world, it was the Greek atomists such as Leucippus (fifth century BC), Democritus (ca. 460–ca. 370 BC) and later Epicurus (341–270 BC) and subsequent Roman Epicureans, in particular Lucretius (95–55 BC), who adopted an early form of atomist determinism rendered compatible with free will by introduction of arbitary swerves in the downward motion of particles. It was only with the rise of modern science in the sixteenth century and the discovery of precisely quantifiable laws of physics, valid for the movements of both supra-lunary stellar objects and for terrestrial bodies, that the modern forms of *nomological determinism* began to emerge. The rediscovery of a single surviving copy of Lucretius by Poggio Bracciolini in 1417 ensured the transmission of Lucretius' ideas to early modern science and philosophy. They had a significant influence on the thought of Galileo (1564–1642), and later Gassendi (1592–1655).

Early modern nomological determinism The scientific advances of the early modern era unified terrestrial physics with the geometry of the heavens, and discovered that spatio-temporal objects and their motion are subject to universal laws. Kepler (1571–1630), unlike Copernicus (1473–1543), was concerned not merely with 'saving appearances', but also with discovering mathematical (algebraic) laws governing the orbital movements of all planets. These were subsequently incorporated as special cases into Newtonian physics. Biology also started to yield its secrets: the field of anatomy developed rapidly, the first comprehensive atlas of human anatomy being compiled by Vesalius (1514–64) and the circulation of the blood being discovered by Harvey (1578–1657). Descartes (1596–1650) produced a metaphysics tailored to the new sciences, and held that biology was reducible to purely mechanical (geometrical) principles and that the whole of nature, animate and inanimate, with the exception of the human mind, was subject to causal law.

Clockwork It is not a coincidence that the gradual emergence of nomological determinism coincided with the development of machines, in particular of ever more sophisticated clockwork, mechanical clocks, and automata of wonderful complexity. The analogy between the clock-maker and his clock and God and the universe did not escape scientists and philosophers (e.g. Kenelm Digby, Gassendi). We are always prone to model our ideas of nature (as well as of ourselves) on our own inventions. It is noteworthy that the planetary system as explained by Newton on the one hand, and mechanical clocks and automata, on the other, were the primary examples available of apparently nomological determinist systems subject to observation and experimental validation. These mesmerized philosophers and scientists alike.

Dualism and Humean compatibilism

While Descartes embraced nomological determinism for the material world, he eschewed causal closure for it, holding that the mind was capable of intervening causally in it. However, he never found a satisfactory explanation of how causal interaction between an immaterial mind and the material pineal gland (which was the alleged locus of interaction between the mind and the body) is possible. Descartes's successors, lumbered with dualism but rejecting causal interaction, struggled to preserve human freedom while acknowledging nomological determinism. They experimented with various apparent possibilities – occasionalism, the two clock analogy, Monadology – but to little avail. The rise of British idealism meant jettisoning the domain of matter as conceived by Cartesians, on the one hand, and differently by Lockeans, on the other. In the hands of Hume, the nomological character of the Cartesian, Lockean, and above all Newtonian universe was projected onto the mind and its ideas and impressions. Hume held that ideas stand in causal relations to each other, so the mental is nomological despite being immaterial. Freedom, Hume insisted, implies neither that one could have done otherwise nor absence of causal determination, but rather absence of constraint. Denial of determinism implies randomness, and to act randomly is not to act freely at all. So freedom actually requires nomological determinism. Accordingly freedom and determinism are perfectly compatible.

Physical determinism: Laplace

The eighteenth century saw the emergence of incompatibilist psychological determinism – in the works of Baron d'Holbach (1723–89) and Denis Diderot (1713–84). Physical determinism was given a vivid representation in the works of the great mathematician and physicist the marquis de Laplace (1749–1827) in the early nineteenth century. According to Laplace,

> We may regard the present state of the universe as the effect of its antecedent state and as the cause of the subsequent one. Given an intelligence which at a certain moment could comprehend all the forces by which nature is set in motion and the respective positions of all things of which nature is composed – an intelligence sufficiently vast to submit these data to analysis – it would embrace in a single formula the

movements of the greatest bodies of the universe and those of the smallest atom; for such an intellect, nothing would be uncertain and the future, like the past, would be present before its eyes. (Pierre-Simon Laplace, *A Philosophical Essay on Probabilities* (1814))

So, any state of the universe is determined by its antecedent state. Accordingly, the history of the cosmos and all things in it has been predetermined 'since the dawn of time'. (Put in more modern terms, the Big Bang already contained within itself the history of the cosmos as well as the history of mankind down to the very last detail, including my writing this sentence.) It is striking that it seems to have gone unnoticed that any antecedent state of the universe is also predetermined by any subsequent state, since the direction of time is irrelevant to Newtonian physics (so, roughly speaking, the omelette you ate for breakfast determines the eggs out of which it was made).

Physical determinism and predictability — Note that we should distinguish explicitly between physical determinism and predictability. Neither implies the other. If things are thus-and-so at time t, then, if the system is determinist, it is impossible for things at time t_n to be different in any way from what they are. In this sense, any given state of things determines any given antecedent or subsequent state. Globally, any given state of the universe determines every other state. It does not follow that any future state is predictable. Conversely, many things may be predictable without the system to which they belong being determinist (in science, it may be quantum mechanical in which the fundamental indeterminacy is ironed out at higher ontological levels; in human affairs, much is reliably predictable (otherwise social existence would be impossible) but that does not imply that individual, social, political, and historical affairs are determinist. The higher intelligence that Laplace introduced to make clear the predictability of the determinist universe he postulated is normally known as God, although in this case it is commonly characterized as Laplace's demon. Laplace's demon would have to know a true state description of the universe and the totality of the laws of nature. Then he would be able to predict or retrodict with certainty any past or future state of the universe. Such predictions would be inferences from the conjunction of the state description and the laws of nature.

D'Holbach and global nomological determinism — Laplace did not concern himself with the consequences of his physical determinism for human freedom of thought and action. But the consequence of such a conception had already been spelled out explicitly long before by Baron d'Holbach:

> Man's life is a line that nature commands him to describe upon the surface of the earth, without his ever being able to swerve from it, even for an instant. He is born without his own consent; his organization does nowise depend upon himself; his ideas come to him involuntarily; his habits are in the power of those who cause him to contract them; he is unceasingly modified by causes, whether visible or concealed, over which he has no control, which necessarily regulate his mode of existence, give the lie to his way of thinking, and determine his manner of acting. ... Nevertheless, in despite of the shackles by which he is bound, it is pretended he is a free agent, or that independent of the causes by which he is moved, he determines his own will, and regulates his own condition.[9]

Global nomological determinism is a *picture*, a preconception of the nature of change in all its forms. It is striking how profoundly appealing it is to many philosophers and scientists, just as divine predestination appealed to many theologians. There could hardly be a greater conflict between theory and practice. For *no one* in practice behaves as if it were true. One might say that Laplacean determinism just is Christian global predestinarianism for the age of science, although, unlike the foreknowledge of God, the Laplacean 'supreme intelligence's' knowledge is inferential.

Problems with states   It is time to submit these ideas to some critical conceptual scrutiny. It is striking that no cogent explanation is provided for what 'a state of the universe' is. To shed some light on *this* idea, we must first bring to mind the notion of a state, then we can move on to examine the idea of a state of the universe. It is noteworthy that the state of the economy, that is of great concern to economists, does not include how many pennies I have in my pocket. The state of the cupboard may include its untidiness or cleanliness, but not the decorations on the mugs or the design of the dinner set. The state of someone's health may be characterized by reference to the condition of his vital organs and to any infections he might have, but would not include the disposition of each hair on his head, let alone of each cell in his body. It seems therefore that the very idea of *a state of the universe* involves subliming the mundane, elastic, vague, and useful notion of a

---

[9] Pierre Henri Thiry, Baron d'Holbach, *The System of Nature* (1770), trans. H. D. Robinson (Batoche Books, Kitchener, Ont., 2001), 98.

state of something. Hence the question arises of whether this subliming has not gone far too far, whether there is any oxygen left for us to breathe among such refined and elevated ideas.

Our scientifically minded adversary might insist that the notion of a state has merely been extended from its common-or-garden use to a systematic scientific one. Then one might say that a state of the Newtonian or Laplacean universe consists in the precise disposition and velocity of every single particle in the universe at a given moment. This is not obviously objectionable. Nor obviously unobjectionable. It certainly detaches determinism from predictability (hence the need for a Laplacean demon or god). But that too is not obviously objectionable. So it asserts merely that, given the laws of nature and given a state of the universe at a certain time, no state of the universe at other times could be different from what it was or will be. It should be noted that there is no *argument* whatsoever for this form of determinism. It is merely a picture that has powerful appeal to scientists and to the intellectual lay public. It is rooted in Newtonian physics and in the Newtonian conceptual scheme. That leaves open the question of whether any of it can be salvaged in contemporary physics and cosmology and the conceptual scheme of relativity theory and quantum mechanics. But this question will not be pursued here, since no philosopher could sensibly wish the question of human freedom to depend on the contingencies of the development of science (as if we might *discover* tomorrow that we and our behaviour are and have always been free, or are and have always been determined).

4. Flaws in reductive determinism

Micro-physical reductive determinism   The form of scientific determinism under consideration is micro-physical. It presupposes bottom-up causation. Hence too it is explanatorily reductionist, that is all objects, events, states, and processes at any given macro-physical level are fully explicable by microphysical events (but not vice versa). It assumes that the domain of causal law is closed or complete, that is that every state or event is fully explicable causally. In principle, all statements concerning macro-physical phenomena are translatable by means of bridge principles into statements at lower levels and ultimately into statements at the final micro-physical level (in the Laplacean case, into the ultimate particles of the universe). None of these doctrines is correct.

- It is microphysical.
- It involves bottom-up causation.
- It involves explanatory reduction.
- It is causally complete (closed).
- It is reductionist.

List 6.1　*Characteristic features of scientific determinism*

It is noteworthy that such forms of scientific (physicalist) determinism are prone to drive their proponents to epiphenomenalism (T. H. Huxley and his followers), supervenience of the mental (Jaegwon Kim), anomalous monism (Donald Davidson), or downright eliminativism (Patricia Churchland). Characteristic of all is a bizarre downgrading or elimination of everything that makes us human and of all the manifold forms of explanation of our thought, feelings, and actions that make us intelligible to each other and to ourselves.

Flaws in the reductive picture　The thought that all causation is bottom-up and the corresponding idea that all causal explanation is reductionist are curious. As Schiller remarked to Goethe, 'That is not an experience, it is an idea.' It is, again, a compelling picture that actually runs contrary to our experience.

(i) Innumerable causings we observe or generate in our daily lives are not 'ultimately' caused by the behaviour of particles. We arise in the morning, make the bed, complete our toilette, go downstairs to breakfast, make a scrambled egg and eat it, and drive to work. We see the postman delivering the mail, observe him closing the gate after having done so, and see him mount his bicycle and ride away. All these are causal transactions. They involve living human beings, that is individual living substances consisting of organic matter of various complex kinds and their interactions with artefacts that consist of inorganic matter of different kinds. But, though the manifold biological constituents of a human being consist of particles (molecules, atoms, ions, or cells of numerous different kinds), no human being is identical with a set of particles. The particles of which we are made constantly change, but we remain the same living organism. Moreover, much of what happens to the particles of which we are made is determined by the way we wend our way through the world: how we move intentionally, what we eat and drink, how what we encounter impacts upon us. Human artefacts are made of various materials, but they are not just a collection of stuffs. They are designed for a purpose and fulfil a function for which they are used. So they have a form in addition to their constitutive matter, and one cannot infer the former from the latter.

(ii) Most of the perfectly correct causal explanations of phenomena in our daily lives do not conform to this picture. The workings of a system of pulleys are not explained by reference to its constituent particles but by the interaction of its gross parts. If it fails to work, however, we might explain that by reference to rust, that is the product of chemical interaction of iron and oxygen. Why is the room so cold? Because the window was left open. Why is the envelope stained? Because it was dropped into a puddle of water. Why are so many plants in the garden broken? Because there was a storm last night. None of these explanations adverts to bottom-up explanation, even though such an explanation could perhaps sometimes be given. The idea that explanation of biological phenomena (physiological and neurophysiological) are essentially and exclusively bottom-up has come under increasing criticism in the last decade. It is of the nature of biological organisms to be multileveled and highly structured, and of the operations of elements at one level to be controlled by the characteristic features of items at the higher level.[10]

(iii) Most human behaviour and the vast majority of explanations of human behaviour do not conform to this model. As we shall see below, most human behaviour, and those aspects of human behaviour that are of most concern to us in daily life are teleological, purposive, and goal directed – not molecular, let alone sub-molecular. Their explanation is by reference not to ions, atoms, molecules, tissues, or organs, but to human beings, their reasons and motives, goals and intentions, customs and conventions.

(iv) The very idea that the cosmos is clockwork, that it is a closed and complete causal system is increasingly being challenged.[11] Indeed, it is striking how natural it is to conceive of clockwork as determinist, forgetting that clockwork rusts, that human beings

---

[10] See, for example, Denis Noble, *Dance to the Tune of Life* (Cambridge University Press, Cambridge, 2017); also John Dupré, *Human Nature and the Limits of Science* (Clarendon Press, Oxford, 2001).

[11] See, for example, Nancy Cartwright, *How the Laws of Physics Lie* (Oxford University Press, Oxford, 1983).

often break their antique clocks when winding them, that anchor escapements wear out and need to be replaced, and that cogwheel teeth sometimes bend or break. In short, we confuse the machine, which is not such that any state of it completely determines any subsequent state, with *the machine as symbol*.[12] For we can derive the movements of the clock from the machine as symbol, but not from the clock itself save with the addition of numerous *ceteris paribus* clauses that exclude the clock's being dropped, metal fatigue, wear and tear, incompetent winding, and so forth. There is no causal closure to clockwork.

(v) The *theory* of modern science is reductionist, although the practice is not. The gap between theory and practice is supposedly closed by admission of ignorance of detail and limitations of human powers. But reductionism was never more than an *a priori* programme (see, for example, the atomism of Epicurus and Lucretius): a particular *picture* of the workings of the universe. In this respect, it was akin to the reductionist programme of sense datum theorists, which was likewise an *a priori* programme: a particular *picture* of the nature of knowledge of the world. In neither case was the programme ever successfully executed. There is no reason for thinking it might be and a multitude of reasons for thinking that it cannot be. Reductionism in biology disregards top-down causation, and assumes the causal closure of the cosmos and the ubiquity of closed causal laws, neither of which assumptions has any warrant. Above all, it disregards all forms of explanation that are distinctive of living phenomena, of biological teleology (*HNCF*, ch. 6), on the one hand, and of everything distinctively human, on the other. This was nicely expressed by Nietzsche apropos mechanistic reduction:

> A 'scientific' interpretation of the world, as you understand it, might therefore still be one of the *most stupid* of all possible interpretations of the world, meaning that it would be one of the poorest in meaning. This thought is intended for the ears and consciences of our mechanists who nowadays like to pass as philosophers and insist that mechanics is the doctrine of the first and last laws on which all existence must be

---

[12] Wittgenstein, *Philosophical Investigations*, §§193–4. For a detailed explanation, see G. P. Baker and P. M. S. Hacker, *Wittgenstein: Rules, Grammar and Necessity*, vol. 2 of *An Analytical Commentary on the Philosophical Investigations: Essays and Exegesis of §§185–242*, 2nd extensively rev. edn by P. M. S. Hacker (Wiley Blackwell, Oxford, 2009), exegesis of §§193–4.

based as on a ground floor. But an essentially mechanical world would be an essentially *meaningless* world. Assuming that one estimated the *value* of a piece of music according to how much of it could be counted, calculated, and expressed in formulas: how absurd would such a 'scientific' estimation of music be! What would one have comprehended, understood, grasped of it? Nothing, really nothing of what is 'music' in it![13]

To be sure, nothing that has meaning or that has a meaning, including human acts, actions and activities, language and its uses, and art in all its manifold forms, can have its meaning disclosed by a mechanist, purportedly determinist, explanation.

5. The random and the determined

It was widely argued by Hume and others that the denial of free will is unavoidable, since if determinism is true we are not free; if determinism is false whatever we do is random; and to act randomly is equally incompatible with acting freely. But this is confused. It is not true that what is random cannot be freely chosen, since the existentialist *acte gratuit* is freely chosen (e.g. by Lafcadio in André Gide's *The Cellars of the Vatican*) and the agent is fully responsible for it. To be sure, it was meant to exemplify freedom at its purist – the unmotivated spontaneous act – and that *is* confused. For the prototype of human freedom does not lie in spontaneity and arbitrariness but in acting for a good reason, in particular such reasons as do not manifest bondage to the passions. Furthermore, it is not true that what is random is the opposite of what is causally determined. There are different grounds for characterizing something as random. Twitches, spasms, epileptic, intoxicated, and drugged behaviour are random inasmuch as they are *not under the agent's control*, but they are certainly caused (although some of these are not actions at all). The roll of a die is *statistically random*, but it is caused and it accords with the laws of mechanics. *What happens by chance is random* but is not uncaused, merely unexpected, accidental, unintended, equiprobable (as in the case of the result of the roll of a true die), or coincidental.[14] In short, the matter of random actions or events is a red herring. The

---

[13] Nietzsche, *The Gay Science*, book 5, §373.
[14] See Mark Hofman, *The Case for Free Will* (Bibliofiles, 2012), chs 4–6.

pertinent contrasts are not between determinism and indeterminism (e.g. between Newtonian physics and quantum mechanics), but between being caused to do something that is no action (e.g. a reflex, slipping, falling, sneezing) and a voluntary action, between one-way powers to do something and two-way powers to do and to refrain from doing something, between reasons and causes of human behaviour, between being made to do something and doing something of one's own free will.

7

Neuroscientific Determinism, Freedom, and Responsibility

1. Neuroscientific determinism

Humans compared to machines The most common form of determinism in the first quarter of the twenty-first century is neuroscientific determinism. It is an heir to psychological determinism, on the one hand, and to physicalist Laplacean determinism, on the other. It typically rests on the supposition that all psychological attributes are identical to states, processes, or events in the brain, and that, given a complete state description of the brain, one can in principle predict what movements the agent will make, and indeed what action the agent will take. 'We are but machines' is a common refrain from neuroscientists and biologists (e.g. Colin Blakemore and Richard Dawkins). Of course, we are *not* machines: we are not made for a purpose, but have purposes of our own; we have a welfare and ill-fare; we are language-using animals with a determinate phased life cycle; we are sexual creatures; we have emotions and emotional attitudes; we are social beings living in complex societies in which we have to work to make a living; we are born to parents and beget children whom we normally love; we enjoy and value friendships; we have the power to act for reasons and are capable of reasoning; we can come to know the difference between good and evil; we have both a mind and a soul (see *HNCF*, 190 n.). None of this is true of machines.

Psychological and brain states and processes compared   Psychological attributes are not states or processes of the brain. As has been made clear through the successive volumes of this study of human nature, it is not the brain that sees or hears, thinks and opines, forms intentions and makes decisions, acts for reasons in pursuit of goals – it is the human being as a whole (which is not to be confused with a whole human being).[1] (i) To engage in reasoning is not for one's brain to undergo a process that yields the result of reasoning, although doubtless, without concurrent brain processes, one would not have arrived at the result. For reasoning is normative, whereas causal connections are not. (ii) To behave thoughtfully is not to be caused to move by thinking, let alone by a thought. It is to act after deliberation in a certain mode warranted by what the agent can sincerely adduce as his reasons. (iii) There could not be a neural correlate of remembering *as opposed to* misremembering, of knowing as opposed to being ignorant, of understanding as opposed to misunderstanding. To know, remember or understand something are complex *powers* of human beings to do certain things: to act for the reason that things are as they are known to be, to correct mistakes, to answer questions correctly, and so on; to retain what one came to know and not forget it; to be able to explain what one came to understand, and to act, think, or feel in the light of what one understood. These are not mental states, *a fortiori* not neural states either. (iv) There could be no distinct neural correlates of implying, insinuating, assuming, or hinting that things are so. For whether one is doing one or the other is not dependent on the state of one's mind or on the state of one's brain, but on the context and antecedents. (v) An emotion felt may be an agitation, an occurrent state with genuine duration, or a persistent state, but it is an agitation or state of a human being, not of his cortex (see *TP*, chs 2 and 3). There could be no neural correlate of feeling remorse as opposed to regret, indignant as opposed to resentful, jealous as opposed to envious.

Brain states and total state of the brain   It is legitimate to talk of the state of the brain, for example of the fact that it is in an oedematous state, or that it is inflamed or short of oxygen. But to speak of a total state of the brain is to stretch the notion of a brain state well beyond breaking point. We no longer know what we are talking about. Furthermore there is no reason at all for supposing that

---

[1] See M. R. Bennett and P. M. S. Hacker, *Philosophical Foundations of Neuroscience* (Blackwell, Oxford, 2003), for comprehensive treatment of these matters.

a 'state of the brain' at time *t* completely determines the 'state of the brain' at every future or past time, that is that it is impossible for the brain to be in any other state at any given moment than the very state it is in. For (i) there is no reason for thinking that the brain is a deterministic system. (ii) Unlike the universe as a totality, the brain is constantly being changed by the impact of external stimuli.

State description of the brain
Just as there is no such thing as a state description of the universe, so too there is no such thing as a state description of the brain. The brain is no less of a dynamic system than the universe. It is in constant flux and in constant response to external and internal stimuli. In the small, each human brain is unique and unlike any other human brain. But there are no neuroscientific laws by reference to which one can deduce from a 'state description of my brain' what I shall do in the next moment or what sentence I shall write next. Might we not discover such laws one day? No, this is not an intelligible hypothesis. We do not know what would count as a state description of the brain, nor do we know what it might be nomologically correlated with. One can imagine comprehensive correlations between cortical and spinal activity and muscular contractions, but not correlations with agential actions as opposed to mere movements, let alone with moves in a language game of a human community at a given stage in human history. Global neuroscientific determinism is a blank cheque on a non-existent bank. If one knows that I am taking dictation or typing from a manuscript, one can predict with tolerable accuracy what I shall write next, given knowledge of the dictation or manuscript. No neuroscientific data about my brain can be used to predict and explain such simple behaviour. There is much that can be predicted from a diseased brain, but little that is of interest in understanding human action and motivation that can be predicted from a normally functioning brain of a human being living in a human society. A large part of the speculative literature on neural prediction of human behaviour rests not on science but on science fiction.

Neuroscientific explanation of movement
However, all this may be cast aside. Neuroscience is able to explain the causation of human movement of limbs, beginning with events in the premotor cortex and ending a fraction of a second later with the contraction of the relevant muscles that are constitutive of moving the relevant limb. This is all that is necessary to raise the bogey of determinism regarding human action. For once the appropriate neural events in the premotor cortex have occurred, is it not causally

determined that, for example, one's arm will rise? Of course, it may not rise if someone holds it still, or if a meteor strikes one – the prediction presupposes a semi-isolated system. Equally, one might change one's mind. However, if Benjamin Libet and his followers are right, one has only 150 milliseconds in which to do so.[2] This seemed to Libet to show that the window of opportunity for the exercise of free will is very small. But that is wrong, since changing one's mind is itself a decision which, on Libet's account, must be identical with a non-conscious neural event that occurred 150 milliseconds before we became aware of changing our mind. So, is neuroscientific explanation compatible with freedom of action? Or does it show that we are not free in our choices and deeds – that we are mere puppets, dancing to the tunes of our brains? This matter will be discussed after a brief recapitulation of the forms and multiplicity of explanations of human action.

2. Explanations of human behaviour: a recapitulation

In *Human Nature: the Categorial Framework* (ch. 7) the diversity of licit explanations of human behaviour was spelled out. It is necessary for our purposes to bring them upon the carpet once more.[3] Human behaviour may be *mere doing* that falls short of action, for example slipping, falling, stumbling, or choking. It may consist of unchosen, involuntary, or only partially voluntary *reactions* such as reflex behaviour and reactive behaviour (e.g. laughing, weeping, sneezing). These may be *partly voluntary* if they cannot be initiated at will, but can be suppressed or terminated at will (e.g. a sneeze, a yawn), or if they can be initiated at will, then once started they cannot be stopped (e.g. blinking). One mark of involuntariness of movement is *lack of control*. This is manifest in uncontrolled utilization behaviour (e.g. compulsively picking up and putting on a pair of glasses even though one is already wearing one's glasses), anarchic

---

[2] They aren't right – see Parashkev Nachev and Peter Hacker, 'The Neural Antecedents to Voluntary Actions: A Conceptual Analysis', *Cognitive Neuroscience* 5:3–4 (2014), 193–208 – but it is remarkable how many scientists and intellectuals have been tempted by their siren songs.

[3] I am much indebted here to the numerous writings of Ilham Dilman, Stuart Hampshire, Anthony Kenny, Bede Rundle, Gilbert Ryle, Georg Henrik von Wright, Alan White, and Ludwig Wittgenstein.

limb behaviour (irresistibly grasping something or someone and not letting go with the hand contralateral to a brain lesion in the pre-supplementary motor area of the brain, or irresistibly and uncontrollably gesturing (hilariously depicted by Peter Sellers in the film *Dr. Strangelove*)) and so forth.

The non-voluntary   Actions may be *voluntary* or *non-voluntary*. Those that are non-voluntary may be involuntary or not voluntary. An individual act is *involuntary* if it is of a kind that the agent can perform at will (voluntarily) but did not on this occasion ('I couldn't help it', one may characteristically say) – as when one smiles involuntarily at something comic or blinks when air is puffed at our eyeball. The act, one might say, is *forced from us*. Acts may be *not voluntary* because they are neither voluntary nor involuntary inasmuch as they are performed unknowingly – as when one unintentionally mispronounces someone's name. Acts may similarly be not voluntary when they are done under duress or when one is obliged by circumstances to perform them. They are not *forced from us*, but *forced on us*. Nevertheless they are intentional.

The voluntary   A criterion of voluntariness of action is that one can do or abstain to order, act earlier or later if there is reason to, hurry up or slow down on request. We are, in general, not surprised by what we do voluntarily (unlike sufferers from utilization behaviour or anarchic limb behaviour), and can say what we are doing if asked. Doing something voluntarily is distinct from doing it willingly, since one can do something voluntarily although reluctantly (as when one offers, without much enthusiasm, to perform an unsavoury task). A person may be reluctant or willing, but not voluntary or compulsory – the former pair concerns the agent's attitude towards what he is doing and admits of degrees, while the latter concerns the agent's control and subjection to obliging or constraining factors. Human actions can be done or omitted at will. So they can be chosen, attempted, and intended. Action may or may not involve deliberation, choice, and decision, and is something one may be asked or ordered to do. It may be voluntary without being intentional, as when one gestures while speaking. In such cases the action is not intentional inasmuch as one is not even aware of making the hand gestures, but it is voluntary inasmuch as it is not forced and one could immediately stop if asked. Action may be intentional without being voluntary (as when one acts under duress). One's action may be intentional without attendant deliberation (as when one eats with one's mind on the conversation). An intention may be formed in advance of action as the upshot of deliberation and decision, just as one may plan in advance of action. But one may act intentionally without advance intention (see fig. 7.1).

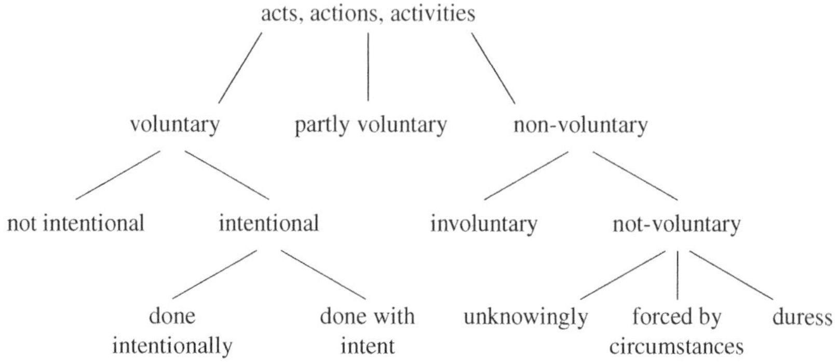

Figure 7.1   *Acts and voluntariness*

Acting for reasons   Intentional actions that interest us in our engagement with others are for the most part done for reasons. But one may act intentionally for no reason (unless doing something just because one felt like it, felt inclined to do so, or just wanted to 'for no particular reason' counts as doing it for a reason). To be sure, this may be irrational or unreasonable, if there are good reasons for refraining. One may act on a sudden urge, craving, or impulse to do what one does, which may have no reason (and indeed may be 'quite mad' and completely irrational – as when Dostoyevsky's Stavrogin bites the governor's ear in *The Possessed*). And one may act intentionally when one responds instantaneously to a split-second opportunity.[4] In such cases one's reason for acting is specified, if one is asked, *after* the event. What *makes it* my reason, as opposed to a reason I had, for doing what I did? The fact that I am willing sincerely to avow it as my reason for action, and to take responsibility for my action thus characterized. Specifying an intention *with* which one acts

---

[4] Cognitive neuroscientists mistakenly suppose that such actions are non-rational, since they involve different neural circuitry from that which characterizes considered deliberate action. This 'short-cut' circuit is allegedly too fast for reasons to be involved. But this manifests a primitive notion of doing something for a reason. Reasons are not antecedent causes, and we have good reasons for much that we do very quickly and unreflectively. They are specified *after* the act has been performed, and to specify them thus *ex post actu* is not to report them but to avow them and take responsibility for what we have done thus characterized.

is not the same as specifying one's intention *in* acting. One may go to London *with* the intention of catching the 10.30 train home, but one's intention *in* going to London (and one's purpose in going) is to go to the theatre, not to catch the 10.30 train home. To specify one's purpose in doing something is to give one's reason for doing it, since one cannot do something with the purpose of doing it, but only with the purpose of doing something further. The consequences of an intentional action may themselves be intentional, unintentional (if not foreseen), or neither intentional nor intended (if foreseen but not wanted and no part of the intention in the act – as when one wakes up one's wife when putting the cat out at night). If an act is not intentional, it may be done knowingly or unknowingly; voluntarily, involuntarily, or not voluntarily; by accident or by mistake.

Movement involving action and description Human behaviour may or may not be movement involving. If it is movement involving, it may be described in terms of the movements of the agent's limbs (his knee jerked) or movements the agent himself makes (he moved his leg), or movements of the agent himself (he turned round and walked away). It may not be movement involving (e.g. observing, listening to a lecture, looking at a painting, thinking up a solution to a problem, waiting motionless, holding the door open, holding tight, or omitting, refraining, or abstaining from doing something). Whether or not an action is movement involving, it can be described in many different ways, for example by reference to its antecedents (e.g. *reply, repeat, re-enter*), to its consequences (e.g. *to break, mend, open, close, wet, dry*) or its circumstances (e.g. *to commit perjury or bigamy, to lie, to steal*), by reference to its manner (e.g. *to mumble, to hurry, to dawdle*), or by reference to social or legal norms (e.g. *trespass, tax, marry, checkmate, score a goal*).

Explanations of behaviour All this bears on the explanations of human behaviour. The character of an action already precludes certain kinds of explanation. Acts that can be performed only once by a person, such as suicide, cannot be explained by reference to frequency explanations, such as habit or custom, even though the prevalence of the act may be so explained (the Hindu practice of sati was once customary in India). Accidents cannot be explained in terms of purpose or motive. Acts that can be performed only intentionally (e.g. *cheating, lying, forging*) cannot be explained in terms of inadvertence or inattention. And so on.

'Why did A V?' is the most general form of a request for an explanation of someone's doing what he did. It includes things that happen to one ('Why did he fall?'), as well as acts and actions ('Why did he jump?').

So such a request can elicit various kinds of explanations, each involving different explanatory factors. More specific kinds of question restrict the answer. 'What did he do that *for?*' asks for the point and purpose of an action and presupposes its being intentional. So it precludes various kinds of explanation in terms of efficient causation, ignorance, or inadvertence. 'What *made* A V?' can be asked even when V-ing is not an act of A's but merely something he did (e.g. stumble, choke, slip), or if it was an act that was patently uninviting, unreasonable, or irrational, that is something that A would not have done but for the explanatory factor. The first is answered by a causal explanation, the second by various forms of *being obliged* to do whatever one did.

Different forms of explanation of action So, with that stage setting, we can distinguish the following different forms of explanations of human behaviour. Their itemization is comprehensive but not exhaustive; moreover, it is not exclusive either, since some of these kinds of explanation may overlap.

(i) Explanations in terms of constitutive redescription: such explanations redescribe an activity (ritualized, semi-ritualized, or otherwise rule governed) in terms that make it clear what the activity in this particular setting amounts to, for example it is a game (or such-and-such a game), a wedding, a political demonstration, a strike, or a parade.

(ii) Social role explanations: he issues orders because he is the leader; he stands up because he is going to speak; he locks the gate because he is the night watchman.

(iii) Explanations by polymorphous redescription: she is rehearsing, practising, obeying an order, or keeping her promise.

(iv) Regularity explanations: explanation by reference to habit, custom, disposition, or personality trait: he is in the habit of going for a walk at three o'clock every afternoon; it is customary here to dress for dinner; she is very shy. These are regularities or regularity implying, hence their instances are to be expected from such a person and no special explanation is called for.

(v) Inclination explanations: an action may be explained as exhibiting a standard preference, a liking or disliking – she loves diving; he hates cold water; they don't like dancing. So no *further reason* need be sought.

(vi) Explanation by reference to expression: she is weeping for joy (not out of grief), laughing with amusement (as opposed to embarrassment), shouting with excitement (not out of anger).

(vii)   Same-level causal explanation: he sneezed because he had just taken snuff; he cried out because he was hit; he jumped because you startled him by shouting in his ear.

(viii) Explanation in terms of reasons: the factor that is cited as a reason is held by the actor to warrant his behaviour irrespective of whether it is an action or omission. The reason cited may be backward looking, citing a past fact as justification, for example I promised; she asked me to; he insulted me; it rained last night. It may be contemporaneous: it is raining; she has arrived; it's time to go. It may be forward looking: for the sake of such-and-such an end; in order to bring about something desired. It may specify the action in normative terms: it is an obligation, a duty, it is his right; or in terms of good or evil.

(ix)   Motive explanations: he did it out of generosity (or kindness, compassion, friendship, or hatred). These do not specify the actual reason for acting but rather the pattern of backward- and forward-looking reasons (B was in need, A V-ed with the intention of alleviating B's need, B's need was satisfied). One kind of motive explanation specifies an action done in order to exemplify a pattern, for example revenge, gratitude (see fig. 7.2).

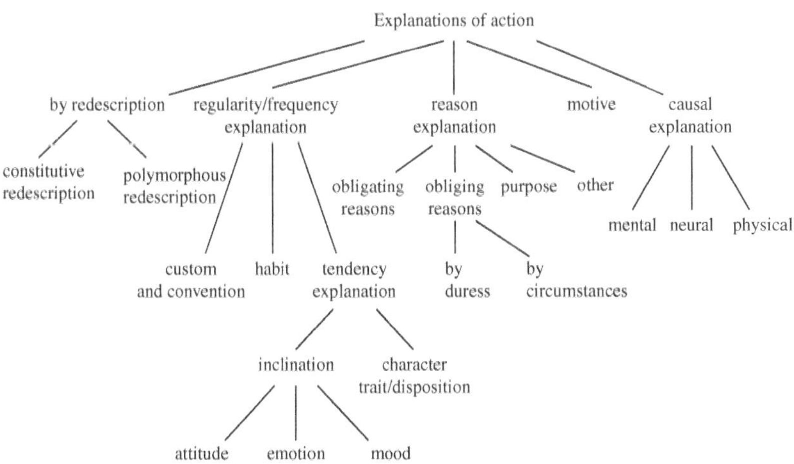

Figure 7.2 *Schema of types of explanation of action (Source: HNCF, 7.2)*

Explaining inaction is often just as important as explaining action. Quite apart from not doing something for a reason that one has or out of a certain motive, there is a variety of forms of explanation distinctive of *not doing* something:

(i)   inability, which may be physical, technical (lack of skill), or intellectual;
(ii) lack of normative power;
(iii) lack of opportunity;
(iv) ignorance of an opportunity;
(v)   prevention, natural or human;
(vi) lack of equipment or instrument;
(vii) forgetfulness;
(viii) thoughtlessness.

With these conceptual maps in place, we should be in a position to locate a neuroscientific explanation, to indicate its limits, and to reflect upon neuroscientific determinism.

3. Neuroscientific explanation and its limits

To explain something is to render intelligible something that is not understood, that is misunderstood, or that is, for one reason or another, baffling. Of course, there are quite different kinds of things that call out for explanation, and the kind of explanation that is appropriate for one kind of *explanandum* may well be completely inappropriate for another. So too it should not be surprising when scientists and philosophers, in their monistic craving, extend one type of explanation beyond its legitimate boundaries. Inductive confirmation, *pace* Quine, has no role in explaining why a mathematical or logical proposition is true. Similarly, *a priori* proof has no role in confirming an empirical proposition, nor, *pace* Kripke, can an analytic proposition be confirmed in experience. What kinds of behaviour can neuroscience explain?

First, let us note the previous eight kinds of explanation of omissions. One may not do something because one cannot physically do it (e.g. reach the box on the top shelf because one is not tall enough, or lift the box because one is not strong enough), because one lacked the necessary equipment (e.g. a ladder), or because one is baffled (e.g. by a set of instructions). In such cases the agent did not act because it was not *possible for* him to act, and, since it was not possible for him to act, it is not *possible that* he acted. One may not do something because one lacks the normative competence (e.g. he did not sign it, indeed could not validly sign it, because he is under-age, she did not vote (indeed could not vote) because she is a member of the House of Lords). In such cases it *was* not *legally possible for* the agent to act, so it *is* not possible that the agent validly signed or voted. One may not do something one wanted to do because one never had an opportunity to do it, or because one did not notice the opportunity until it was too late. One may not do something because one was prevented, either by nature (an avalanche that blocked the road, a flood) or by human intervention (duress). And, of course, one may forget to do something that one wanted to do or that one was under an obligation to do. Or one may fail to do something one should have done through thoughtlessness – it simply never crossed one's mind.

Omitting, refraining, abstaining   We distinguish between *not doing something* (*simpliciter*), *omitting or failing to do something*, *refraining*, and *abstaining*. There are indefinitely many things one does not do, which are not omitting, failing, neglecting, refraining, or abstaining from acting. I didn't go to indefinitely many places today, didn't eat indefinitely many foods or dishes, didn't meet indefinitely many people. These 'inactions' require no explanation nor need there be any reason for them (I had no reason for not going to Berlin, not meeting the prime minister, and not eating ice cream at midnight). One may omit doing something deliberately, intentionally, unintentionally, inadvertently, by accident, or by mistake. So some omissions may be explained as done by mistake (if one thought the meeting was tomorrow rather than today) or by accident (if one missed the train). Failing or neglecting to do something, unlike omitting, always involves recognition of a standard that one failed to live up to and is always a matter of mistake, thoughtlessness, or accident. One may omit the usual preliminaries and go straight to the point, which may or may not be a fault, but if one fails to give one's condolences to a friend on his or her loss, that is a fault. One cannot refrain from doing something unintentionally (accidentally or by mistake). So refraining from doing something is explained in terms of reasons or motives (e.g. if someone gave offence, one may snub him at a party and refrain (not abstain) from talking to him because of what he had done, and out of contempt). Abstaining is similarly intentional, but it has a more limited scope. To abstain from voting (unlike forgetting to vote) is to take a deliberate stand; to abstain from alcohol or smoking is to restrain oneself for reasons (e.g. of health, or out of moral considerations). What is noteworthy about all this for present purposes is that it is inconceivable that a neuroscientific explanation could be relevant to understanding why someone omitted doing something, save in the case of inability due to motor abnormality or paralysis – and that would explain why he did not do it, not why he omitted doing it. No description of neural events could explain one's refraining or abstaining, let alone replace familiar kinds of explanation.

Explanations of behaviour beyond the purview of neuroscience

Putting omissions, failings, refrainings, and abstainings aside, do the neuroscientific tales displace the various forms of explanation that render the puzzling behaviour intelligible? They don't, and they couldn't:

(i) If a foreigner is puzzled to see hundreds of thousands of people converging on Parliament Square, no tale about neural events is going to render the puzzling phenomenon intelligible. But to redescribe it as a demonstration against Brexit will do so immediately.

(ii) If a foreigner is sitting in the Visitors' Gallery in the House of Commons and is puzzled to see people springing to their feet on the floor of the house, neuroscientific descriptions of the complex events that are involved in getting up will go no way to explaining that the MPs are trying to catch the speaker's eye and to get permission to speak. A neuroscientific story may explain *how it was possible* for them to rise to their feet. It will not explain that they *could not but* rise to their feet, since it is evident that they were not *made* to rise. If they had been made to rise by cortical events that necessitated their rising to their feet, they would all have been dismayed and perplexed, perhaps terrified, and would have apologized to the house. To be sure, they have to rise if they are to catch the Speaker's eye.

(iii) To explain someone's playing the piano in the next room as practising for tomorrow's concert, to explain someone writing their name as signing a cheque, a letter, a declaration; trying out a pen; or practising a new style of signature is to make clear what is going on. But to describe neural events in the motor cortex that make the movement of the hand possible will go no way to rendering the observed phenomena intelligible. One may ask someone why they are telephoning, and the reply may be that they promised to call a friend to arrange a meeting. That will render the conduct transparent. But whatever neural events are described to explain the innervation of the arm, hand, and fingers cannot explain the difference between V-ing and keeping a promise to V, between playing and practising, between signing a letter and signing a cheque.

Knowing what one is doing   But surely, the neuroscientific reductionist may say, the agent knows what he is doing, knows that he is signing a letter as opposed to signing a cheque, that he is practising the piano as opposed to giving a performance, and that knowledge must be represented in his brain. So a proper description of the brain *will* differentiate between these. However, as we have seen, knowledge is not a mental state, *a fortiori* not a neural state (*IP*, ch. 4). It is ability-like, and there is no such thing as storing, as opposed to retaining, abilities or ability-like powers. If we come to know something, we may retain what we have learnt (i.e. not forget it). But retention of knowledge is not the same as storage of knowledge, and while knowledge can be stored in notebooks, books, and computers, it cannot be stored in a brain. So whatever neural changes are involved in remembering something to be so, they are not such as to enable anyone to read off what is retained but not stored. So no description of the state of the brain or of any segment of the brain could possibly explain what is being done in cases of polymorphous descriptions of action. The most it can explain in regard to human acts, activities, and actions is muscular contractions.

(iv)–(v) Regularity explanations are similarly irreplaceable, since no neuroscientific description of a brain state or of a sequence of neural events can contain the information that Herr Professor Dr Kant goes for a walk at three o'clock every afternoon – that it is a habit and so his going for a walk today calls for no *special* explanation. Nor could it explain why the Prince Regent wore a kilt on his famous visit to Scotland in 1822 – that he thought it was a custom, and that doing so would endear him to the Scots. Similarly, an explanation of behaviour in terms of a non-pathological personality trait is beyond the reach of neuroscience, since that someone is exhibiting sentimentality or manifesting shyness or touchiness can have no neural imprint from which it might be read. For character traits are tendencies and pronenesses, not one-off affairs.

(vi) There is all the difference in the world between crying with joy and crying in grief, or between laughing with amusement, laughing sympathetically without knowing why others are laughing, laughing ironically, and laughing cruelly. The physiology and neural activity may be the same, but the differentiation that renders the laughter perspicuous is not to be found in the brain but in the context, in what happens before and after the laughter. To understand a person's laughter one has to know its context (someone just told a joke; everyone else was laughing), its cause (he was being tickled) or its object (what he was laughing at), the attitude that accompanies it, the temperament of the subject, and so on. No doubt a neuroscientific tale will explain the movement of the muscles, the stimulation of the lacrymal glands, the release of air from the lungs, and the chuckle, shriek, or roar. It may explain why the agent couldn't help laughing, although, for the most part, laughter can be suppressed. But it could not explain the character of the laughter, let alone the joke.

(vii) There is, of course, room for causal explanation of behaviour. But it is typically same-level explanation. One may cry out with alarm at a sudden noise, for example when someone shouts in one's ear in order to make one jump. One may sneeze if one's nostrils are being irritated. One may weep when cutting onions. Such explanations are not defective for lack of neural descriptions, and the neural descriptions do not stretch to the context that renders the subsequent response perspicuous (e.g. laughter, irritation, wiping one's eyes, and opening the kitchen window).

(viii)–(ix) Explanations in terms of reasons and motives are the explanations that are of most interest to us in our enquiries into the behaviour of our fellow human beings, and in the explanations of our own behaviour that we may offer others. This has a number of reasons: it indicates the reasoning of the agent; it commonly explains action in terms of its purpose or goal – what it is or was *for*; it reveals much about the character of the agent for it shows what kinds of things he is willing or eager to do, and what considerations move him. No neuroscientific explanation can do the service of an explanation in terms of the reason for which an agent acted, let alone show that when one acts for a reason what one does is inevitable. On the contrary, if someone V-s for such-and-such a reason then, with qualifications regarding the perceptual powers, he must have a two-way ability and an opportunity, and there must have been a possibility of omission.

4. How possible, not why necessary

It is evident that neuroscientific explanations of movements, in particular of muscular contractions, cannot replace explanations of the manifold forms of action or omission. It cannot render perspicuous human behaviour that may be opaque to us in many different ways. But that is exactly what our normal forms of correct explanation do. Now we may turn to the crucial question of whether the theoretical possibility of neuroscientific explanations of behaviour demonstrates that we are not free. Does it show that, in all cases, what we do is predetermined by neural events? Does it show that our explanations of action and omission in terms of reasons and motives are misguided, mere epiphenomenal rationalizations? Does it show that our form of life, which presupposes freedom and responsibility, rests on illusion, that we are no more than puppets dancing to the music of our neurons?

Do action potentials demonstrate determinism? Neuroscientists have discovered the character of the neural activity in the premotor cortex immediately antecedent to movement, and the nature of the neural impulses from the brain to the muscles in the relevant limb that will make them severally contract or relax. Does this not mean that, once the neural activity commences, the muscles *have to* contract and the limb *has to* move accordingly? And, if that story is correct, has freedom of action (or 'freedom of the will') not evaporated? It is tempting to think so (just as it tempting to think that the world in which we live is actually, 'really', colourless and soundless). Notoriously, Benjamin Libet and his numerous later followers succumbed to this temptation. It is not necessary to repeat in detail our criticisms.[5] The conceptual confusions are manifold: decisions and intentions are not causes of the actions they explain; psychological predicates have no intelligible application to the brain or its parts; the so-called action potential (*Bereitschaftspotential*) cannot in principle be specific to the particular act (act individual, as it is sometimes called) that ensues, but is more akin to a neural response to affordances (apprehension of possibilities); willing and intending are not feelings; trying to time decisions and intention formation to the millisecond is incoherent (akin to 'Time to the millisecond the moment dinner began!').

This much seems clear:

Making possible differs from making necessary (i) Unless some appropriate neural events and processes take place in advance of a movement-involving action, the agent will not be able to move his limb, but will, at least for a moment, be paralysed. So appropriate neural activity is *required* for an agent's voluntary or intentional action. The neural activity involved in a voluntary movement or non-voluntary intentional movement makes it *possible for* the agent to do what he does: it is a condition of the actuality of a

---

[5] Bennett and Hacker, *Philosophical Foundations of Neuroscience*, 228–31; see also n. 2 above.

possibility. It does not make it *necessary* or *inevitable*. Given the appropriate neural activity, the relevant muscles *will* contract, but that does not imply that the agent could not have changed his mind at the last moment. To be sure, the last moment is not 100 milliseconds. Otherwise our voluntary and intentional actions would be like our blinking or our withdrawal reflex. But they patently are not. Nevertheless, is it not inevitable that the muscles will contract once the efferent neural impulses have occurred? One is tempted to affirm this, forgetting that someone might hold one's arm down (so it is not inevitable) and that one might bethink oneself and change one's mind once one has started to raise one's arm and draw it back on second thoughts.

(ii) If the antecedent neural activity made the subsequent movement *inevitable* or *unavoidable*, then there would be no voluntary action or behaviour to predict or explain. We often do things voluntarily but inattentively (such as knitting while watching television). What makes it voluntary is that the agent can stop if asked, can slow down or speed up on request, and can say what she is doing. Moreover, she can explain why she is knitting. There is nothing unavoidable or inevitable about what she is doing – unlike reflex movements that, given the stimulus (e.g. a tap on the patella) are inevitable unless forcibly prevented. And there can be no current or antecedent cortical events or processes to be correlated with these unactualized abilities.

(iii) The neuroscientific account can explain the contraction or relaxation of the muscles consequent on the transmission of a neural impulse from brain to limb. When that happens in the course of my voluntarily raising my arm, nothing *makes* me raise my arm – I raise my arm freely. I don't make my arm rise or bring it about that my arm rises – that would make voluntary and intentional action into a form of telekinesis. (I can make my arm rise, for example, by attaching a rope and pulley to it, and pulling on the rope with my other hand.)

Arm rising and arm raising   But doesn't the neural impulse cause my arm to rise? Does it not *make* my arm go up? No, when I raise my arm, it would be at best misleading to say that my arm rises. My arm rises when someone else raises it, or when I push it against the wall for a minute and then step back and *allow* my arm to rise. When I nod my head in assent, my head does not nod – that happens when I nod off. Nor do I *make* it nod – which is something I might do were I to seize my head in my hands and move it up and down. When I rise to my feet my body does not rise, let alone rise to *its* feet and when I turn around, my body does not turn around (*HNCF*, 153–60). Wittgenstein's notorious question, 'What is left over when you subtract your arm's rising from your raising your arm?', is not to be answered but deconstructed – it is a malformed, illegitimate question, akin to 'What is left over when you subtract seeming to see from seeing?'[6]

(iv) No neural events can show voluntary, intentional, or deliberate omissions, refrainings, or abstainings to be inevitable or predetermined.[7] For it is not neural events that make such acts of omission possible, but rather the fact that the agent *could have done otherwise* – that he had the appropriate two-way power and that there was an opportunity. What makes an omission a case of *refraining* from acting is that the agent intended thereby to achieve a certain end (snub the person who had offended him) – but intending something is not itself an event or phenomenon that could have a neural correlate. Similarly, the decision to snub NN whenever one meets him may have been made weeks or years before, and decisions have no phenomenal characteristics that might be correlated with neural events. So no neural events could show that we are not free to omit, refrain, or abstain from acting as we please. Were any cortical tale to show that no movement was *possible*, that absence of *Bereitschaftspotential* shows that the agent *could not move*, then, of course, he would not have omitted, refrained, or abstained since he would have been paralysed, and were he paralysed there would be no omission, refraining, or abstaining at all, let alone an inevitable omission, refraining, or abstaining.

Explanations in terms of reasons and motives

(v) It was noted above that explanation in terms of reasons and motives are the most important explanations to render non-pathological human behaviour that is opaque to us intelligible (see *HNCF*, ch. 7). Such explanations presuppose mastery of a language inasmuch as someone who acts or refrains from acting for a reason must *be able to reason*, that is to apprehend a consideration as a warrant for a given conclusion and as a justification for acting or refraining. He must be able to answer the question 'Why?' by citing his reason, which explains and purports to justify his doing what he did *for* that reason.

---

[6] Larry Weiskrantz would suggest that what is left is blindsight. But that is mistaken (see John Hyman, 'Visual Experience and Blindsight', in John Hyman (ed.), *Investigating Psychology: Sciences of the Mind after Wittgenstein* (Routledge, London, 1991), 166–200).

[7] Save for the *absence* of any neural events characteristic of Gilles de la Tourette syndrome. But the absence of such abnormal neural activity does not show that my omissions and refrainings are inevitable.

In the case of others whose behaviour he understands, he will explain it by reference to what he takes to be *their* reason or reasons for doing what they do. If we want to know why Hannibal did not besiege Rome after Cannae, or why Raphael painted the central figure of Heraclitus as the only figure *with boots* in *The School of Athens*, or why Elizabeth I of England etceterated her letters to Philip II of Spain, we do not want to know anything about concurrent cortical events in their heads. Reasons may be specified or given in various grammatical forms. 'A's reason for V-ing was that things were so' is common (A took an umbrella because it was raining); so is 'A's reason for V-ing was in order to do some further thing' (A went to London because he wanted to go to the theatre) and 'A V-ed for B's sake'. In some cases we cancel the factivity of the specification of A's reason by inserting 'as far as he knew' or 'because he believed' in the sentence 'A V-ed because things were so', (e.g. 'Jack went to see Jill because he believed that she wanted to see him'). But in such cases it is not A's believing that is his reason for V-ing, but what he believed, namely that things were so (which may be false).

In none of these cases is the explanatory factor a cortical event, state, or process. Nor is anything that is cited as a reason (e.g. *that it was going to rain*; *for B's sake*; *in order to go to the theatre*) something that might be *correlated* with a cortical event, state, or process. The incongruity is even greater when one recollects that what is commonly a criterion for such-and-such's being A's reason is nothing he said or did antecedently to V-ing, but rather what he *would* have said had he been asked, or what he said or would have said *after* V-ing.

5. Varieties of responsibility

Rationality and responsibility Being rational (capacity rationality), being free, and being responsible for our actions and omissions are essential species characteristics of mankind. Other animals have two-way powers, pursue their goals, and have desires and appetites but they are not rational, that is they are not sensitive to reasons, do not do things for reasons, and cannot reason from premises to conclusions. Some are intelligent, but they are not language users and they do not have a will – a capacity for practical reasoning. They are not moral agents even though they may have attractive or unpleasant dispositions. They lack knowledge of good and evil, they are not answerable for what they do, and we do not hold them responsible for their deeds (see Appendix 1). We are unique in our world in being responsible for our choices and for our actions and omissions. We are not responsible for chance and fortune, but we are responsible for our lives. We do not deal the pack, but we must play our hand as best we can.

Forms of responsibility The concept of responsibility is neither clear nor distinct. Before we proceed further, some mapping of the conceptual terrain here will shed light on the concept.[8] Various forms of responsibility have been distinguished by philosophers and lawyers:

(i) *Causal responsibility*: The floods are responsible for damage to the local housing. The cold snap was responsible for the decline in the bird population. Here the responsible agent can be a natural object or event. But it can also be a human being or his action – as when we say that the prime minister's resignation is responsible for the fall in the value of shares on the stock exchange. Whether the human agent in such cases is *also* morally or legally responsible for such causal consequences is a further question.

(ii) *Capacity responsibility*: The insane are not responsible for some of the things they do and say. They sometimes do not know what they are saying or why, and they are sometimes not in control or in full control of their actions. Here we are speaking of general rational capacities for acquisition, possession, and utilization of knowledge, for understanding, thought, and decision, as well as capacities for control of impulse, desire, and for movement, action, and restraint that are generally presupposed for someone to be *held responsible* for their deeds. Various forms of mental impairment undermine normal human rational capacities. Continental legal codes characterize this general precondition for assignment of responsibility as 'imputability'.

(iii) *Act responsibility*: People are generally *held responsible* for their specific actions and, sometimes, for their specific omissions on a given occasion. Over a wide range of kinds of actions and in a wide variety of circumstances human beings are normally competent to exercise their rational capacities and their two-way powers of action. They can control their movements, their actions, and the mode of the performance of their actions. But the question arises, in both law and morals, whether

---

[8] In what follows I am indebted to H. L. A. Hart, especially to his essays in *Punishment and Responsibility* (Clarendon Press, Oxford, 1968).

an agent, *on a specific occasion*, was in possession of such rational powers, and whether, though generally in possession of such powers, the
agent was able to exercise them on the occasion in question. Continental
legal systems characterize failure to do so on an occasion in which the
law is transgressed as 'fault' (*Schuld, faute, dolo*). English law amalgam-
ates imputability and fault under the general heading of *mens rea*.

Such act responsibility is typically a condition of *legal* or *moral
liability*, that is likewise commonly referred to as responsibility – as
when we speak of 'strict responsibility' or 'vicarious responsibility' in
the law (and also 'strict liability' and 'vicarious liability'), and of
moral responsibility. Strict liability is a feature of penal laws in which
mental conditions of responsibility (*mens rea*) do not apply and the
offender is liable for punishment irrespective of whether he did or did
not take all possible precautions to avoid the delict.

(iv) *Liability responsibility*: Agents, human beings and their institutions ('artificial persons'), are held responsible, that is liable, for
their deeds and for their sphere of responsibility, which may stretch
beyond their own deeds – as in the case of vicarious liability responsibility. Liability may be moral, social, or legal. Moral liability
responsibility is liability for blame. (Of course, one may be morally
deserving of praise if one is responsible for good deeds or for acting
over and above the call of duty.) Social liability responsibility is for
blame and social stigma. (And, of course, one may be deserving of
public praise or reward for meritorious actions.) Legal liability is for
punishment or enforced compensation. In general, both capacity
responsibility and act responsibility are conditions of legal liability
responsibility – but there are exceptions, as in cases of strict liability.

(v) *Role responsibility*: We speak of a person's responsibilities *as* a
ship's captain, *as* a nightwatchman, *as* a doctor. The duties and obligations of the role are the responsibilities of the person who occupies
that role. The occupier of the role is appointed to take care of precisely those matters, and it is for them that he is *accountable*.

(vi) *Responsibility as a character trait*: We distinguish between
responsible people and irresponsible people. Here we identify character traits. Responsible people are those who can be relied on. They
will faithfully fulfil their responsibilities. They can be left in charge
and can be relied upon to act reasonably – not to be careless or reckless, negligent or inattentive.

(vii) *Other*: We also speak of *collective responsibility* when a group
or even each member of a group is held liable for the deeds of an individual member of the group. We speak of *corporate responsibility*,

when an 'artificial person' such as a corporation is held to be responsible for the activities of the corporation and liable for them. And we speak of people *taking responsibility for their lives* when we urge people to stop blaming others, fate, and fortune for their plight and to face up to reality and discipline themselves to take charge of their lives. These are not relevant to our concerns.

Many writers on this subject refer to these distinctions, or ones akin to them, as differentiating *senses* of the words 'responsible' and 'responsibility'. But there is no need to multiply polysemy beyond necessity. There are connecting links between these different applications, and the concept of responsibility is best viewed as a focal concept in the Aristotelian sense. The moot question is: what is the focus of the concept of responsibility? The clue is furnished by etymology. The English word 'responsible' (a surprisingly late usage for a very early distinction) is derived from the Latin *respondere* – meaning 'to answer'. But the notion of answering here is not that of merely answering a question, but answering or rebutting charges or accusations. The focal point of this cluster of usages is the idea of *being answerable for* or *being accountable for* an action (see fig. 7.3).

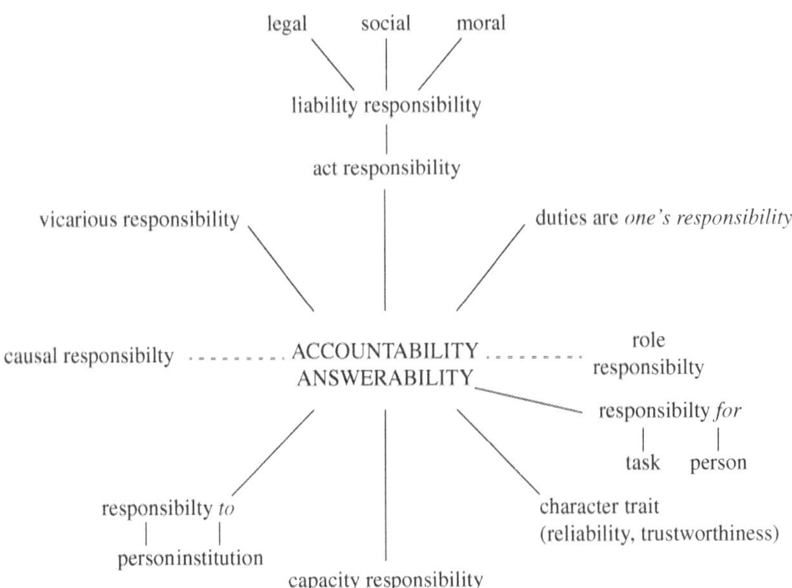

Figure 7.3 *Schematic representation of the focal concept of responsibility*

Normal human competence   At the heart of the very idea of responsibility lies the notion of a normal rational human agent with normal human faculties. Such a person generally knows what he is doing, and knows his own situation and the relevant circumstances with which he is confronted in his actions. He is responsive to reason, and capable of deliberation and intention formation and execution. He has a behavioural repertoire at his behest any item of which he can perform at will, given an opportunity. This is the sphere of normal human competence. Particularized to an individual human being, it is the domain of the individual's competence. It is with this range of kinds of action under his control, his responsiveness to reasons, and his capacities for knowledge and understanding that a rational agent confronts the situations and opportunities for action that he encounters. If he sets himself to perform any of these actions in normal circumstances, then, other things being equal, he will do them.[9] It is paradigmatically in the execution of actions that lie within his sphere of competence that a human being can legitimately be held accountable, and challenged accordingly.

Accountability   The agent may refuse the challenge with the reply 'I'm not responsible for that'. Depending upon context and elaboration, this may amount to the denial that the act was his responsibility – he is not in charge, or this does not fall within his remit. Or it may be a denial that he actually did the deed – someone else did it. However, he may admit responsibility for a nefarious deed or claim responsibility for a meritorious act. If he did the deed, he is *accountable* – unless he can show that he fails to satisfy conditions of accountability, that is conditions of *mens rea*. If he is accountable, he may nevertheless plead justification or excuse as exculpatory or mitigating conditions. So, given that the agent did the deed, assignment of responsibility for it presupposes, other things being equal, that the agent possesses normal human rational capacities and control. Capacity responsibility and act responsibility are generally conditions for moral or legal liability responsibility. Then, in the absence of excuse or justification, a person can be called to account. Then they are liable for punishment, or, extra-legally, deserve criticism and blame.

Role responsibility   Role responsibility is obviously associated with the idea of answerability for the performance of the duties of one's role. One's duties and obligations are 'one's responsibilities' (irrespective of whether they are duties of a role or not). A responsible person is one who can be relied on to fulfil his

---

[9] The idea of a 'domain of secure competence' is nicely discussed in Joseph Raz, 'Being in the World', *Ratio* 23 (2010), 433–52.

responsibilities. Role responsibility is in turn linked to the idea of being responsible *for* a person or institution, or *for* a specific task assigned one – and these too are connected to the notion of accountability. For one is responsible for a person who *is in one's charge* and for an institution in which one has appropriate managerial role responsibilities and which is *within one's control*. To be responsible *to* a person or institution is a matter of *to whom* or *to which institution* one is answerable and accountable.

Causal responsibility Causal responsibility is evidently linked to the idea of answerability in the thin sense of being to blame for something – as when the weather is responsible for the failure of the crops or the winter floods for the state of the roads, or when the explanation of some benefit is laid at the door of (explained by reference to) some inanimate agent that brought it about, or some event that caused it. And similar considerations apply to assignment of causal responsibility to human agents.

Vicarious responsibility The notion of vicarious responsibility or vicarious liability is also linked to the idea of being answerable – but not for one's own actions; rather for the actions of others, for example one's employees or one's pet, or for events that may even involve no human action or blame (e.g. adulteration of food that one manufactures, despite having taken all possible precautions; collapse of buildings). Here one is liable for punishment (or for enforceable compensation) for things that are neither one's acts nor one's omissions. In such cases the reason for holding the person (or company) responsible is that because of their position (e.g. as employer, pet owner) they bear the risk of such harmful events.

What can one be responsible *for*? One can be responsible for one's acts and 'inactions' of various kinds (omissions, failings to act, refrainings, abstentions), for the acts and inactions of others, for the consequences of one's acts and inactions, and for the consequences of the acts and inactions of others. One may be responsible for non-acts, such as illegal possession. One may be responsible for tasks, for people, and for the care of things associated with one's role responsibility. The phrase 'responsible for' is strikingly polysemic.

6. Elaboration

So, central to the idea of responsibility and to our practices of holding people responsible in law, in social practices, and in morality are the notions of voluntariness and of control, of intention formation and the execution of intentions in action, of reasons and motives, and of acting intentionally. It is in these terms that we characterize our freedom as rational agents. But, against this background, many qualifications and modifications are necessary. For we are responsible for both more and less than our actions, for more and for less than our voluntary and intentional behaviour, and for more and for less than the sphere of our control. And we are responsible for items in different senses.

Scope of responsibility We are responsible for things that are not our acts (or 'inactions'). Among the things for which we are legally liable are not only the acts of others for which we are answerable, but also our own unexecuted plans and intentions (which are not acts), when these are plans or intentions to transgress the law. We are also responsible for illegal possession – which is not an act (or inaction), for example being in possession (in charge) of a car while intoxicated (even if one is quietly sleeping it off in the back seat of the parked car), being in possession of stolen goods, or being in possession of drugs or explosives.

We are responsible for our acts. But we are not responsible for *all* our acts.[10] One is not responsible for things done in one's sleep, while sleepwalking, during epileptic seizure, or when in a fugue or under hypnotic suggestion. Act responsibility is limited by requirements of knowledge, intention, and control, which are not satisfied in these kinds of case. Nevertheless, we are sometimes responsible (liable) for acts that are not intentional, and we are sometimes not responsible for acts that are intentional. Negligent acts are an obvious case in which absence of intention does not relieve one of responsibility. So too, in certain circumstances, are unintentional omissions when performance is legally or morally required. We are commonly held responsible for some accidents and mistakes that are consequences of our action, both intentional and unintentional. Conversely, there are actions that we perform intentionally for which we are not held responsible, for example those done under duress. For duress is an excusing condition (save in cases of homicide), since one did not do the act of one's own free will – it was forced on one by threats of harm that are sufficient to exculpate. Provocation (in cases of homicide), though not an excusing condition, is a mitigating one.

---

[10] Nor are we responsible for movements that are not acts of ours, such as reflex behaviour, for example reactions to a bee-sting, which may result in harm (e.g. while driving). So too for lack of muscular control due to disease (e.g. Parkinson's disease), and pathological mental conditions involving uncontrollable behaviour.

Just as the sphere of the intentional does not coincide with the sphere of responsibility, so too the sphere of responsibility does not coincide with the sphere of actions over which one has control. As was already pointed out by Aristotle, voluntarily disabling oneself does not relieve one of responsibilities for what one does when not in control of oneself, for example when drunk. And, as we just pointed out, accidents are events beyond our control, are neither intended nor foreseen, and yet we can be held responsible for them. So too we may be held responsible for things that are not voluntary – namely for acts or omissions that are unknowingly or inattentively done.[11]

7. Irresistible impulse and temptation

The M'Naghten Rules in English law limited any insanity defence to considerations of cognitive conditions. For the accused to plead the insanity defence in cases of homicide, it was required that he be shown to have been, at the time of the commission of the offence, 'labouring under such a defect of reason, as not to know the nature and quality of the act he was doing, or, if he did know it, that he did not know that what he was doing was wrong'.[12] Even at the time (1843), considerable unease was expressed by judges, jurists, and psychologists alike at the exclusion from the consideration of the courts of any evidence pertaining to defects of will and of control of the emotions. For it was recognized, by judges such as Lord Cockburn, as well as by psychiatrists such as Isaac Ray, that powers of self-control, no less than cognitive powers, may be impaired by mental disease, damage, and dysfunction. Lord Cockburn argued that, 'though the patient is quite aware that he is about to do wrong, the will becomes overpowered by the force of irresistible impulse; the power of self-control, when destroyed, or suspended by mental disease, becomes, I think, an essential element of responsibility'.[13] According to Hart, Belgium, Germany, and France had all recognized defects of will as well as of understanding in the nineteenth century. England had to wait until the Homicide Act of 1957, which recognized the

---

[11] For an illuminating discussion of voluntariness, see Alan R. White, *Grounds of Liability: An Introduction to the Philosophy of Law* (Clarendon Press, Oxford, 1985), ch. 5.

[12] *R. v. M'Naughten*, 9 Eng. Rep. 718 (1843).

[13] B.P.P., 1874, vol. ix, p. 549.

category of diminished responsibility as a mitigating condition for reducing the accusation of homicide to one of manslaughter.

Irresistible and unresisted impulse The standard legal objection to the acknowledgement of the relevance of diminished powers of self-control and to the very idea of 'irresistible impulse' is that there is no way to distinguish between an irresistible impulse and an unresisted impulse. We cannot distinguish, it is argued, between a person who was irresistibly tempted to perform an illegal act and a person who did not resist the temptation. Moreover, it is held, no psychiatric or other medical evidence can definitively show that, on the occasion of the *actus reus*, the accused actually lacked the ability to resist the temptation, as opposed to merely not resisting it.[14] Indeed, the Butler Committee (1975) held that in practice 'it is fair to say that the only evidence of incapacity to conform to law is the act itself'.[15] In discussing the issue in 1984, Anthony Kenny went even further: 'the difficulty in telling the difference between unresisted and irresistible impulses is not a contemporary and contingent one which progress in science may remove. The notion of irresistible impulse is an incoherent piece of nonsense.'[16] Since the 1970s there has been extensive debate, in both England and Wales and the USA on what modifications of the M'Naghten Rules are desirable in the light of modern psychiatry. My concern is not with the current state of the law, so I shall not trace the developments and debates that ensued. My sole concern is with Kenny's claim that the very idea of irresistible impulse is incoherent.

His arguments were two. First, that there is no way, even in principle, of deciding whether the accused is someone of normal strength of will succumbing to abnormally strong impulses, or someone of abnormally weak will giving way to normal impulses. The very same behaviour can be adduced with equal justice as evidence in support of either of the conflicting hypotheses. This in itself suggests that the distinction is misconceived ('a metaphysical fiction'). Secondly, if somebody *acts* on impulse, his act must be voluntary. If an act is to be voluntary, it must be possible for the agent to act otherwise. If so the impulse cannot be 'irresistible', on pain of losing the name of action.

---

[14] e.g. *R. v Byrne* [1960], 2 QB 396, 404, *per* Lord Parker CJ.

[15] *Report of the Committee on Mentally Abnormal Offenders*, Cmnd 6244 (1975), ch. 18, quoted by Anthony Kenny, 'The Psychiatric Expert in Court', *Psychological Medicine*, 14 (1984), 291–302, at 299.

[16] Kenny, 'The Psychiatric Expert in Court', 299.

Both arguments are forceful. But it is by no means clear that we cannot and do not make similar judgements with cogency. We can observe someone over time in various circumstances, and compare him with others. We can and do judge that a person has the greatest of difficulties in resisting temptations, for example to indulge in excessive eating or consumption of alcohol. Someone who goes so far as to have his jaws wired together to ensure that he will desist from food consumption is fighting against temptations that are, if not impossible for him to resist, at least unbearably difficult. Someone who opts for aversion therapy for alcoholism is taking desperate measures against recurrent failure to resist temptations. In such cases, and in many other less abnormal ones, there is a recurrent refrain of 'I really did try as hard as I could'. It would be amiss to dismiss such claims of inability to resist the temptations as nonsensical and incoherent.

Acting on impulse We may grant that to act on impulse is not to be caused to act by a mysterious mental cause denominated an 'impulse'. To act on impulse is to act spontaneously, without reflection. The action itself may be meritorious, indifferent, or culpable. An impulse is no more a mechanical cause than a whim. To act impulsively is not to be *caused* to move – if it were, then impulsiveness would indeed fall short of *action*. It would be no more than movement that is not under one's immediate control, such as movements of those with some form of Tourette syndrome or anarchic limb behaviour.[17] But this alone does not show that there are no defects of will that should not be taken into account in morality and law alike.

Loss of self-control To lose control of oneself as a result of an overwhelming emotion is a common eventuality, especially in cases of indignation, rage, or fear (e.g. Billy Budd's response to Claggart's malicious and false accusation, or of Winston's action in the face of the terror of rats eating his eyes in Room 101 in Orwell's *1984*). Is it always the case that an agent, in the grip of such an emotion, *could* have controlled himself if he had tried hard enough? That seems anything but obvious. (Recourse to the 'objective standard' of what the reasonable man would have done or foreseen – as in *DPP v*

---

[17] Such pathological neural conditions as anarchic limb behaviour are *not* cases of action, at least in so far as the agent can prevent the anarchic limb movement only by seizing hold of the limb with his other hand. If so, then these are *involuntary movements* rather than non-voluntary actions. That is why the terms under which they are discussed by neuroscientists, who speak here of 'impulsivity', are misleading, since these are not really cases of *acting* on impulse at all.

Smith [1960] – is surely beside the point, for the question should surely have been what the reasonable man would have done if he were frightened out of his wits.) But such behaviour *is* action, not mere uncontrolled bodily movement.

Similarly, various forms of addiction exhibit intense craving, powerful 'delay aversion' (preferring a short-term desideratum rather than a delayed benefit), and an overwhelming inclination to act in order to satisfy the craving. It is not clear that it is never correct to say that the person tried as hard as they could, but in the end succumbed. Should the law give no recognition to such cases? One should bear in mind the fact that the law recognizes the plea of provocation in cases of manslaughter, thereby recognizing degrees of loss of control that at least sometimes play a mitigating role, reducing the crime to one of manslaughter.

Such cases may incline us to qualify Kenny's judgement. These kinds of act, manifesting loss of control under the pressure of immediate intense emotions, or succumbing to temptation under conditions of addiction and intense craving, are not (in the cases that concern us) involuntary bodily movements. They are indeed acts. But they are surely only *partly voluntary*. Should such a category not be recognized? For in all the kinds of case in question, the agent's ability to refrain from the delict is impaired.

PART III

Of Pleasure and Happiness

8

Pleasure and Enjoyment

1. Varieties of hedonism

Doing things for the sake of pleasure has a significant role in most people's lives. Entertainments and celebrations are meant to give audiences and participants pleasure. Those fortunate enough to have time for recreations usually enjoy the activities that they engage in for pleasure and are pleased when they are successful in them. Pleasure and enjoyment are an integral part of flourishing human life, and the desire for pleasure and enjoyment is a distinctive aspect of human nature. It is therefore appropriate that we should investigate the field of hedonic concepts here.

An eloquent paean to hedonism was made by Jeremy Bentham (1748–1832):

> Nature has placed mankind under the governance of two sovereign masters, *pain* and *pleasure*. It is for them alone to point out what we ought to do, as well as determine what we shall do. On the one hand the standard of right and wrong, on the other the chain of causes and effects, are fastened to their throne. They govern us in all we do, in all we say, in all we think: every effort we make to throw off our subjection, will serve but to demonstrate and confirm it. ... The *principle of*

utility recognizes this subjection, and assumes it for the foundation of that system, the object of which is to rear the fabric of felicity by the hands of reason and of law.[1]

This is an impressive clarion call. But the trumpets are too noisy. Like Bentham himself, we must distinguish. We must distinguish between *psychological hedonism*, which is a descriptive account of the essential nature of human motivation, and *ethical hedonism*, which is a prescriptive doctrine that lays down how we ought rationally to conduct our lives. These different doctrinal views are in turn to be distinguished from *practical hedonism* – a variety of ideologically motivated forms of life pursuing pleasure ranging from Epicureans to sybarites, advocated by some and passionately rejected by many religious and secular moralists alike. This will be discussed in sections 7 and 8 below. These are all doctrinal distinctions. We shall also have to draw many conceptual, non-doctrinal, distinctions: between active and passive pleasures, between different logical objects of pleasure, as well as between pleasure, enjoyment, and being pleased.

Psychological hedonism
Psychological hedonism is a descriptive doctrine concerned with giving an account of actual human motivation. Pleasure and pain, Bentham declared, govern us in all we do. They are our masters and they determine what we do. So, although it may seem that we do things for the sake of a multitude of goals, such as the achievement of knowledge or the pursuit of justice and fairness, appearances deceive. For those who spend their lives in pursuit of knowledge do so only because gaining knowledge gives them pleasure, and that is really what they aim at. And, although the pursuit of justice and fairness may be meritorious, those who dedicate themselves to such ends do so only because achieving justice and fairness pleases them. As a matter of fact, all human action is designed to avoid pain and to obtain pleasure for the agent of the action. We are all, in fact, egoistic hedonists, although some, like Bentham, actually find their greatest pleasure in ameliorating the lot of others.

To be sure, descriptive hedonism, when carefully formulated, must restrict its claims a little. Acts that are voluntary but not intentional (such as laughing at a joke) and unintentional acts (such as spilling one's tea) are obviously not done for the sake of pleasure. The doctrine, like ethical hedonism, is designed for what we choose to do and do on purpose.

---

[1] *An Introduction to the Principles of Morals and Legislation* [1780/1789], ed. J. H. Burns and H. L. A. Hart (Athlone Press, London, 1970), ch. 1.1.

Ethical hedonism Ethical hedonism is a prescriptive doctrine that advances the view that human beings *ought* to pursue pleasure and avoid pain, that prospective pleasure and pain are severally the *only* good reasons for action. Those benighted by certain forms of religious asceticism may eschew the pursuit of pleasure, but it is irrational to do so. (It is unclear how this is consistent with descriptive hedonism save by reference to the view that we should eschew pleasure in this life for the sake of pleasure in an afterlife.) With these views Bentham and his utilitarian followers made contact (but no more than contact) with the ancient hedonism of the Cyrenaics and Epicureans. Epicurus (341–270 BC) had written:

> For the end of all our actions is to be free from pain and fear, and when once we have attained all this, the tempest of the soul is laid. … For this reason we call pleasure the alpha and omega of a happy life. Pleasure is our primary and natural good. It is the starting point of every choice and every aversion, and to it we revert and make feeling the criterion to judge of every good thing …
>
> He who has a clear and certain understanding of these things will direct every preference and aversion toward securing health of body and tranquility of mind, seeing that this is the sum and end of a happy life. (*Letter to Menoeceus*)

Egoistic and universal hedonism We must draw further distinctions within prescriptive hedonism: between egoistic hedonism and universal hedonism. Egoistic hedonism is the doctrine that the only rational actions to be taken by a human agent are those that the agent does to further his own pleasure or happiness. This attempts to reconcile descriptive hedonism with prescriptive hedonism: we cannot but pursue pleasure, but it often takes careful thought to choose the correct course of action that will maximize the pleasure and its likelihood. Universal hedonism is the utilitarian prescriptive doctrine that right action is action that produces the greatest happiness for the greatest number. This can be reconciled with descriptive hedonism only if we find our own greatest happiness or pleasure in maximally benefiting all. Where this is not so, the only remedy is to enact such laws that by means of threats and punishments make it in the interest of citizens of an enlightened utilitarian state to maximize the pleasure or happiness of all.[2] However, our

---

[2] There is an ineliminable conflict in utilitarianism between the aggregative principle of maximizing quantity (greatest pleasure) and the distributive principle (for the greatest number). This is irrelevant to our investigations.

concern is not with utilitarianism in all its axiological and normative varieties but only with the hedonism of classical utilitarianism. The fundamental motivational and prescriptive doctrines that underpin it are flawed.

Against psychological hedonism
Nature has not placed mankind under the two sovereign masters of pleasure and pain. They do not govern us in all we do, in all we say, and in all we think. We perform or avoid actions and seek or avoid the prospective results of action for a multitude of reasons: for the sake of those we love or cherish, of fairness and justice, of solidarity, of truth, or of the attainment of knowledge, as well as out of revenge, vindictiveness, and malice. It is true that we sometimes do things for pleasure. It is also true that there are many things we enjoy doing, enjoy having done to us, and we engage in or indulge in for the sake of pleasure. And it is true that many things please us. But it is not true that we do everything we do, either immediately or proximately, for the sake of pleasure and enjoyment – even if it is true that we get pleasure out of, or are at least pleased by, doing justice, acting fairly, indulging our curiosity, achieving our ambitions, getting our revenge, or injuring those we hate. Nor is it true that, whenever we fulfil our desires or attain our goals, we get pleasure from so doing or even are pleased. Sometimes the upshot of doing what we want to do is very disappointing. Contrary to Epicurus, it is by no means obvious that pleasure as such is always good and is rendered bad only by its consequences. For, as we shall see, pleasure is always pleasure of, or in, something, and to take pleasure in tormenting or torturing another or in killing others is evil. The pleasure a sadist gets from torturing another is not a causal consequence of torturing his victims but intrinsically related to the heinous activity. We shall explore these counter-hedonistic claims below. Our first task is to gain an overview of the field of hedonic concepts and the disorderly relations between them.

2. Pleasure, enjoyment, and being pleased

We are concerned in the first instance with the concept of pleasure and its manifold siblings and cousins. Given the role of pleasure and enjoyment in human life, it is unsurprising that the available vocabulary is rich and diverse. Part of it is displayed in Table 8.1.

Nouns	*Verbs*	*Adjectives*
a pleasure, a delight, a joy, glee	to take/derive/obtain/ feel/ pleasure in/from, to delight in	delightful, joyous, gleeful
delectation, relish, refreshment	to indulge in, to relish, to treat oneself to	delectable, delicious, relishing, refreshing
enjoyment, amusement, entertainment, fun, zest, enchantment	to enjoy, to be enthralled by, to be charmed by, to be delighted with, to revel in, to be enchanted/ entertained by	enjoyable, overjoyed, charming, amusing, entertaining, enchanting
gladness, pleasantness, satisfaction, gratification, fulfilment, contentment, well-being, felicity, happiness	to please, to rejoice in, to gratify, to satisfy	pleased, pleasant, pleasing, agreeable, satisfying, gratifying, glad, charmed, captivated
sexual satisfaction/bliss/ gratification, rapture, ecstasy	to pleasure, to slake one's appetite, to gratify	rapt, enraptured, transported, ecstatic, blissful

Table 8.1 *A part of the hedonic vocabulary*

It is evident that expressions for the pleasures of taste and sex are prominent in our vocabulary, and these terms are used derivatively in much wider contexts too. Other centres of variation are hedonic reactions to what is pleasing, and the enjoyment of activities.

Pleasure, enjoyment, and being pleased To a first approximation, let us distinguish between pleasure, enjoyment, and being pleased.[3] First, when one enjoys something, one *enjoys oneself*, but when one is pleased at, by, or with something, one does not please oneself, and when one takes pleasure in something, one does not pleasure oneself. Secondly, what one enjoys 'subjectively' is *an activity or passivity of one's own*, that is one is doing or perceiving something enjoyable to do or perceive, but what one enjoys 'objectively'

---

[3] I am here indebted to Elizabeth Telfer's *Happiness* (Macmillan, London, 1980), 12–18.

is *whatever it is that gives one pleasure to do or perceive*. One might distinguish here between the primary object and the secondary object of pleasure or enjoyment. If one enjoys watching the children play, the primary object of one's enjoyment is watching them play, while the secondary object is their playing – which is enjoyable *to watch*. Thirdly, one may be pleased with, at, or by something because of the way *it fits one's plans, preferences, or prejudices* rather than because of any of its intrinsic characteristics, but one enjoys something because of its *intrinsic features*. One may be pleased at the death of evil men, but to take pleasure in their death would be inappropriate *Schadenfreude*, and to enjoy watching their death (as Madame Lafarge enjoyed watching the guillotining of aristocrats in *A Tale of Two Cities*) would be repulsive and morally degrading. Enjoyment is a species of pleasure, but although taking pleasure in something and enjoying something are usually pleasing, being pleased is not a form of pleasure. Being pleased is a responsive attitude involving a favourable judgement and a preference. It may accompany enjoying oneself and taking pleasure in something. One is usually pleased at either.

'Pleasure', 'enjoyment', and 'being pleased' are three graces that stand at the intersection of psychology, axiology, and practical reasoning.

Pleasures of the senses   (i) *Psychology*: Sensual pleasures are internally related to the senses and their use. For each of our five senses, Aristotle pointed out, there is a corresponding form of pleasure derived from its optimal exercise with respect to an appropriate object. The senses of taste and touch are primal. We take pleasure in the taste of good food and drink. We enjoy eating the food, and good food gives pleasure to those with discriminating taste. This, together with the social significance of dining with family, friends, and households is the root of the culinary arts. The indulgences of food and drink are refined as one acquires a discriminating palate through experience and instruction. We enjoy touching or stroking various things, such as fur, silk, and velvet, which are soft to the touch, hence the pleasures of dressing in clothes made of such expensive materials. We take pleasure in stroking the limbs and face of a lover, and we enjoy sensations of warmth when it is cold and of coolness when it is hot. The olfactory pleasures of scents are fleeting but distinctive, and artificial scents are used to enhance sexual attraction. The pleasures of sight and sound are the most complex and allow of indefinite sophistication in the enjoyment of music (from the primal beat of drums to Beethoven's late quartets) and the pleasures of vision, ranging from attractive or beautiful human beings or landscapes to the visual arts of painting or sculpture. The more refined the object of sense, the more the subjective pleasure and enjoyment has a cogitative and cognitive component. The appreciation of great music and the visual arts requires a natural sensitivity, an educated eye and ear, and extensive knowledge and exposure. This is less true of the pleasures of taste, smell, and touch, although even here the wine connoisseur's and gourmet's pleasures are rich in association and in refined distinctions. To be sure, many sensual pleasures and enjoyments involve the stimulation and exercise of more than one sense. The gourmet's pleasures of food involve taste, texture, smell, and appearance; the pleasures of the garden and countryside involve scent, hearing, and sight. The pleasures of the opera and ballet involve both sight and hearing. Erotic pleasures encompass all our senses and a variety of emotions and attitudes. Further psychological connecting links among the pleasures of the senses are the notions of wanting and desiring, liking and craving, being tempted by and wishing for.

Pleasures of desire and strength of will   The latter psychological concepts provide a connection between the pleasures of the senses and the pleasures of desire and of strength of will. Many of the pleasures of desire are shared with animals, but what might be called *the competitive pleasures of the will* are distinctive of mankind (although not without analogues among animals). We not only have immediate desires to satisfy our appetites or to indulge our curiosity; we have long-term goals and purposes that we pursue for reasons. We form intentions to act on the basis of such reasons and can plan in advance of action how best to achieve our goals. We often enjoy planning, and take pleasure in the successful exercise of skills put to use in the execution of our plans. The creative pleasures of arts and crafts involve not only the pleasures of the senses but also skills, knowledge, and planning. We sometimes enjoy the pursuit of our goals, in particular the single-minded concentration this demands and the intensity of effort. The struggle against odds is occasionally pleasing, the struggle against others is sometimes satisfying, the testing of our strength of will is often fulfilling, and the challenge gratifying. Hence the pleasures of competitive sports and the delights of mountain climbing where the goal is, in a sense, trivial and the endeavour is all. Hence too the pleasure some men take in mortal combat the intensity of which may elevate them to abnormal levels of consciousness, as was the case with beserkers, for whom subjective time slows down in battle frenzy and they see the movements of their adversaries in slow motion. The pleasure these Norsemen took in slaughter was such that their conception of an afterlife in Valhalla for a warrior who died with sword in hand was of perpetual battle interspersed by evenings of comradely drinking and song.

Pleasures of the intellect   Apart from the pleasures of the senses and the competitive pleasures of the will, we are also blessed with the pleasures of the intellect. Many of us enjoy learning, most of us take pleasure in understanding, and the possession of knowledge is not only useful but pleasing and often gratifying. Our recreations may include reading and studying, which we enjoy. We find satisfaction in indulging our curiosity, in careful examination and observation of what catches our attention, in finding things out, and in experimenting. We play thought-demanding games such as chess or bridge, in which we pit our wits against others for the pleasure of it. Other intellectual faculties too are sources of pleasure. Memories of past joys and delights are pleasant to dwell on. We take pleasure in the reproductive imagination as well as in the exercise of the creative imagination.

Sociable pleasures   In addition to these three sources of pleasure, enjoyment, and being pleased that are rooted in our nature is a fourth, which can incorporate all of them, namely sociable pleasures. We are by nature social animals and we find pleasure in various social activities. We take pleasure in dining and drinking together – this being one of the primal forms of social bonding. Similarly, we enjoy parties and celebrations, both on a familial level and on a tribal or national level. The latter forms serve to mould a tribal or national identity. We enjoy playing games together, and playing team games with others, and we enjoy watching such intensely competitive endeavours. This too contributes to a sense of group, tribal, or national identity (e.g. chariot racing between Blues and Greens in Byzantium, and soccer in current times). We may enjoy giving others pleasure, as when we invite others to dine with us, or when a person higher in a social hierarchy gives parties for employees or followers. Furthermore, we may take pleasure in the successes of others (e.g. of our children or pupils) and indeed in their enjoyment. However, one should not forget the negative pleasures of *Schadenfreude*, and even the pleasure the evil take in torturing, tormenting, and humiliating others. There is little doubt that some take pleasure in killing their adversaries or arbitrary victims.

To be sure, this fourfold classification is not exclusive. Some pleasures of the senses, such as music or the visual arts, are also intellectual pleasures, and the pleasures of playing intellectual games are also challenges to our competitive instincts. The sociable pleasures include the sensory pleasures of taste at a dinner party, as well as the intellectual pleasures of discourse, and the pleasures of game playing and of war involve the pleasures of the will and of the exercise of skills, as well as the pleasures of malevolence. So the four are interconnected centres of variation that correspond to psychological faculties and powers and to our social nature and status.

Pleasure not a 'simple natural quality' (ii) *Axiology*: Pleasure, enjoyment, and being pleased are *axiological* concepts, that is they essentially involve evaluations. It was a grievous error of G. E. Moore to suppose that pleasure was a simple natural quality, logically comparable to colour qualities. Something's being green involves no value, but something's being a pleasure does. To take pleasure in something; to be pleased at, with, or by something; or to find something pleasant is to value it positively – to find it good.[4] To be pleased to be given, to have, or to do something is to think highly of it, to like it, and to show a preference for it. What pleases one is welcome, be it good news, a visit from a friend whose company one values, or an achievement (of one's own, of members of one's family, of a compatriot). One's emotional response to what pleases one is positive – good news may put one in a good mood; a visit from a friend is valued for the delight one takes in his or her company; an achievement is pleasing and sometimes a cause of joy inasmuch as it has gained a valued goal. The pleasures of benevolence, on the one hand, and of malevolence, on the other, are patently value laden, for their characterization presupposes standards of good and evil.

Hedonic goodness Von Wright distinguished, among the varieties of goodness, the category of hedonic goodness. The primary objects of hedonic goodness, he suggested, are *experiences* of those things that give us pleasure and enjoyment and that please us. Exclamations such as 'That's good', 'That's delicious', 'How wonderful', and 'Oh, yes' are characteristic verbal expressions and manifestations of sensory pleasure derived from our pleasures, recreations, and amusements. The goodness of the items and activities that give us such experiences are secondary forms of hedonic goodness. We take

---

[4] Of course one may be shocked and perturbed to find oneself taking pleasure in something (e.g. in watching a bullfight), see p. 230.

pleasure in food and drink the consuming of which gives us pleasure. We say of such foods and drinks that they are good inasmuch as they are the source of our pleasure. In this sense, their goodness is hedonic. Of course, they may be the subjects of other forms of goodness too. A good apple may be an apple with a good taste and so pleasant to eat, but it may also be an apple the consumption of which is healthy (and so beneficial), or an apple that is good for cooking (utile goodness), or an apple in good condition (good of its kind)

Moore's open-question technique

These considerations show how misleading Moore's 'open-question' technique in *Principia Ethica* was. His suggestion that the question 'Is pleasure good?' makes sense and, moreover, is open, is confused. For, in the absence of a specification of the mode or variety of goodness, it makes no sense. Furthermore, the specified question 'Is pleasure hedonically good?' is no more a genuine question than asking whether pleasure is pleasant. Of course, one may ask whether certain pleasures, amusements, and entertainments are beneficial. Some are, and others are not and indeed may be positively harmful (e.g. various forms of addiction). Indulging in pleasures and amusements may be morally bad if they are at the expense of other people or self-destructive. Some pleasures are intrinsically bad, such as the pleasures of malevolence.

For the sake of pleasure

(iii) *Practical reasoning*: Pleasure and enjoyment are essentially connected with practical reasoning inasmuch as the fact that one enjoys doing something or takes pleasure in something is a reason for doing it or getting it. Pleasure and enjoyment are not, so to say, practically inert. In one's own case, they may not be very weighty reasons unless one is a committed hedonist, and they may be overridden by other considerations and claims upon one, such as one's obligations, duties, or other pressing goals. In the case of giving pleasure to another person, in particular someone one loves, it may, other things being equal, be a weighty consideration, save in ascetic and puritanical cultures. If one is asked why one is doing something and one answers, 'Because I enjoy it', or 'Because it gives my dear friend much pleasure', or 'In order to please my spouse', such answers terminate the enquiry, as long as what one is doing is or imparts an intelligible pleasure or enjoyment. There is nothing puzzling about doing something for the sake of pleasure or enjoyment, or doing something in order to give pleasure or enjoyment to someone one loves or is fond of. What may be puzzling is how people can take pleasure in, or get enjoyment out of, some specific thing that gives them pleasure or enjoyment, for example out of playing certain games that seem boring to one or out of watching some games (cricket mystifies those not brought up on it), bird-watching, collecting beetles, or studying the behaviour of fleas. Of course, they will give reasons why they find such things enjoyable, but they may well not convert one. To be sure, some of the things human beings enjoy are harmful to themselves or others, some are repulsive or wicked. In such cases the fact that they are enjoyable should be overridden by their repulsiveness and by their wickedness or evil. Other enjoyable activities or indulgences are harmless in moderation but may be harmful in excess.

3. Pleasure, pain, and the pleasures of sensation

Subjective and objective pleasures We have distinguished between objective and subjective pleasure or enjoyment. We speak of the pleasures or enjoyments of food, drink, or sex (the objective sensual pleasures of the appetites), and *taking* pleasure in or enjoying eating, drinking, and sex (someone's subjective pleasure of sating these appetites). Failure to note this ambiguity leads to conceptual confusions, for what can be attributed to pleasure in one sense is sometimes unintelligible in the other. Something is *a* pleasure, or *an* entertainment, if some people normally take pleasure in it or find it entertaining. Some cultures may license evil pleasures, such as watching gladiatorial games or the slow torture of captured braves. We judge such subjective pleasures to be morally repulsive and reprehensible, and the objective pleasures to be evil. The pleasures that human beings pursue vary from person to person. One person's main objective pleasures may be walking in the countryside, camping, and travelling; another's may be gardening, listening to music, and visiting museums. That something is one's main pleasure does not imply that one always enjoys it – one may not be in the mood, or the particular instance may not meet one's standards of excellence.

Hedonic opposites In English it is natural to assume that pleasure and pain are opposites, our two 'sovereign masters', as Bentham put it. But this is mistaken. There is no strict contrary to 'pleasure', since we lack an expression signifying absence of pleasure. 'Unpleasant' is indeed available, but it is not the contrary of 'pleasure', but of 'pleasant' and of 'pleasing'. But so is 'painful'. 'Displeasure' is a form of annoyance or anger, and not an opposite of 'pleasure'.

'Being displeased', however, is perspicuously the contrary of 'being pleased'. So, curiously, the hedonic conceptual baseline, so to speak, in English is a source of potential confusion (see Table 8.2).

pleasure	no pleasure
pleasant	unpleasant/painful
pleased	displeased, disappointed
approbation	displeasure
no pain	pain

Table 8.2 *A medley of confusing opposites*

Pleasure and pain are not opposites – neither contraries nor contradictories. One may be neither in pain nor enjoying a pleasure – indeed that is normally how things are with one. Moreover, they do not exclude each other, for one may find certain pains pleasurable, as when one scratches an itch to excess and has a sensation that is a blend of pleasant and painful constituents. Equally, during copulation human beings sometimes bite and scratch each other in their sexual passion – here we have a pain which, in the circumstances, is patently both desired and enjoyed. So a 'pleasant pain' or 'pleasurable pain' is not a contradiction in terms. However, one cannot take pleasure in something one finds unpleasant, displeasing, disagreeable, distasteful, or unwelcome in the circumstances. One cannot enjoy something one finds irritating, tiresome, repulsive, or nasty. One cannot be pleased at something one takes to be dismal, distressing, deplorable, vexatious, or horrible.

Epicurean pleasures

Epicurean hedonism held that pleasure, or at any rate *natural pleasure*, is largely privative. Epicurus himself was more stoical than sybaritic, advocating a life of restraint:

> He who understands the limits of life knows how easy it is to procure enough to remove the pain of want and make the whole of life complete and perfect. Hence he no longer has any need of things which are not to be won save by labour and conflict. (*Principal Doctrines*, §21)

Natural pleasure, he averred, consists primarily in the absence of pain and suffering, in particular the absence of hunger, thirst, and cold, which are natural needs. Satisfying these is (dynamically) pleasurable, and having satisfied them – having quenched one's thirst, satisfied one's hunger, and warmed up – is a (static) pleasure. These are natural needs, to which correspond natural desires the satisfaction of which is a natural pleasure. However,

> The just person enjoys the greatest peace of mind, while the unjust is full of the utmost disquietude. (*Principal Doctrines*, §17)

> Justice consists in abiding by the social contract, not harming others and not threatening them. In addition, the pleasures of friendship are necessary for a good life. Of all the means which are procured by wisdom to ensure happiness throughout the whole of life, by far the most important is the acquisition of friends. (*Principal Doctrines*, §27)

According to Epicurus, the satisfaction of natural needs, together with justice and friendship suffice for living a happy life. Other natural pleasures do not correspond to natural needs and are not necessary. Such are eating or drinking to excess (the gourmand) or the pleasures of luxurious food or drink (the gourmet). Sexual pleasure, Epicurus averred, is natural but it produces no good and often produces harm. Other pleasures, for example the pleasures of power, wealth, public honours, and the intellect, are not natural and not necessary either. They are unnecessary either because their absence involves no pain or alternatively because they are futile inasmuch as they are likely to bring pain in their wake. The wise will not pursue such pleasures.

Against Epicurus, it should be pointed out that absence of somatic pain is not a form of enjoyment. Of course, one may be pleased, feel relieved, and find it most pleasant *not to be* hungry, thirsty, cold, and weary. Here the pleasure is indeed privative. It is also 'pure', bringing no suffering in its wake. By contrast, friendship, which Epicurus recognized as a form of pleasure, is patently not privative. However, it can be the source of great psychological suffering in the form of grief at the loss of a loved friend or sorrow at the break-up of a friendship. So, although it is a positive rather than a privative pleasure, it is not a pure pleasure.

Pain and sensation   Somatic pain is a sensation. As we have seen,[5] it typically has a location, which is the place the sufferer assuages and at which he points when asked where it hurts. There are also overall somatic sensations, such as weariness, aching all over, being sunburnt all over. Somatic sensation has degrees of intensity

---

[5] *IP*, ch. 7.

which range from the slight to the unbearable. It admits of judgements of comparison, as when we aver that the pain in our wrist is worse than the pain in our ankle, or that someone's pain today is worse than it was yesterday. Pain has causes but no objects. It has phenomenal qualities the characterization of which is partly causal and analogical, for example a burning sensation, a stabbing sensation, or a dull, nagging sensation. Pain is largely undesired inasmuch as one normally wishes it to cease. It is attention demanding, forcing itself, as it were, on our attention. One cannot have a pain of which one is unaware, although in certain cases one can be distracted from one's pain.[6] Pain tends to impede both thought and action for it makes it difficult for the sufferer to concentrate on other things.

Pleasure and sensation The contrast with pleasure in its various forms is striking. This makes it all the more curious that pleasure, since the decline of scholasticism and the rise of early modern philosophy, has for the most part been conceived to be a sensation. In the late eighteenth and the nineteenth centuries the utilitarians took pleasures and pains to be opposites – as is patent in the quotation from Bentham above. Since somatic pains are evidently sensations, it was taken for granted that pleasures must be sensations too. This supposition seemed to be confirmed by the fact that some sensations are undoubtedly pleasant and give pleasure – paradigmatically the pleasures of satisfying the appetites. The pleasures of sex in particular are of unparalleled intensity. This categorial confusion led to egregious distortions and oversimplifications in what is a complex conceptual field, with non-trivial deleterious consequences in early psychology and economics.

If pleasure is thought to be a sensation, it seems that it must either be a sensation *of* pleasure or a pleasurable sensation. If it is a sensation of pleasure (as somatic pain is a sensation of pain) then it would have to be a special sensation caused by a stimulus or by an act, action, or activity. The stimuli that occasion the sensations of pleasure would be the pleasures one enjoys. The pleasure one feels would be the sensations of pleasure one has. The sensation of pleasure would accordingly be externally related to what occasions it. So on any given occasion on which one felt pleasure, one would have to check what it was that gave one pleasure, as one sometimes checks to see what hurt

---

[6] See Norman Malcolm, 'Consciousness and Causality', in D. M. Armstrong and Norman Malcolm, *Consciousness and Causality* (Blackwell, Oxford, 1984), 15–16.

one or gave one pain. On taking pleasure in going to the opera, one would have to check to see whether it was *Così fan tutti* or the seat in which one was sitting that gave one pleasure.[7] But that is absurd.

If pleasures were sensations, then 'quantities of pleasure being the same', innumerable pleasures would be interchangeable. One could get exactly the same pleasure from a bottle of wine as from a play by Oscar Wilde, and one needn't bother getting tickets for *Così fan tutti* since one could get the same pleasure from playing bridge. As Bentham put it succinctly, 'Prejudice apart, the game of push-pin is of equal value with the arts and sciences of music and poetry',[8] which Mill paraphrased as 'If the quantity of pleasure be the same, pushpin is as good as poetry'.[9] What is objectionable about this is not merely its philistinism, but its incoherence. Poetry is not as good a game as pushpin, since poetry is not a game. A game of pushpin is not as good a poem as 'Fern Hill', since a game of pushpin is no poem. The value of a poem does not consist in how much pleasure it gives anyone. 'The Wasteland' is not a better poem than 'The Highwayman' because readers get more pleasure from it. It makes little sense to compare how much one enjoyed reading a poem and how much one enjoyed a game of pushpin (see below), and if one enjoyed pushpin more there can be no answer to the question 'How much more?'

Of course, one may take pleasure in playing games and one may enjoy declaiming poetry. Aristotle pointed out that such activities are generally better, that is more skilfully performed, if they are enjoyed.[10] But if pleasure were merely a sensation that *accompanies* the exercise of a skill, that would imply that one would philosophize better while listening to music one enjoyed. But evidently enjoyable music might well distract one from philosophizing rather than improving one's performance.[11]

---

[7] See A. J. P. Kenny, *Action, Emotion and Will* (Routledge & Kegan Paul, London, 1963), 128–34.

[8] *The Rationale of Reward* (Robert Howard, London, 1830), 206.

[9] J. S. Mill, 'Bentham', in *Dissertations and Discussions, Political, Philosophical, and Historical reprinted chiefly from Edinburgh and Westminster Reviews* (John W. Parker & Son, London, 1859), 389. Mill was highly critical of Bentham's view, and wished to distinguish between higher and lower pleasures. One may indeed draw such a distinction, but only if one acknowledges that there are values other than pleasure.

[10] *Pace* Aristotle, there are exceptions. A good philosopher may enjoy writing easy popular articles more than the hard grind of writing original philosophical works. The latter may be fulfilling without being enjoyable.

[11] Aristotle, *Nicomachean Ethics*, X. 1175^{a}21–^{b}7.

Actually, pleasure is not a sensation at all, although there are pleasant sensations. But pleasant sensations are not sensations *of* pleasure on the model of painful sensations being sensations of pain. A sensation of pleasure, one might say, is not a pleasure sensation. Rather, it is a sensation, otherwise unspecified, that one enjoyed having, liked to have. Such are sexual pleasures, the pleasures of tactile sensations; perhaps, on appropriate occasions, the sensations of floating or of speed. It is therefore not surprising that pleasure felt has no somatic location. One cannot ask 'Where is the pleasure?' as one can ask 'Where is the pain?', but only 'Is that pleasant?', 'Is that enjoyable?', or 'Does that give you pleasure?', where the 'that' is a sensation. Nor is pleasure an overall bodily sensation like weariness. One may ache all over, but one can't feel pleasure all over, only take pleasure in one's overall somatic condition of relaxation or sexual ecstasy.

Pleasure and awareness　　The relation between pleasure and awareness or consciousness of pleasure is subtle and unlike the corresponding relation to having a pain. It is possible to take pleasure in, or enjoy, V-ing without being aware that one is doing so until after the event. This is particularly true of competitive activities that demand complete concentration. The greater the concentration in an enjoyable activity the more likely it is that it is so engrossing that one is not even aware of enjoying oneself. It is only afterwards that one may realize what pleasure it was, how much fun one got from it, and how much one enjoyed it.

Intentionality of pleasure　　A fundamental difference between pleasure and sensation is that pleasure has an object and not merely a cause. One takes pleasure *in* V-ing, derives pleasure *from* V-ing, enjoys V-ing, and is pleased *that things are so*. In some cases, most obviously in being pleased that something is so, the object is intentional – for one may be pleased at something which is not the case, as long as one thinks it is the case. A woman lying on the beach with her lover may take pleasure in his rubbing her back, only to discover that it is a stranger rubbing her back – she will not only be offended, but cease to take any pleasure.

4. Enjoyment and the pleasures of activities

Active and passive pleasures　　One may classify pleasures along a scale ranging from active to passive. Active pleasures are those one engages in – they are activities rather than receptivities (playing a game as opposed to watching a game being played). Playing games, playing musical instruments, making things (arts and crafts), travelling on holiday, mountain climbing, hill walking, and gardening are active pleasures. Observing games being played, watching a craftsman at work, watching a travelogue, and sunbathing are passive pleasures. It is evident that there is a spectrum ranging from patent activities to patent passivities (see fig. 8.1). Between the two ends of the spectrum lie pleasurable activities that are neither clearly passive nor clearly active, such as the pleasures of eating and drinking, of going to a museum, of reading. Active pleasures evidently characterize the active life, on the one hand, and recreations, on the other. Active pleasures take time, as in playing a game, travelling abroad, going for a walk, enjoying a party, as well as enjoying one's work or one's creative activities as a painter, writer, composer, or craftsman. As already noted, in the case of creative activities, as one's skills increase, one's enjoyment *tends* to increase too. The more one enjoys the activity, the better one's performance of it is *likely* to be. Generally

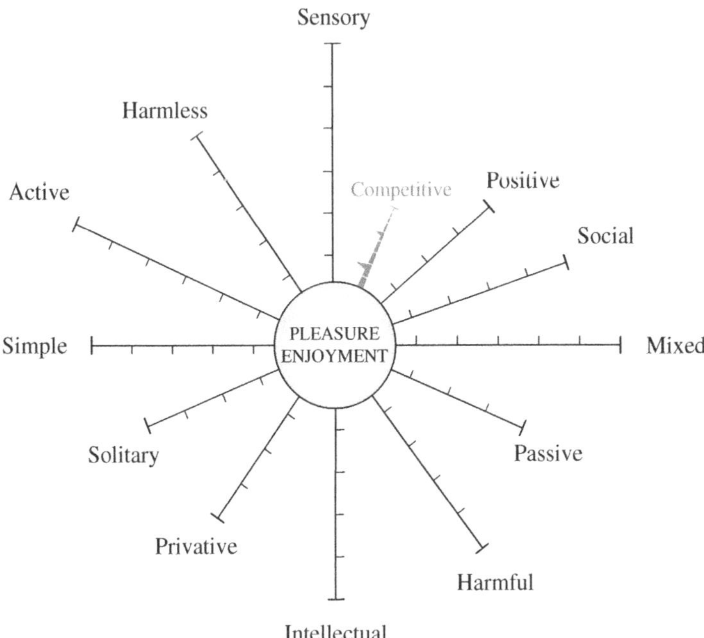

Figure 8.1 *Continua of pleasure and enjoyment in different dimensions*

speaking, if one enjoys V-ing, one tends to like V-ing, to be keen on V-ing and to V with zest and enthusiasm.

Passive pleasures are, roughly speaking, the pleasures of receptivity. Right at the end of the spectrum, so to speak, are the pleasures of idleness (*dolce far niente*): of sunbathing on the beach or sitting in a deckchair in the garden. In close proximity lie the pleasures of the senses, in particular the pleasures of perceiving something and, derivatively, the pleasure taken in what one perceives. It is clear here that there is a gradation. The pleasure of listening to good music well played is a passive pleasure relative to playing good music well. But unlike sunbathing, it demands attention and concentration, as well as knowledge of the musical tradition that gives one a trained ear, and familiarity with other performances of the same piece that enable one to appreciate and evaluate alternative interpretations. The pleasures of taste may range from the crude enjoyment of a glutton or gourmand to the refined pleasures of the trained palate of the gourmet. The pleasures of sight often demand careful looking and concentrated observation, as well as a trained eye, that is to say: knowing *how to look at* things. So, such intermediate passive pleasures usually imply some form and degree of activity, although they are not aspects of the active life. But they are integral to the life of the practical hedonists and also elements in the lives of most human beings.

Pleasure as unimpeded activity

While the utilitarians took the prototype of pleasure to be sensations and pleasures of the senses, Aristotle had taken the enjoyment of activities as prototypical. In the twentieth century he was followed in this respect by Ryle and subsequently by Anscombe and Kenny, who developed the idea of 'adverbial' pleasures. If pleasure is not a sensation or experience that accompanies an activity that gives us pleasure and from which we derive pleasure, and engagement in which we enjoy, is it then identical with the activity? It cannot be identical with the enjoyable V-ing, for it is perfectly possible to engage in the normally enjoyable V-ing *without* enjoying it (perhaps one is in pain, or is preoccupied with some other pressing matter). But it can't be different from the enjoyable activity either, since then the pleasure would either always be the same no matter what one was doing, which is absurd, since then the pleasure of playing football could be obtained from reading a book or listening to music. Alternatively, the pleasure would always be different depending on what one was doing, which is absurd, for one would then have to find out exactly what it was that one was enjoying. One way out of the impasse is to change the question. Instead of asking what happens when one enjoys an activity that does not happen when one does not, perhaps one should ask: 'What does not happen when one enjoys an activity which does happen when one does not?' Well, one might venture, one is not bored. One is not normally reluctant to engage in the activity. Indeed, one is not easily distracted from the enjoyable activity. So one does not, for example, feel a headache (or other distracting pain or discomfort) that might divert one's attention from the pleasurable activity. Further, the more skilful and successful one is at a given activity the more likely one is to enjoy it. It is, therefore, not surprising that Aristotle should have suggested that pleasure (or, better, enjoyment) is *unimpeded activity*. The interpretation of Aristotle's various writings on the theme is notoriously controversial and will not be ventured here. But the suggestion that enjoyment is unimpeded activity is of interest, diverting attention from the putative result of an activity to the zest with which it is it is engaged in.

Criticism of Aristotle   Nevertheless, this proposal will not do. First, being unimpeded is not sufficient for an activity to be enjoyable. There are many activities that one may engage in without distraction or impediment, but which are hedonically indifferent or patently not enjoyable, in particular routine activities at work, walking through a minefield, or defusing a live bomb. One may be very good at certain activities (e.g. breaking bad news to patients or pupils), but nevertheless not enjoy doing them. So the indisputable correlation between skilful and successful V-ing and enjoyable and pleasurable V-ing is not straightforward. Secondly, unimpeded activity is not necessary for taking pleasure in something, for innumerable pleasures are not activities at all, both wholly passive pleasures (such as sunbathing) and pleasures of anticipation. Thirdly, the simple formulation does not link pleasure and enjoyment *qua* unimpeded activity with reasons for action. Some feature of pleasure and enjoyment must provide a (defeasible) warrant or justification. It must be perspicuous why 'for pleasure', 'for fun', and 'because I enjoy it' are terminating answers to enquiries why one is doing something. 'Because it is an unimpeded skilful activity' is not an adequate answer.

But this much can be gleaned from the conceptual link between being enjoyable and activities: V-ing for pleasure, if it is an activity, is done willingly, not reluctantly. Other things being equal, it is not something that one would prefer not to be doing. It involves a favourable attitude (it is something one likes doing) and positive mode of attention. So it provides one with a justification for and/or explanation of action. One's enjoyment is manifest in the zestful manner of performance and the visible inclination to engage in it. It is both circumstance dependent and person relative, dependent upon individual taste.

To these features we may add further marks. One may do something *for pleasure* and *not enjoy* doing it, because one does not do it well. At the same time, one *may* enjoy an activity despite not doing it well, for example when one enjoys a competitive game despite losing. However, it is true that if one V-s for pleasure and succeeds in V-ing, then other things being equal, one will enjoy V-ing and take pleasure in it. Anything we enjoy gives us pleasure, and vice versa. But we may engage in V-ing for non-hedonic reasons (it may be obligatory or part of our duty) and nevertheless enjoy it. However, this conceptual link between pleasure and enjoyment on the one hand, and activities on the other, does not illuminate enjoyable passivities, such as sunbathing or lying in a hot bath, where no skill is involved, and various pleasurable sensations, which are not something one does.

Pleasure and prolongation   Although there are some pleasures one normally wishes to prolong (e.g. various passive pleasures), there are many pleasures and enjoyments of which that is not true. One may greatly enjoy listening to a piano sonata or watching a good play, but that does not imply that one would like it to go on; one may enjoy a good meal, but only a glutton would want to prolong the eating. Knowing that V-ing is enjoyable does not mean that one would like to V: it may not be to one's taste. Moreover, one may have more important things to do.

Pleasure as restorative   It is in general true, and a very important truth, that pleasure and enjoyment are restorative. Doing enjoyable things alleviates unpleasant mental states such as boredom, which is ameliorated by doing or watching something exciting, fascinating, or amusing. Indeed, even exhausting pleasures of hyperactivity may be restorative. Jangled nerves are alleviated by soothing enjoyments; disaffection for the present may be ameliorated by nostalgic pleasures, on the one hand, or the pleasures of anticipation, on the other.

The conceptual diversity of the field of hedonic concepts is so great that it is futile to search for definitions of pleasure, enjoyment, and being pleased. It is, it seems, a conceptual domain with different centres of variation. On reflection, that is exactly what one should expect from creatures like us whose individual histories are unique and whose modes of engagement with each other and with our environment are so complex and whose abilities are so diverse. What is needed to resolve puzzlement is to lay bare the logical geography by connective analysis.

5. Pleasure, desire, and satisfaction

Logical and psychological satisfaction distinguished If something, some passivity or activity is a pleasure to have, to undergo, or to do, then it is a possible object of desire. Whether one wants to obtain it, undergo it, or do it is, as we have seen, person relative – dependent upon individual taste and occurrent mood. It is also circumstance dependent. Getting something one wants or wishes for, passively engaging in something that generally gives one pleasure, or engaging in some activity that one usually enjoys is called *satisfying one's desire* or *fulfilling one's wishes*. This is a logical relation, not a psychological one. It is akin to the relation between a command and fulfilling it. 'The command to V' $=^{df}$ 'the command that is fulfilled by V-ing'. V-ing, one might say, is the fulfilment condition (or obedience condition) of the command to V, as the obtaining of the state of affairs that p is said to be the truth condition of the proposition or statement that p.

Whether the satisfaction of the desire for something or of wanting to do something that normally pleases one gives one *a feeling* of satisfaction, gratification, or pleasure is a further, empirical, psychological, question. Often getting what one desires or wants turns to ashes in one's mouth. Confusing or conflating the logical and the psychological senses of *satisfaction* is a root of the misconceived doctrine of psychological hedonism. Two complementary errors are common. First, supposing that because obtaining what one desires is called satisfying one's desire, therefore what one *really* wants is not to have the object one wants to have or not to engage in the activity one wants to engage in, but rather the satisfaction of the want for that object or activity. Secondly, supposing that the satisfaction of one's wants or desires is psychological rather than logical – inferring that the logical satisfaction is itself a feeling of pleasure or enjoyment, and that is really all one ever wants. But it is evident that what one objectively wants when one wants O or to V or be V-ed, is O or to V or be V-ed. To say that one wants one's desire for O or desire to V to be satisfied means no more than that one wants to have O or to V. (This is precisely parallel to the assertion that what one is commanded to do when one is commanded to V is the fulfilment of the command to V, which means no more than that what the command to V commands is to V.)

What one does for pleasure is not coextensive with what one enjoys. One may be fortunate and enjoy one's work, but, for all that, one does it to earn a living not for pleasure – even if one would not have chosen this kind of work had one not enjoyed it. Those things one does for recreation are pleasures, although there are many things one enjoys and many things that give one pleasure that are not recreations (such as satisfaction of appetites). 'He enjoys V-ing' and 'V-ing gives him much pleasure' explains his V-ing when the activity is purposive (like playing chess). Otherwise it merely explains why someone *continues* V-ing. One generally likes and likes to do whatever gives one pleasure, but not everything one likes to do (such as rising early to get things done) is done for pleasure.

Being pleased   One may be pleased by all manner of things that one views with favour or hopes for and the occurrence of which is preferred to its non-occurrence. We are generally pleased to be enjoying or to have enjoyed ourselves or to have taken pleasure in something. But it is possible to take pleasure in something and to be disturbed, shocked, or horrified to find oneself doing so, as when one finds oneself taking pleasure in doing or in watching something morally objectionable, demeaning, or wicked (e.g. bull-fighting, hunting big game, hare-coursing). Being pleased is conceptually related to what is pleasant inasmuch as experiencing something pleasant is pleasing. It is further internally related to a group of positive responsive attitudes, such as feeling delighted, overjoyed, triumphant, thrilled, grateful, or relieved. Being pleased is an intentional attitude: we are pleased *by* something. What pleases us need not be so, although if one finds out that it is not so one's pleasure is replaced by disappointment. We may be pleased by something without its giving us pleasure or enjoyment – as when we are pleased and relieved not to have caught cold. Conversely, we may enjoy something, such as rock climbing, without being pleased at anything. Being pleased, in particular being pleased by what is not privative, is stronger than being satisfied, indicating more positive feelings, more favourable attitude, and a moderate degree of enthusiasm.

Avowals and criteria for taking pleasure   Does one invariably know whether one is pleased by something, takes pleasure in something, or enjoys doing something? Is the subject's word final and authoritative? It is tempting to think so. The Lockean introspectionist tradition claims that the subject knows whether or not he is enjoying himself by inner sense, by 'peering into his breast' to see whether there is an idea of pleasure within. The Cartesian will contend that one cannot be pleased by something or take pleasure in something, and at the same time doubt whether one is. So when one takes pleasure in something, one knows one does. But investigation is needed. The first step to investigate is not introspective, but observational. What are the criteria for whether people are enjoying themselves, pleased by something, or taking pleasure in something? Obviously the criteria are behavioural, including verbal behaviour or avowal. Equally obviously, the behaviour that is apt for being pleased by or with something or someone differs profoundly from the behaviour that is apt for enjoying oneself. So too the hedonic behaviour will vary greatly according to what it is that one is enjoying, taking pleasure in, finding pleasant, or being pleased by. It will also vary greatly depending on the age, gender, and personality of the subject, as well as upon the circumstances and the social mores – it may be inappropriate or undignified to show how pleased one is to have achieved such-and-such an accolade. In all cases, however, it is the responsive behaviour and utterance that provide the criteria for the subject's hedonic response. The behaviour may be smiling, exclamations of delight, jumping for joy, and intense concentration on playing the game, followed by expressions of satisfaction, complete absorption in the book one is reading or in listening to the music to which one is attending, and so forth. Deception is possible – one may be loath to disappoint one's donor or host. Truthfulness does not always guarantee truth, for self-deception is also possible. One may only slowly realize that one has not enjoyed the last half a dozen parties one attended, and loath to admit that the enjoyment of V-ing is waning. So the criteria are defeasible. But, if not defeated, the agent's word goes, not because he is an 'authority' on his own pleasures, nor because he has 'privileged access' by means of introspection to the contents of his mind, but because his behaviour and utterances are *expressions* of pleasure, enjoyment, and being pleased. The exclamation 'Oh, that was good' is no less an avowal of pleasure than 'Ow, that hurt' is an avowal of pain, although pain is not intentional and pleasure is.

6. Comparability and quantification

Bentham's hedonic calculus   Bentham noted an array of dimensions of pleasure: intensity, duration, certainty, propinquity or remoteness, fecundity, and purity. At first blush this may well seem plausible. *Intensity* of pleasure is supposedly parallel to the intensity of pain and adverts to the fact that pleasure, like pain, can vary in intensity. Some experiences may be pleasant or enjoyable, enjoyable or overwhelming, to the point of rapture or ecstasy. Pleasures may be *momentary*, like smelling a whiff of scent or sampling a passing taste, or they may *endure* for a while like the pleasure of listening to a concert or reading a good book. The *certainty* of a prospective pleasure can be estimated as possible, probable

to one degree or another, or certain. If one course of action will certainly result in pleasure and another only probably, then, other things
being equal, the former is to be preferred. *Propinquity* or *remoteness
in time* speaks for itself – an immediate pleasure is preferable to a
deferred one. By *fecundity* Bentham meant the degree to which one
pleasure is likely to produce further experiences of pleasure or not.
Pleasures of understanding tend to be fecund: the more one now
enjoys gaining an understanding of something, the more one is likely
to enjoy deepening one's understanding and knowledge. By *impurity*,
he meant the degree to which a given pleasurable experience is mixed
with various kinds of unpleasantnesses. The purer the pleasure the
more desirable it is. It is noteworthy that intensity, duration, and
purity determine how much pleasure is to be derived from V-ing or
being V-ed, whereas fecundity, propinquity, and certainty determine
rational choices between different pleasure-yielding acts or activities.
These six dimensions concern the individual human being and his
pursuit of his own pleasure. To it must be added the dimension of
extent, that is, the number of people who will derive pleasure from
the event or act in question – the more, the better the act.

Bentham assumed that any pleasure one takes can be evaluated in
these seven dimensions, which can then be added up to yield a measure of the rational desirability of pursuing one course of action or
another. This he hoped would lay the foundations for a calculus of
utility. The pleasure in question can then be compared with other
courses of action taken in the past in order to judge which was preferable and to guide one in future choices. Similarly, prospective pleasures can be assessed in the seven dimensions and ordered cardinally
to determine what should be done. This conception laid the foundations for marginalist economics.[12] This theoretical form of analysis
subsequently abandoned cardinal utility in favour of mere ordinal
utility, and later relinquished the very notion of utility and replaced it
first by desire satisfaction and later by preference fulfilment.

*Comparability
of pleasures*
A question that a conceptual investigation into pleasure must confront is whether it makes sense to evaluate
the quantity of pleasure to be obtained from an event,
act, activity, or passivity and subsequently to compare pleasures in

---

[12] It is noteworthy that the principle of diminishing marginal utility that lay at the
heart of Marshallian economics may have some plausibility when applied to consumer
goods, but is sheer nonsense when applied to morally good action. The third act of
kindness or justice in a row is not less good or less desirable than the first.

order to choose the most pleasure-yielding course of action. To the modern ear, accustomed to being asked at the hospital to rank pains on a scale of 1 to 5, or asked by social surveys to evaluate the happiness of one's life on a scale of 1 to 10, Bentham's felicific calculus may seem intelligible. But this is something to be investigated.

Two things are surely clear – First, the thought is geared to the presupposition that pleasure is a sensation. But, as we have seen, it isn't, even though some sensations are pleasant or enjoyable. Secondly, various rough criteria that may be invoked apropos pleasure and enjoyment in special contexts or circumstances are being sublimed and idealized. One may be asked whether one enjoyed the soup – and it is easy to give a reply: it's excellent/very good/not bad/ not very good. But these are evaluations of the gustatory merits of the soup, not of the intensity of one's pleasure in eating it, although there is a presumption that the better the soup the more pleasure one gets from eating it (but not if one is ill, or at a luncheon interview). But is the pleasure more *intense*? Limited comparisons make sense here. 'Was today's soup better than yesterday's?' If one answers affirmatively, could one further answer the question 'How much better?' Could one answer the question 'Was the soup better than the pudding?' Only if one was good and the other awful. But suppose that both were good? Given that one enjoyed the soup and that one also enjoyed the Mozart sonata that was being played, could one answer the question 'Did you enjoy the soup more or less than the sonata?' We are prevailed upon to think that comparisons of pleasures (and their intensities) make more sense over a wider range than they do, because of the frequency with which one can unhesitatingly say, 'Oh, I enjoyed A much more than B'. One may say that one enjoyed the soup much more than the pudding because the pudding was disgusting. One may confidently assert that one enjoys Mozart more than Wagner, not because one has compared a performance of *The Magic Flute* with a performance of *Das Rheingold*, but because, like Nietzsche, one has come to detest Wagner. We may indeed compare two operas, or two performances of the same opera, but the basis of comparison is not the amount of pleasure one derived or derives from hearing them. We constantly make choices between alternative entertainments for a given occasion. But our choices do not turn on elements of a felicific calculus, but on our current mood, on how tired we are, on the inclinations of our companion, on whether we want a change, whether we are curious about the entertainment (if it is new), how interesting/

amusing/exciting/soothing/engaging the prospective form of pleasure or recreational activity is likely to be.

Pleasure and duration   The relation of a pleasure to duration has similarly been idealized. Not every enjoyable activity can be simply related to duration. Can one always say that the pleasure one derives from V-ing is temporally coextensive with V-ing? Does the pleasure of reading a good novel last as long as the reading? Does it begin on the first page? Does it continue after the last page or does it stop abruptly? Is it the same on the second evening when one is halfway through? One enjoyed playing a game of football – all the time or only when one was actively engaged? Other things being equal, is a prolonged pleasure therefore better than a brief one? So are great symphonies always more enjoyable than great sonatas *because they are longer*, or novels than short stories? These questions are surely absurd, and it is not a coincidence that we do not raise them. They might approximate sense if pleasure were an accompaniment of the pleasures and enjoyments – but it isn't. Of course, one *can* say that one enjoyed the first half of a book, a play, a concert, but not the second – because the second half tailed off, became boring, was too vicious, and so forth. But the notion that one's pleasure or enjoyment is clockable is, in the case of adverbial pleasures, as misconceived as the idea of measuring the duration of the feeling with which a piece of music is played.

Similar considerations apply to propinquity and remoteness. Is jam today really always better than jam tomorrow? Perhaps one has no time today, being busy on other matters. Isn't that because the other matters are more enjoyable? Only if one has accepted descriptive and egoistic hedonism. There are endless reasons, which have nothing to do with pleasure, that one might have for postponing an entertainment or indulgence. Moreover, one may well postpone the best until the end.

Limits of hedonistic axiology   It is not necessary to go any further, since it should be obvious that this hedonistic axiology and its associated norms are profoundly misguided. It is monist, recognizing only one single value, whereas no axiology for social beings with the power to reason and sensitivity to reasons, as well as the ability to feel sympathy, can be anything other than pluralist. For justice and fairness are not forms of pleasure or enjoyment. We do not value integrity, autonomy, and human dignity for hedonic reasons. The value of transcending one's own concerns, including one's hedonic desires, for the sake of love for another is not measured in the currency of pleasures. We recognize a multitude of different values, and we know full well that they often conflict. Such conflicts are rarely reconcilable by felicific calculations. Hedonistic axiology is psychological, whereas the forms of goodness extend far further than the value of mental states or conditions. In particular, the goodness of morally good deeds is not a form of hedonic goodness (of maximizing pleasure for the greatest number). Moreover, pleasure, contrary to both Epicurus and Bentham, is not good in itself. *Schadenfreude* is not good at all, but typically mean and nasty. So too are the pleasures of malevolence and evil-doing.

7. First-person judgements of pleasure

Pleasure and first-person authority   The utilitarians were prone to aver that each person has final authority on whether something gives him pleasure (or, as they put it, *happiness* – since they misguidedly assimilated the two). 'Don't you tell me that I am enjoying this – I'm not!' is an exclamation that has a use. 'Only you can say whether you like it or not!' seems a grammatical statement that amounts to the grammatical proposition that your word goes, that you have the last word on the matter. How can this be? The utilitarians were in broad agreement with the Cartesian and Lockean conception of the inner and the outer. Only the subject can perceive, by inner sense, the lively theatre of ideas and impressions in his mind. And what he perceives there he cannot doubt. He knows, with certainty, whether what he is experiencing is pleasure and enjoyment or not. Only he can thus introspect and report back for others what he there finds (the mind as a private peep show). So the subject has first-person authority on his own pleasure and enjoyment. This is a powerful and persuasive picture. As first cardinal utility, and then ordinal utility, were rejected in favour of desire satisfaction, this picture was retained. That the subject has the last word on what he wants, coupled with the notions that giving people what they say they want is good for them and hence morally good, are presuppositions of much demotic and democratic ideology.

Hedonic utterances   We have already seen ample reason to reject the venerable picture of the inner and the outer. First-person present-tense utterances of pleasure, enjoyment, and being pleased are not knowledge claims. They do not rest on evidence provided by inner sense or introspection. They are not known immediately, or yet mediately, since the idiom of knowledge is here inept. First-person present-tense utterances are *expressions* of pleasure, enjoyment, or delight. They belong together with exclamations of pleasure such as, 'This is delicious', 'How beautiful it is', 'To hear that [e.g. Maria Callas singing Tosca] was an experience of a life-time', 'That was most enjoyable/exquisite/wonderful', as well as first-person utterances such as 'I'm so pleased to hear that' and 'I *did* enjoy that', or 'I find it wonderful'. There is a gradation from exclamations to reports, but such utterances are not descriptions of the subject's hedonic responses. A token of that lies in the fact (as pointed out by Wittgenstein in another context) that here truthfulness guarantees truth. By contrast the corresponding third-person statements are indeed descriptions, for example 'He is pleased', 'She takes pleasure in ...', 'Look at him. He *is* enjoying himself', 'She thinks it's wonderful'. These are true or false statements the criteria for which lie in the agent's behaviour, including exclamations, facial expressions, and demeanour. These criteria are defeasible, since the agent's sincerity may be called into question in some circumstances – his hedonic avowal may be no more than politeness, self-deception is possible, and there may be a clash between the agent's behaviour and avowal. First-person expressions of pleasure or enjoyment are valuations, whereas third-person statements are not.

Pleasure and the good of man   It was part of utilitarian doctrine and later of the various forms of consequentialism, to argue that pleasure and enjoyment, and later desire satisfaction, are constitutive of the good of man. Right action is action that is conducive to the greatest happiness, pleasure, utility, or desire satisfaction for the greatest number. This mesmerizing doctrine, coupled with the view that each person is the best judge of what gives him pleasure, yields political demoticism (masquerading under the title of *democracy*) and the economics of the marketplace (which allegedly maximizes *choice* under the guise of satisfying cognitively self-transparent preferences). Giving people what gives them pleasure, what they want, or what they prefer is held to be the high road to human happiness and a good society. Would that the path to felicity were so straight.

Intemperance   While a degree of pleasure and enjoyment are normal constituent elements in a good and flourishing life, excess of pleasure has generally been held to be self-indulgent and intemperate. Intemperance is a vice associated with the domain of the hedonic, in particular with pleasures of the senses and specifically with pleasures associated with the senses of touch and taste. It is the gourmand and gourmet, the glutton and the drunkard, who are the primary bearers of the vices of intemperance. Concupiscent indulgence is not normally viewed as intemperate, but rather as lechery or venery.

It is striking that indulging in excessive reading is commonly viewed adversely, especially if done at the expense of taking exercise or at the cost of one's responsibilities, but is not held to be intemperate. Similarly, someone who listens to music to excess may indeed be criticized, but not as intemperate. Hedonic vices are forms of self-indulgence.

8. The hedonic life

Cyrenaics and Epicureans   To live the life of a *practical* hedonist is to live a life guided by the pursuit of pleasure. Some Western philosophers in the ancient world advocated such a form of life and put it into practice.[13] Most famous among these was Aristippus the

---

[13] To be sure, the ideal of a life of pleasure was articulated and advocated long before, for example in the advice tendered to Gilgamesh (Sumerian version, ca. 2100 BC, Akkadian version ca. 1300 BC):

> Gilgamesh, where are you running?
> You won't find the immortal life you are seeking.
> When the gods created mankind
> They ordained death for humankind
> And retained immortality for themselves.
> So Gilgamesh, let your belly be full.
> Be merry every day and night.
> Make each day a day of joy.
> Dance, play, by day and by night.
> Wear clean clothes.
> Let your hair be washed and your body bathed with water.
> Cherish the little child who grasps your hand.
> Let your wife rejoice in your arms
> For this is the destiny of mankind.

The same attitude of 'eat, sing and be merry for tomorrow we die' is evident in the following ancient Egyptian song dating from the Twelfth Dynasty (1991–1802 BC):

> Let thy desire flourish,
> In order to let thy heart forget the beatifications* for thee.
> Follow thy desire, as long as thou shalt live.
> Put myrrh upon thy head and clothing of fine linen upon thee,
> Being anointed with genuine marvels of the gods' property.
> Set an increase to thy good things;
> Let not thy heart flag.
> Follow thy desire and thy good.
> Fulfil thy needs upon earth, after the command of thy heart,
> Until there come for thee that day of mourning.

*Funerary preparations of the corpse.

Elder (ca. 435–356 BC), a pupil of Socrates who established the Cyrenaic school. He advocated the pursuit of sensory and sensual pleasure as constitutive of a good life for human beings, provided one does not become enthralled to pleasure. He was not an amoralist, in so far as he insisted that one should not harm others. Nevertheless he incurred the disapproval of both Plato and Xenophon for his hedonic life.

Epicurus, as already noted, was the founder of the most renowned school of hedonism in the ancient world, but it was a hedonism of a very different brand from the Cyrenaics'. Pleasure was viewed primarily as privative and the goal of this form of hedonic life was *apatheia* (equanimity) and *ataraxia* (tranquility) – very far removed from 'eat, drink, and be merry for tomorrow we die', *a fortiori* from 'wine, women, and song'. Epicurus was renowned for living a sparse and restrained life of pleasure as he understood it. His hedonism was not merely philosophical theory but very much a practical hedonism according to which he and many of his followers lived.

Stoics   It is striking that the Stoics, although they too aimed to achieve *apatheia* and *ataraxia*, were averse even to the very mild hedonism of the Epicureans. They adamantly rejected the idea of the desirability of a life of pleasure as contemptible. In his dialogue *Old Age*, Cicero, a Stoic, puts into the mouth of Archytas of Tarentum the following remark:

> This much can be made clear if you think of someone enjoying the most delightful sensual pleasure imaginable. It will be generally agreed that during the process of enjoyment he is incapable of any rational, logical, or cerebral process. The consequence is that such pleasures are exceptionally repulsive and harmful. Indeed their substantial, prolonged indulgence will plunge the whole light of the spirit into darkness.(*On Old Age*, 12.40)

Comparably critical views are to be found in Seneca:

> Now go, question yourself:

> If you are never downcast, if your mind is not harassed by any apprehension, through anticipation of what is to come, if day and night your soul keeps on its even and unswerving course, upright and content with itself, then you have attained to the greatest good that mortals can possess.

> If, however, you seek pleasures of all kinds in all directions, you must know that you are as far short of wisdom as you are short of joy. (*Letters from a Stoic*, letter 59)

In general, Stoics distinguished sharply between joy and pleasure. Pleasure was identified with sensual and sensory pleasures, essentially fleeting and detrimentally pursued in disregard of the demands of virtue (Ovid's *Ars Amatoria* was definitely *infra dig*). Joy, however, was the upshot of living a virtuous life and was thought (quite wrongly) to be impervious to fortune.

Christianity and the hedonic life [39] As sketched in *The Passions: a Study of Human Nature*, Christianity, from its Pauline inception, was ill disposed towards sensory pleasures in general, and towards sexual pleasure in particular. Such pleasures were associated with capital vices, for example gluttony, on the one hand, and *luxuria* or concupiscence, on the other. Practical hedonism was officially condemned even more adamantly than by the Stoics, and held no place in the Christian conception of the good life. Indeed, it was held to distract the sinner from the required service to God and from the exclusive love of God. To be sure, this did not determine the lives of all as is evident in Le Roy Ladurie's *Montaillou*, which recounts the beliefs and form of life of a Pyrenean village in the thirteenth century. The lay literature and poetry of the high and late Middle Ages (ranging from the *Roman de la Rose* (1230, and especially the 1275 version) on the one hand, to Boccaccio (1313–75) and Chaucer (ca. 1343–1400) on the other, exhibit the uninhibited bawdy of the courtly aristocracy as well as of the vulgar (François Villon (ca. 1431–ca. 1463). Nevertheless, the most common popular moralizing attitude towards the hedonic life was unequivocally expressed in Sebastian Brandt's *Ship of Fools* (1494), chapter 50:

> Oh, worldly pleasure's likened to
> A meretricious woman who
> Traverses streets, a common whore,
> Bids one and all approach her door;
> ...
> Oh, fool, you should consider well:
> Before you plunge yourself in hell,
> In Heaven rather store up treasure
> Than cultivate such earthly pleasure
> ...
> Who concentrates upon delights
> Knows not the pain which he invites;
> Some quaff the sweet and never think
> Upon the gall they'll sometime drink,
> For joys on this terrestrial waste

> At last impart a bitter taste,
> Despite the fact that Epicurus
> Sought pleasure first, as some assure us.[14]

However, as Europe became more civilized and cultivated, and as classical learning revived throughout the courts of Europe, the pursuit of refined pleasures, restrained by reason and morality, became a thinkable form of life again among the ruling classes. This did not involve the advocacy of a practical hedonistic form of life on the model of either the Cyrenaics or the Epicureans. But it did involve the pleasures of literature and the arts, of elegant dress and politesse, of palaces and gardens, and, in due course, of music and theatre, over and above the perennial venery, the pleasures of the senses and of hunting that had characterized medieval courts and continued to characterize aristocratic life. By the seventeenth century it was no longer so much the Catholic Church that attempted to curb such pleasures, but the various Protestant and Puritan sects. Rakes and libertines doubtless abounded, but no new practical hedonist ideologies to speak of were advanced.

Utilitarians were not practical hedonists but committed political radicals devoted to the public good. They assimilated pleasure to happiness, and both to utility, and adopted the principle of utility as their guide to right action. The numerous distinguished nineteenth century utilitarians were great social reformers, but were not noted bon viveurs.

Aestheticism   A curious form of practical hedonism emerged in the late nineteenth century, concurrent with late Romanticism. Aestheticism was an avowedly literary movement in England, consisting of a group of literary and artistic figures advocating art for art's sake. It was inspired by Walter Pater (1839–94), art critic and novelist. Pater's novel *Marius the Epicurean* (1885) is less of an accurate representation of ancient Roman Epicureanism in the Antonine age than a vision of a nineteenth-century academic art critic who had lost his religious faith and sought to replace it with aesthetic experience. It is, he held, only in the fleeting artistic experience of beauty that any meaning can be found in life. His most renowned declaration of principles is in the conclusion of his *Studies in the History of the Renaissance* (1873):

---

[14] Sebastian Brant, *The Ship of Fools*, trans. William Gillis (Folio Society, London, 1971).

The service of philosophy, of speculative culture, towards the human spirit is to rouse, to startle it into sharp and eager observation. Every moment some form grows perfect in hand or face; some tone on the hills or the sea is choicer than the rest; some mood of passion or insight or intellectual excitement is irresistibly real and attractive for us, – for that moment only. Not the fruit of experience, but experience itself, is the end …

To burn always with this hard, gemlike flame, to maintain this ecstasy, is success in life. In a sense it might even be said that our failure is to form habits: for, after all, habit is relative to a stereotyped world, and meantime it is only the roughness of the eye that makes any two persons, things, situations, seem alike. While all melts under our feet, we may well catch at any exquisite passion, or any contribution to knowledge that seems by a lifted horizon to set the spirit free for a moment, or any stirring of the senses, strange dyes, strange colours, and curious odours, or work of the artist's hands, or the face of one's friend. Not to discriminate every moment some passionate attitude in those about us, and in the brilliancy of their gifts some tragic dividing of forces on their ways, is, on this short day of frost and sun, to sleep before evening. With this sense of the splendour of our experience and of its awful brevity, gathering all we are into one desperate effort to see and touch, we shall hardly have time to make theories about the things we see and touch. …

Well! we are all condamnés, as Victor Hugo says: we are all under sentence of death but with a sort of indefinite reprieve – les hommes sont tous condamnés a mort avec des sursis indefinis: we have an interval, and then our place knows us no more. Some spend this interval in listlessness, some in high passions, the wisest, at least among 'the children of this world', in art and song. For our one chance lies in expanding that interval, in getting as many pulsations as possible into the given time. Great passions may give us this quickened sense of life, ecstasy and sorrow of love, the various forms of enthusiastic activity, disinterested or otherwise, which come naturally to many of us. Only be sure it is passion—that it does yield you this fruit of a quickened, multiplied consciousness. Of this wisdom, the poetic passion, the desire of beauty, the love of art for art's sake, has most; for art comes to you professing frankly to give nothing but the highest quality to your moments as they pass, and simply for those moments' sake.[15]

Pater's impact was extensive, both upon his pupils (e.g. Oscar Wilde, Gerald Manley Hopkins) and among the following generation of art

---

[15] Walter Pater, *Studies in the History of the Renaissance* (1873), ed. Matthew Beaumont (Oxford University Press, Oxford, 2010), conclusion.

critics, such as Bernard Berenson, Roger Fry, and Kenneth Clark. His counterpart in France was the critic and author Théophile Gautier. The hedonism of the Aesthetes had a darker, decadent side, as is evident in Oscar Wilde's *The Picture of Dorian Grey* and *Salomé*, in Aubrey Beardsley's illustrations, as it is among the Pre-Raphaelites, such as Swinburne and Rossetti. An heir to the Aesthetes was the Bloomsbury Group in the inter-war years, who averred a deep commitment to the intrinsic value of personal relationships and aesthetic experience – a set of beliefs derived in part from G. E. Moore. The idea that the *only* intrinsic good consists of *experiences* should strike anyone who has reflected upon the varieties of goodness (see Chapters 1 and 2 above) as altogether bizarre.[16]

---

[16] For a brief overview of the representation of pleasure in post-medieval European visual art, see Appendix 4.

9

Happiness

1. The linguistic terrain

Happiness: systematic disagreement Happiness has been at the centre of philosophical reflection ever since Plato and Aristotle. It is commonly held to be an end, if not *the* end, of life. All of us, if asked whether we should like to be happy, would answer without hesitation 'Yes', or 'Of course', for no one would wish to lead an *unhappy* life. Nevertheless, it is unclear what happiness is, and what counts as a happy life. Epicureans thought of happiness as the satisfaction of one's minimal needs and the absence of further desires. Stoics thought that happiness must be so robust as to inure one from all misfortune and render one imperturbable. Some Christians have held that true happiness is not to be found in this life but only in heaven, in the vision of the glory of God. Utilitarians held that happiness is pleasure and the absence of pain. With the recent emergence of a putative 'science of happiness' (to be discussed in Chapter 10), it has become urgent to elucidate the concept of happiness lest we succumb to pseudo-scientific theories and scientistic advice about how to lead a happy life. We must get a *distinct* idea of what happiness is, i.e. distinguish it from related concepts with which it might be confused, such as contentment, pleasure, enjoyment, joy and bliss. It is equally necessary to get a *clear* idea of what happiness is, that is specify what features characterize happiness. We shall begin

The Moral Powers: A Study of Human Nature, First Edition. P. M. S. Hacker.
© 2021 John Wiley & Sons Ltd. Published 2021 by John Wiley & Sons Ltd.

with some linguistic botanizing, as Grice called it – namely with an investigation into the meaning and etymology of the English expressions 'happy' and 'happiness', and a brief glance at comparable terms in other languages.

Etymologies   Both the adjective 'happy' and the abstract noun 'happiness' are derived from the late medieval English expression 'hap' (late fourteenth century). Hap is the chance or fortune that befalls one, one's luck or lot in life. Accordingly the word can be modified by such prefixes as 'good', 'bad', 'evil', and 'ill', and by possessive pronouns. Someone lucky has good hap, whereas the unlucky have bad or ill hap. The association of 'hap' with good fortune came to dominate usage and was transformed into the adjective 'happy'. This association is not accidental. Most Indo-European languages link the notion of happiness to luck and chance. French *bonheur* (happiness) and *heureux* (happy) have their root in the Old French *heur*, which meant luck or chance. The German *Glück* is to this day ambiguous, meaning both luck and happiness, and the Germanic *geluk* (Dutch) and *lykke* (Norwegian) retain the nexus. The Italian *felicità*, the Spanish *felicidad*, and the Portuguese *felicidade* are all derived from the Latin *felix* (fortunate) and *felicitas* (luck, fortune). Of especial interest to any philosophical investigation into happiness is the ancient Greek etymology and synonyms. The most common term is *eudaimonia*, combining the prefix *eu-*, meaning good or well-disposed, and *daimon*, meaning a spirit, genius, or divine power. To be favoured by a divine power (or guardian spirit) is to prosper, hence the term *olbos*, signifying prosperity granted by a god, and *olbios* meaning prosperous, blessed (in the archaic sense), happy. In Hesiod *eudaimonia* and *olbios* are interchangeable, and human happiness is conceived to be, as it were, a plaything of the gods. A life free of reverses of fortune is not given to man (a prominent theme in Pindar). Consequently *eudaimonia* is linked to *tyche* (luck) and a happy man is *eutyches*, that is fortunate or lucky. *Eudaimonia* is a lucky gift of the gods, liable to arbitrary reversal and exposed to the ravages of time and hazard.[1] *Makar* and *makarios* mean bliss or blessed, and *eu zên* signifies living well or a good life. It is noteworthy that most of these expressions apply only to a person's whole life – not to afternoons, days or nights.

––––––––––––––––

[1] I am indebted to Rosanna Lauriola's 'From Eudaimonia to Happiness: Overview of the Concept of Happiness in Ancient Greek Culture', *Revista Éspaço Acadêmico* 59 (2006).

It is therefore striking that in Hebrew, which is not an Indo-European language, there is no etymological link between happiness and the notion of luck or good fortune. Ancient (and modern) Hebrew is rich in eudaimonic vocabulary, since joy, happiness and rejoicing played so central a part in ancient Jewish religion and celebration. But there is no connection either between the numerous close synonyms (e.g. *simcha* - happiness, celebration; *osher* - deep and lasting happiness; *orah* - light, happiness; *gila* - exuberant outburst of joy; *rina* refreshing happiness; *ditza* - sublime joy; *sasson* - unexpected joy or happiness; *tzahala* - happiness of dancing; *chedva* - happiness of togetherness).

The evolution of the English eudaimonic vocabulary is striking: 'happy' could apply to a human being to signify a person favoured by good fortune, who is lucky, successful, and fortunate (first occurrence identified by the *Oxford English Dictionary* in 1387). It could also be applied to an event or period in which someone enjoyed good fortune or good luck. Hence one could speak of a happy accident, a happy coincidence, or a happy position to be in (1393). A further familiar usage followed rapidly, namely a predication of action or speech to signify being felicitous or apt, as in 'a happy thought', 'a happy reply', or 'a happy situation' (1400). Usage associated with contentment enters the stage surprisingly late. The *OED* records 'happiness'/'happy' being used to signify a deep sense of pleasure or being pleased, glad or satisfied with one's condition or circumstances, content, only in 1477, and its currently common application to an event or period as characterized by contentment or as being joyous, as in 'happy occasion', 'happy times', or 'happy childhood', only as late as 1547. 'Happy memories' and 'happy ending' followed on, as did the use of 'and they lived happily ever after' in storytelling. It must have seemed altogether natural to extend its use to the expression of good wishes, as in 'Happy birthday/New Year/Christmas'. The distinctive use of 'happy' to signify willingness, readiness, or eagerness to do something, as in 'happy to do/make/ deliver/show/pay' is recorded in 1633. The extension of the application of 'happy' to a group or community as in 'happy ship', 'happy crew', 'happy society', 'happy home', or 'happy team' apparently came only in 1717. The opposite of 'happy' is 'unhappy', the two being exclusive but not exhaustive, as one may be neither happy nor unhappy. They are contraries rather than contradictories, and not being unhappy does not imply being happy.

Unhappiness   Being sad, miserable, depressed, unfortunate or hapless, ill-fated or doomed, wretched, and so on are incompatible with being happy. Being weighed down by a sense of the meaninglessness and pointlessness of life likewise precludes happiness (see Chapter 11). Indeed, there are indefinitely many ways of being unhappy. One cannot spend a happy evening at a delightful party if one is bored, if one is feeling very ill or is in great pain, if one is there against one's will, and so on. One cannot spend a happy fortnight's holiday if one is sunk in depression or distraught with grief. One cannot have a happy childhood if one is constantly maltreated by one's parents or sibling, or if one is beset with incessant anxiety. One cannot enjoy a happy youth if one is subject to the ill will of another or suffers intolerable deprivation of the necessities of life. One cannot have a happy marriage if one shares no interest with one's spouse, or finds one's spouse repulsive, or loves another (the fate of many an imperial, royal, or aristocratic marriage throughout history). So unhappiness is asymmetric with happiness. There is a limited number of ways in which one may enjoy happiness (in its various very different forms), but indefinitely many ways of being unhappy.[2] This is a corollary of the very large number of preconditions of happiness, the absence of any one of which *may* (but need not) be the cause of a person's unhappiness in so far as it is a source of discontent, grief, and frustration. We shall explore this below.

Examining such linguistic data enables one to put some preliminary order in such a complex field of currently related and interwoven expressions (see table 9.1).

Scrutinizing this ordering makes it possible to obtain an overview of the network of concepts in which that of happiness is interwoven (see fig. 9.1).

2. A distinct idea of happiness

Objective and subjective happiness   With this before us, we should at least be able to recognize the terrain in which we are walking and begin to orient ourselves. It is already evident that, just as we distinguished 'subjective' from 'objective'

---

[2] A thought implicit in Tolstoy's opening remark in *Anna Karenina*: 'All happy families resemble one another, but each unhappy family is unhappy in its own way.'

Happiness (n.)	Happy (adj.)	Adjectives excluding happiness in one of its forms or another
satisfaction, contentment, well-being	Carefree, untroubled, pleased, satisfied, gratified, contented	dissatisfied, unsatisfied, troubled, displeased, discontented, unfulfilled, disgruntled, frustrated
good spirits, cheerfulness, light-heartedness	cheerful, jovial, blithe	cheerless, heavy-hearted, low-spirited, dejected, dolorous, hapless
merriment, gaiety, joy, enjoyment, pleasure	merry, gay, joyful	miserable, sad, depressed, anxious
glee, delight	gleeful, delighted, joyous, overjoyed	bored, joyless, vexed
exuberance, exhilaration, elation, jubilation, euphoria	exuberant, exhilarated, elated, jubilant, euphoric	wretched, desolate, disconsolate
ecstasy, rapture, bliss	ecstatic, rapturous, blissful	desperate, doomed, accursed
	fortunate, lucky, favourable, advantageous, convenient, timely	unfortunate, unlucky, unfavourable, disadvantageous, inconvenient, untimely
	apt, felicitous, successful	unhappy, infelicitous, inept, inappropriate
	willing, ready, eager to do	unwilling, reluctant, loath to do

Table 9.1 *A possible ordering of some felicific and contra-felicific nouns and adjectives*

pleasures (Chapter 8.1), so too we must distinguish between something's being a kind of happiness (objectively) and someone's being happy (subjectively) when fortunate enough to attain it. True happiness may be the love of another, or successful and virtuous public service recognized by society, or successful engagement in a

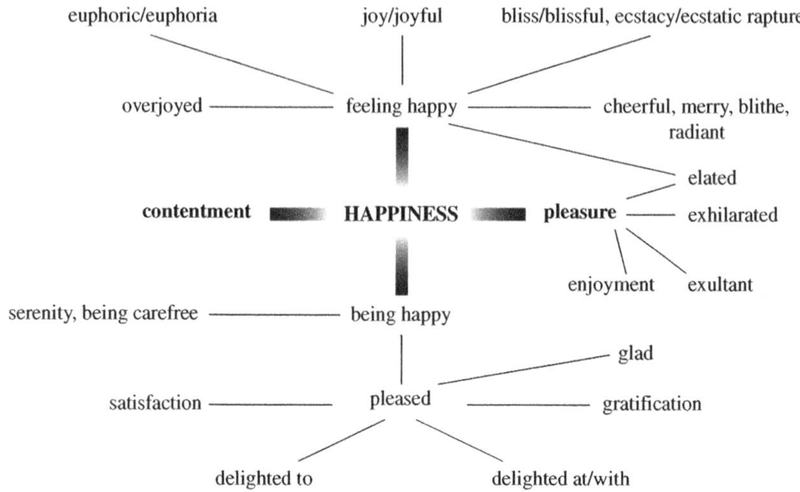

Figure 9.1   *The web of subjective happiness: a network of related attributes/
concepts*

favoured activity.[3] When such happinesses are given, then one is
happy – one is blessed with happiness.

Now we may turn to drawing further distinctions so that we can
obtain a distinct idea of happiness. Subsequently we shall endeavour
to attain as clear an idea as possible.

Pleasure and contentment   First, relying upon the conceptual geography of pleasure delineated in the previous chapter, we need to clarify the relation between pleasure and contentment. This is a preliminary to investigating the relation between happiness and contentment, and subsequently the relation between happiness and pleasure.

Pleasure may be momentary, as when one takes pleasure in a taste
or scent, but there is no such thing as being content (as opposed to
taking pleasure in something or being pleased) *for only a moment*. It
is precisely because pleasure may be momentary that taking pleasure
in something is compatible with depression or dejection, for even
someone who is sad and unhappy may enjoy the taste of something
they like or the wafting scent of some sweet-smelling blossoms.

---

[3] I once heard someone charmingly say: 'Sitting on the edge of the Grand Canyon
on a beautiful day with a good pair of binoculars and watching the the flight of the
condors – that's as good as it gets!'

Conversely, contentment is perfectly compatible with absence of current pleasure or enjoyment. Again, for similar reasons, one can feel pleasure knowing that it will shortly cease, but by and large one cannot feel contentment knowing it will shortly cease.[4] Contentment is intrinsically related to the satisfaction of needs and wants, but although one may take pleasure in satisfying needs (e.g. for food and drink) many things that are not linked to needs and wants may nevertheless give one pleasure (e.g. a walk in the park, a good book). Hence too, contentment is related to one's judgement on how things are going for one, for it implies holding that things are on balance all right. But no such judgement is involved in a brief pleasure or enjoyment.

Happiness and contentment   How then is happiness related to contentment? Long-term happiness implies the absence of abiding discontent. One cannot be leading a happy life, and yet be discontented with how things are with one, feel dissatisfied with one's lot in life, or have a feeling of haplessness. Does this form of happiness imply contentment? It seems that overall contentment with life is a necessary condition of such happiness, but not a sufficient one. One may be content with one's lot without being happy, since, although things are all right with one and could be much worse, one is nevertheless lacking something vital, such as the deep love of another, or the liberty and dignity of a free man. (A fortunate slave of a considerate and fair slave-owner may be content but not happy (e.g. Tiro and Cicero, as imagined by Robert Harris in his Cicero Trilogy), for he is not in control of his life). One may be content with how things are with one in one's current life, fully aware of how much worse they could be, although sad at the loss of one's creative powers

---

[4] An exception is to be found in cases like Hume's, who was content despite knowing that his contentment would shortly cease. He wrote in his autobiographical essay 'My Life': 'I now reckon upon a speedy dissolution. I have suffered very little pain from my disorder; and what is more strange, have, notwithstanding the great decline of my person, never suffered a moment's abatement of my spirits; inasmuch, that were I to name the period of my life which I must choose to pass over again, I might be tempted to point to this later period. I possess the same ardour as ever in study, and the same gaiety in company. I consider besides that a man of sixty five, by dying, cuts off only a few years of infirmities; and though I see many symptoms of my literary reputation's breaking out at last with additional lustre, I knew that I could have but few years to enjoy it. It is difficult to be more detached from life than I am at present' (*Essays: Moral, Political and Literary* (Oxford University Press, Oxford, 1963), 614–15).

and physical prowess. One may do various things for the happiness of another whom one loves or admires, which one would not do merely for their contentment (one can hardly imagine Sidney Carton going to the guillotine in order to make Lucie Manette contented with her lot in life in *A Tale of Two Cities*).

Happiness, pleasure, and enjoyment How is happiness related to pleasure and enjoyment? One *takes* pleasure in certain things, but one *finds* happiness. One may pursue pleasure, but one does not pursue happiness in the same way – although one may want to be happy. The forms of pleasure, for example sensual, physical, aesthetic, and intellectual pleasure, are not forms of happiness, even though they may contribute to it. One can have a momentary pleasure, but one cannot have momentary happiness. A lifetime of pleasure is the aim of the hedonist. But it is not the same as a happy life – as is evident from fact (e.g. the histories of roués such as Casanova), from art (e.g. Hogarth's *Rake's Progress*), and from the literature of libertinage (e.g. Laclos's *Les Liaisons dangereuses*, Wilde's *The Picture of Dorian Gray*). One can find enjoyment in a pleasure one has never wanted – it may be a complete surprise to find how enjoyable it is to do or undergo something novel (e.g. a sauna, a parachute jump). But happiness involves the satisfaction of at least some of one's important undertakings or goals, of which one is perfectly cognizant and which one has long wanted. To be sure, a happy life, or a happy period in one's life, does indeed imply *some* pleasures and enjoyments (see below). One may live a *fulfilled* life without much if anything by way of pleasures and enjoyments, but it would not have been a happy one. An engineer may build wonderful bridges and railways, design great ocean liners, and yet lead a miserable and lonely life. An architect may fulfil his genius in the great buildings he designs, but have an unhappy marriage and worthless children.

3. A clear idea of happiness

The previous survey has shed some light upon our concept of happiness. It has made the idea more or less distinct. Now we can proceed to analysis in order to render the idea clear.

Happiness unique to humanity (i) *What are the primary subjects of attributions of happiness?* They are, first and foremost, human beings or persons, signified by personal names, definite descriptions, or personal pronouns. It is important to note that non-human animals cannot be happy. They may have better or worse lives, and they may take pleasure in things, but lasting human happiness involves a degree of self-reflection and judgement, an evaluation of one's past and present, and hence an apprehension of time that is possible only for language users. Even the happy child can look back to the green and golden day he has spent, and look forward to a carefree, glorious morrow among the haystacks and the horses, master of the cattle and lord of the fields. No animal lacking a language can do this. However, one may say of one's pet dog that it led a happy life: it was well fed and cared for, enjoyed good health, received and reciprocated its owner's affection, and enjoyed its daily run in the meadow. Surely, this is what it is for a pet dog to lead a happy life. One may indeed say so. The price one pays for this is to transform 'happiness' into more of a family resemblance concept than it would otherwise be, and to mask crucially important distinctions between human nature and the nature of all other animals.[5]

It can also be said of social groups (family, team or crew, village, society) that they are happy, meaning that they enjoy a harmonious life, in which there are few if any animosities and no serious conflicts of any kind, and in which good social relations dominate the lives of the members of the group. In such societies mutual aid and concern are the norm.

Happiness and time   Secondly, times may be said to be happy, in one of two different senses. One may spend a happy hour looking at an admirable painting or listening to a wonderful concerto, just as one may spend a few happy hours walking on the downs with one's lover or watching condors at the Grand Canyon. One may spend a blissfully happy night with one's lover, as is celebrated by innumerable poets, for example Goethe:

---

[5] It also follows that human babies cannot be happy. But we do say of a neonate, 'He's a happy baby'. That means no more than that he chortles and smiles more often than he cries, and is responsive to the smiles of adults. One may go along with this, insisting that this is what it is for a baby to be happy. Again, the price is to make 'happiness' more of a family resemblance concept than it would otherwise be, and to obscure what is distinctive about human nature. In a study of human nature one may prefer to view this idiom as natural slippage that, for current purposes, may be disregarded. (It is comparable to speaking of a horse as *having a will of its own*, meaning that it is stubborn and refuses to do what its rider demands. But a horse no more has a will than it has an intellect.) Animals and babies can suffer, be in pain, and feel pleasure – it does not follow that they can be happy.

Ein rosenfarbes Frühlingswetter
Umgab das liebliche Gesicht,
Und Zärtlichkeit für mich – ihr
Götter!
Ich hofft es, ich verdient es nicht!
Doch ach, schon mit der
Morgensonne
Verengt der Abschied mir das Herz.
In deinen Küßen, welche Wonne,
In deinem Auge, welcher Schmerz!
Ich ging, du standst, und sahst zur
Erden
Und sahst mir dir nach mit naßem
Blick:
Und doch, welch Glück! geliebt zu
werden
Und lieben, Götter, welch ein
Glück!

   (Goethe, *Willkommen und
     Abschied*)

All of spring's freshness rich with
roses
Was gathered in your lovely face,
And tenderness, ye Gods! I hoped
for,
But never could deserve such grace
But oh, with dawn the time to part
Constricted my heart;
What bliss in your kisses
What pain in your eyes!
I went, you stood looking down
Looking after me with tearful gaze;
And yet, what happiness it is to be
loved!
And to love, ye Gods, what
happiness!

   (Goethe, *Welcome and
     Departure*)

So too one may think that the years one spent in such-and-such a country were the happiest years of one's life, for while one was there one lived a happy life. But it can also be said that such-and-such years were happy times, as Gibbon said of the Antonine age, meaning that those were times in which the fundamental preconditions for personal and social happiness were satisfied (absence of war and insecurity, an appropriate degree of relative prosperity, sound and stable government, justice, liberty, and broad social harmony).

Happiness and temperament/mood Thirdly, people are said to have happy temperaments. One who, for the most part, suits his wishes to circumstances more readily than most people do may be said to have a happy temperament. A person with a happy temperament is someone who is usually in good cheer and who does not let himself be cast down by misfortune for very long. He will be prone to see a silver lining in most clouds (like Jacobowsky in Franz Werfel's *Jacobowsky and the Colonel*). He may, for example, be debonair and cheerful.

Associated with a happy temperament, which is a proneness or disposition of character, is the notion of a happy mood. When in a happy mood, one *feels* happy. One views the world through pink spectacles, is hyper-sensitive to what is good and relatively oblivious to what is bad, perceives the good in human beings, and overlooks or is oblivious to the wickednesses of mankind. One inclines to an optimistic view of the future, and is tolerant of human foibles. One can be in a happy mood without having a happy temperament, although someone with a happy temperament is bound to be in a happy mood with some frequency. Similarly, one may be in a happy mood without leading a happy life. One may be in a happy mood without knowing the reason why, since one *need* have no reasons for one's moods. If asked 'Why are you in such a happy mood today?', one may well answer 'I have no idea'. But if one is aware of leading a happy life and is asked what makes one's life so happy, one could not answer 'I have no idea'.

So much, then, for the primary subjects of predication relevant to a philosophical analysis of happiness.

Degrees of happiness   It is noteworthy that, like pleasure and enjoyment, happiness comes in degrees. One may be happier now than one was ten years ago, and one person may be happier than another. So comparative judgements of happiness are possible. It does not follow that they are always possible. Just as one may say that Jack is taking more pleasure than Jill in dinner, when Jack tucks in with gusto and Jill just picks at her food desultorily, so too one may say that Jill has a happier temperament than Jack. But with lives, qualitative comparisons are possible only where there are many similarities in the lives and careers. However, where what makes the one happy differs fundamentally from what makes the other happy, comparison is meaningless. One cannot qualitatively compare the happiness of a great creative artist with that of a great leader of a nation (it makes little sense to aver that Titian led a happier life than Caesar) or the happiness of a flamboyant adventurer like Sir Richard Burton with that of a quiet, retiring farmer living on his land.[6] We labour under the illusion that more

---

[6] Celebrated by numerous poets on the model of Horace's second epode: 'Beatus ille qui procul negotiis.' Thus Andrew Marvel's 'The Garden' and Pope in his poem 'The Quiet Life':

> Happy the man, whose wish and care
> A few paternal acres bound
> Content to breathe his native air
> In his own ground

comparisons are possible than this. The reason is that, when we compare someone who is happy with someone else who is manifestly unhappy, we can indeed say that the first is happier than the second. But this is deceptive. It does not follow that, when two people are, in their different ways, happy, one can always intelligibly say that one is happier than the other.

(ii) *Happiness and wants*: We have already noted the link between happiness and wants, needs, and desires. This must now be examined in more detail. In one use of 'want', to *want for something* is to lack something needed either in itself or for the sake of some subjective end. One may also want something, want to do something, want to have something done to one, or want to be something or someone. The use of 'to want' has so evolved as to become among the most nebulous verbs in our language.

The Categorial Framework (pp. 134–7) distinguished different kinds of wanting. These are useful to bear in mind:

(a) wishing and hoping for something (as in 'I want the Blues to win');
(b) felt desire: appetites, cravings, obsessive preoccupations with lack of something;
(c) inclination ('I wanted to do it.' 'Why?' 'I just felt like it.');

> Whose herds with milk, whose fields with bread
> Whose flocks supply him with attire;
> Whose trees in summer yield him shade
> In winter fire.
>
> Blest, who can unconcern'dly find
> Hours, days and years, slide softly away
> In health of body, peace of mind,
> Quiet by day
>
> Sound sleep by night; study and ease
> Together mix'd; sweet recreation,
> And innocence, which most does please
> With meditation.
>
> Thus let me live, unseen, unknown;
> Thus unlamented let me die;
> Steal from the world, and not a stone
> Tell where I lie.

(d) endeavour and purpose ('He wanted to do it.' 'Why?' 'In order
     to ...');
(e) reasoned wanting ('The reason he wanted to do it was that ...')[7]

Apart from wishes and hopes, wants are essentially linked with
acting – given an opportunity, and with trying – where endeavour is
required. Similarly, they are linked with creating or trying to create
opportunities for acting in the desired way. Only creatures with the
power to V and the power to refrain from V-ing can be said to want
to V. A being can be said to V *because* it wants to, if (a) it V-s for its
own sake (i.e. *that* is what it wanted to do, either for no reason (it just
felt so inclined), or because it is, for example, pleasant or enjoyable or
right and good). It may also V (b) for the sake of some further goal or
with an ulterior motive, for example, to attain something it is aware
of as lacking, and cares about sufficiently to take action. Apart from
felt desires such as appetites and addictions, wanting is not a mental
state, event, activity, or condition of any kind.

One cannot be said to be happy in the long run or to feel happy in
the short run if one is lacking something important to one, the lack of
which disturbs and distresses one. For being distressed is incompatible with both being happy and feeling happy. However, one may
obtain things important to one without therefore being happy – they
may turn to dust in one's mouth, or may be displaced by other pressing goals (Alexander's desire for conquest was, it seems, insatiable).
The pursuit of one's most important desires may involve doing evil,
which may then prey upon one's conscience, making one fearful and
suspicious of all, like Macbeth. Nevertheless, one aspect of adult happiness is that at least some of one's most important wants are being
achieved to a satisfactory degree – that one holds what one is doing
in one part or another of one's life to be worthwhile, that one is

---

[7] The absence of so-called second-order wants is not an omission. For it may be
doubted whether this much discussed category introduced by Harry Frankfurt, is
intelligible. I may wish that I wanted to do something (e.g. to stop smoking, to go for
a run every day), but not want to want to do something, let alone want that I wanted.
If I want something that is difficult to attain, I may try to attain it. But if I could want
to want something, then I should be able to try to want. But what would count as *trying to want*? What would be the criteria for my wanting to want as opposed to wishing that I wanted? What action would show that I wanted to want? I may want to
bring it about that I want something, but that is not to want to want something.

reasonably successful in doing it, and that one finds meaning in one's life, in one's relationships, and in one's activities.

Objects of wants   In considering the relation between one's wants and happiness we must bear in mind the distinctions between wanting

(a)   to get something;
(b)   to have something;
(c)   to do something;
(d)   to be engaged in doing something;
(e)   to become something;
(f)   to be something;
(g)   another person to do one or other of the above.

In the case of wanting another to get, have, do something, and so on, one must have some authority, or persuasive power over him or her, otherwise it is merely a matter of wishing or hoping.

There is much that human beings want to have, ranging from basic needs such as food and drink, clothing and shelter (housing), to the various items considered to be necessary for a tolerable human existence relative to the society of which one is a member. Their lack is likely to be a cause of concern and anxiety, and often of discontent or misery. Beyond biological needs and social necessities lie the manifold possessions human beings covet and the numerous social positions and relationships they aspire to attain. Failure to achieve these may be a source of frustration, indignation, resentment, and a sense of failure incompatible with happiness.

Terminative and regulative wants   Among the things people aim at we must distinguish, as at a fairground, between the targets and the merry-go-rounds: between terminative and regulative goals. Many human goals take the form of a target to be hit, often with a resultant prize to be possessed. One may want an article of clothing, to eat a meal at a certain restaurant, a certain consumer durable, or a specific valuable artefact, a certain qualification such as a diploma or doctorate, or a social and political position. One wants the object *for something*, for example to wear, to eat, to use, or to display for one's own pleasure or for prestige. Other goals involve continuous activities with no immediate terminus or no terminus at all. One may want to be a good spouse, a good parent, a good farmer or beekeeper, a good doctor or author, and so on. One may also want to be a good human being and to live a happy life. These may be called 'regulative ends'.

Happiness and desire satisfaction   Although overall happiness is related to what a person wants, to his aims and purposes, goals and ends, it can surely not be said to be a consequence only of getting *all* one desires. For, quite apart from the practical improbability of this ever happening, except perhaps in the case of such deliberately curtailed wants advocated by Epicureans, far too many of one's diurnal wants are too trivial or misguided for their satisfaction to play any role in achieving happiness. One may even desperately want some goal, for example, to achieve acceptability in appropriate social circles, like Trollope's Phineas Finn who craved the brilliance and glitter of aristocratic parliamentary society, only to discover the vanity of his desire as he achieved it. Moreover, it is probable that among one's many wants are some that are logically, and others that are practically, inconsistent. So one could never obtain all that one desires. One's desire to fulfil one's destiny, or to serve one's country, may conflict with one's want for a felicitous family life with a person one loves (e.g. Nelson and Lady Hamilton), and one has to be sacrificed for the other. Furthermore, one may, like Don Juan, cease to want what one wanted as soon as one attains it. For the Don craved the love and embraces of a virtuous woman, which once achieved was discarded in favour of a further seduction.

It has been suggested that happiness is not so much a matter of getting what one wants, but rather of being satisfied with and liking what one has.[8] This is half-true. Being *dissatisfied* with one's condition in life and disliking aspects of it is indeed a source of unhappiness, for it is a cause of frustration and vexation. But merely being satisfied and liking what one has is insufficient for a person's happiness if some important goal is unfulfilled, or if one lacks what is necessary for self-respect and self-esteem and for the respect of others, or if one is lonely and unloved and craves love. The epsilons (the drawers of water and hewers of wood, so to speak) in Huxley's *Brave New World* are satisfied with their lot and like it, but they cannot be said to be happy.

(iii) *Different kinds of happiness*: We noted above the varieties of happiness with regard to duration. To be sure, there is no momentary happiness, but one may be happy for a few hours, as Natasha was at

---

[8] Jean Austin, 'Pleasure and Happiness', *Philosophy* 43 (1968), 51–62.

her first ball. Here the epithet 'bliss' finds a natural place. Such passing happiness is akin both to a mood and to an experience.

> [Count Rostov] asked his daughter whether she was enjoying herself. Natasha did not answer at once: she only looked up with a smile that said reproachfully: 'How can you ask such a question?'
>
> 'It's the loveliest time I ever had in my life!' said she, and Prince Andrei noticed how her thin arms rose quickly as though to embrace her father, and then dropped again. Natasha had never been so happy. She was at the highest pitch of bliss when one becomes completely good and kind, and cannot believe in the existence or possibility of evil, unhappiness and sorrow.[9]

It is not this kind of happiness that is held to be the *summum bonum*. It is not relatively abiding and it is excessively fragile. It is not sufficient for a happy life or phase of life, since one may experience brief times of blissful happiness without leading a happy life or phase of life. But it may well be a characteristic element in the happiness of youth.

One may be happy for a span of days, weeks, or months, as Yuri Zhivago and Lara were in Varykino, isolated for a brief time from the horrors of the Russian Revolution and civil war. Here too they enjoyed hours of blissful happiness when they did not think of the dangers surrounding them. But this likewise is not the form of happiness that may be constitutive of a happy life or of happy years, for it is happiness in a bubble, detached from the rest of society and from social life, with its commitments and ties. It involves no friendships, no filial and parental obligations, and no social responsibilities that are part of the lives of intelligent, social, language-using creatures and that are possible sources of happiness.

Happiness and phases of life One may be happy for a stage or phase of one's life. One may be fortunate to enjoy a happy childhood, wonderfully depicted by Dylan Thomas in 'Fern Hill', the two opening verses of which give the nostalgic flavour of this great poem:

> Now as I was young and easy under the apple boughs
> About the lilting house and happy as the grass was green,
>   The night above the dingle starry,

---

9 Tolstoy, *War and Peace*, II. iii. 17.

> Time let me hail and climb
> Golden in the heydays of his eyes,
> And honoured among wagons I was prince of the apple towns
> And once below a time I lordly had the trees and leaves
> Trail with daisies and barley
> Down the rivers of the windfall light.
>
> And as I was green and carefree, famous among the barns
> About the happy yard and singing as the farm was home,
> In the sun that is young once only,
> Time let me play and be
> Golden in the mercy of his means,
> And green and golden I was huntsman and herdsman, the calves
> Sang to my horn, the foxes on the hills barked clear and cold,
> And the sabbath rang slowly
> In the pebbles of the holy streams.

So too, one may be happy in one's youthful maturity, when the world is one's oyster and all may go swimmingly – as Wordsworth celebrated:

> Earth was then
> To me, what an inheritance, new-fallen,
> Seems, when the first time visited, to one
> Who thither comes to find in it his home.
> He walks about and looks upon the spot
> With cordial transport, moulds it and remoulds,
> And is half pleased with things that are amiss,
> 'Twill be such joy to see them disappear.
> (*Prelude* (1850), 11. 145–52)

Youthful happiness involves intensity of feeling, engagement with the passing moment, the discovery of first love and of sexuality (see Goethe's 'Welcome and Departure' above), and the joys of dedication to a beloved – depicted in numerous plays (such as *Romeo and Juliet*) and novels (such as *Tom Jones*). Youthful adventure, the excitement of youthful travel to see the world, as well as the thrill of strife in warrior and aristocratic society and the triumph of victory, play no small part in forms of happiness distinctive of young adulthood.

One may have a happy maturity, perhaps with a loving family and friends. One will by then have chosen determinate paths and goals in life, and may be successful and happy in their pursuit. The hot passion of youthful erotic love will have passed, but, if fortune favours one, it will be replaced by mature erotic love, which was similarly celebrated by Goethe in his *Roman Elegies* no. 19:

> Faustine gives me so much happiness: she gladly
> Shares a bed with me, and preserves mutual faithfulness!
> Coy resistance appeals to impetuous youth; I love
> To rejoice languidly, secure in a treasure.
> What bliss it is! We exchange trusting kisses,
> Breath and life we confidently suck and gush.
> Thus we enjoy the long nights, bosom pressed to bosom,
> Listening to storms and rain and downpour.
> And so dawn breaks; the hours bring
> Fresh flowers festively to adorn our day.
> Do not begrudge me, Oh Romans, this happiness, and may the god
> Grant to all the first and last of the world's good things.[10]

And one may, with Browning's 'Rabbi ben Ezra', enjoy the wisdom and detachment of old age:

> Grow old along with me!
> The best is yet to be,
> The last of life, for which the first was made:
> Our times are in His hand
> Who saith 'A whole I planned,
> Youth shows but half; trust God: see all, be not afraid!'

Now free from the arduous efforts of striving, detached from professional and most parental responsibilities, one may find happiness in the role of *pater familias* or *mater familias*, derive joy from grandchildren,

---

[10] Darum macht Faustine mein Glück; sie teilet das Lager
>> Gerne mit mir, und bewahrt Treue dem Treuen genau!
> Reizendes Hindernis will die rasche Jugend; ich liebe,
>> Mich des versicherten Guts lange bequem zu erfreuen.
> Welche Seligkeit ists! Wir wechseln sichere Küsse,
>> Atem und Leben getrost saugen und flößen wir ein.
> So erfreuen wir uns der langen Nächte, wir lauschen,
>> Busen an Busen gedrängt, Stürmen und Regen und Guß.
> Und so dämmert der Morgen heran; es bringen die Stunden
>> Neue Blumen herbei, schmücken uns festlich den Tag.
> Gönnet mir, o Quiriten! Das Glück, und jedem gewähre
>> Aller Güter der Welt erstes und letztes der Gott!
>> (Goethe, *Römischen Elegien*, 19)

and take pleasure in conversation with friends and with the young as one views the world pass by.

The different forms of human engagement involve different kinds of happiness. The happinesses of the active life in its manifold forms differ from those of the creative life, which in turn are unlike those of the scholarly or contemplative life, or of the professional or artisanal life. Different kinds of happiness are accessible at any given phase of life, at different historical times, in different societies, and in different strata of society. What makes for a happy childhood is importantly different from what makes for a happy youth and early adulthood, or what forms constitute happy maturity, let alone the happinesses of old age.

Herodotus reports Solon's remark that one may not judge a man *eudaimon* until after his death, since calamity may strike at any time. This was said in response to King Croesus of Lydia, a successful and very wealthy monarch who asked who was the happiest of men, expecting to be told that he was. Solon replied that Tollus of Athens, who had two beautiful sons, and grandsons, and had died a gallant death in battle was the happiest of men. Croesus, chagrined at not being cited himself, asked again, and Solon cited the two brothers Cleobis and Biton, who, when the family oxen had not returned home, harnessed themselves to the yoke of a cart and brought their aged mother, Cydippe, priestess of Hera, to a festival in Argos in honour of Hera. Thereupon Cydippe prayed to the goddess to grant her sons the best possible gift. The sons fell asleep in Hera's temple and never awoke. The moral is that these youths had been granted great concrete gifts – health, beauty, physical strength, filial piety – and only death could ensure that they were not deprived of these goods by fickle fortune. This is not true of our concept or conception of happiness: one can rightly say of a person that he or she is happy, has led a happy life until now, even though disaster may still ensue.

A happy life Apart from the happiness of phases or periods of someone's life, one may also speak of someone's having led a happy life as a whole (here our conception of happiness approximates *eudaimonia*). Such a life is indeed blessed – a rare good fortune and a rare achievement. It involves reciprocated loves and friendships; a preponderance of happy times; a reasonable degree of pleasure and enjoyment; the adequate achievement of one's meaningful goals and purposes in life; whole-hearted engagement in one's endeavours; and the maintenance of good relations with one's fellow human beings. It requires, among other negative exclusionary features, that one not be dogged by guilt and remorse, by hypocrisy and self-deception, or by a sense of the futility and meaninglessness of one's life or of human life in general.

The concept of happiness   It seems then that there are different kinds of happiness, most obviously those linked to each phase in human life. It is noteworthy just how different they are. The happinesses of childhood would count for little in maturity, and the joys of happy youth: going on the razzle with friends, enjoying adventure and the excitements of novelty, and the obsessive passion of first love ill befit the aged. In addition, we recognize the happiness of a life at any stage the subject has reached. For we may judge of someone who dies before reaching a grand old age that, though their life was cut short, they did (or did not) have a happy life, and we may say of ourselves at any stage in our lives that we have been fortunate and have enjoyed a happy life thus far. In addition, we have noted other happinesses, of social groups, of the passing hour or month, of temperament, and of mood. Is happiness then a family resemblance concept? It seems not, since it is characteristic of family concepts that they constitute an open class, that new members may be added to the family (as signed integers, rationals and irrationals, etc. were added to the concept of number, or as ballet and opera were added to the arts). But it does not seem that new forms of happiness can be added to a pre-established family.

Are the kinds of happiness that we have been surveying *varieties of happiness* on the model of von Wright's varieties of goodness (and perhaps of truth and existence too)? Perhaps there is an analogy between the relation between the different forms of goodness and the good of man, and the relation between the different forms of happiness and the happiness of a human life. In some uses of 'happy' it evidently functions as an attributive rather than a predicative adjective (like 'good X') – as in 'happy childhood', 'happy youth', and 'happy old age'. But in others it seems not to, for example in the case of 'being happy (eager or willing) to do something', or 'happy (fortunate) coincidence', and 'happy (apt) thought'. So too the dense intricacy of conceptual relations that von Wright disclosed in the case of goodness is not to be found in the case of happiness. So maybe one should rest satisfied with characterizing the pattern as yet another instance of a concept with multiple centres of variation (see fig. 9.2). Within this multiplicity, the notion of a happy life enjoys philosophical prominence. For it plays a significant role in philosophical anthropology and in our conception of human nature.

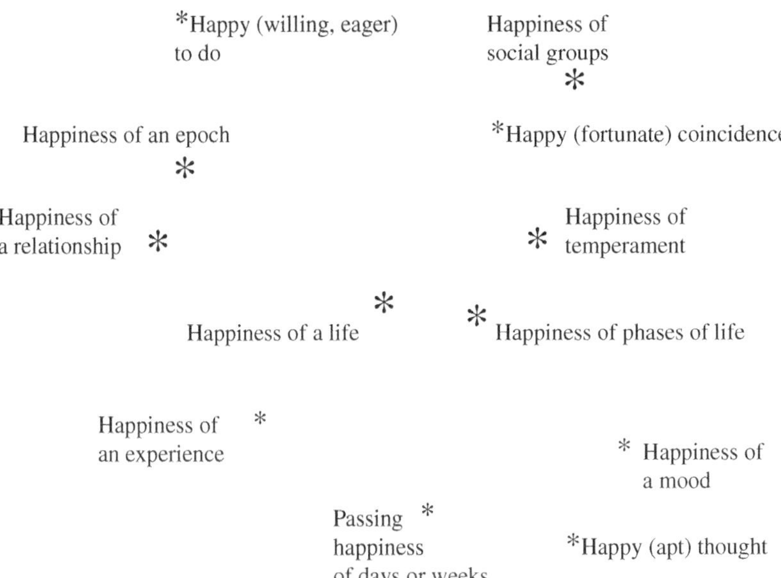

Figure 9.2 *Centres of variation of the concept of happiness*

4. Preconditions of happiness

Granted that long-term happiness is a rare blessing, that it is partly in the hands of fickle fortune, that it is both precious and destructible, one may raise the question of the preconditions of leading a happy life or a happy phase of life. By 'preconditions' is meant neither necessary nor sufficient conditions of happiness, but conditions of the possibility of happiness. They are conditions the non-satisfaction of which makes the achievement of happiness more difficult, but not impossible. They are facilitative rather than necessary conditions. Some such preconditions are personal, bodily, and psychological.[11]

Happiness and health Physical normality and good health are patently preconditions of happiness inasmuch as physical abnormalities and persistent diseases are barriers to leading a healthy and normal human life. For physical abnormalities that limit one's mobility or dexterity prevent one from engaging in the activities of healthy normal human beings, and diminish or deprive

---

[11] I am indebted in this section to the reflections of Anthony and Charles Kenny, in *Life, Liberty and the Pursuit of Utility* (Imprint Academic, Exeter, 2006).

one of the vital pleasures and enjoyments, especially those of normal childhood, youth, and early maturity. So too do deficiencies of sense faculties, such as eyesight and hearing, as well as functions such as speech. Such deficiencies *need not* exclude happiness, but they surely make its achievement much more difficult in so far as they are grounds of suffering, distress, alienation, bitterness, and resentment, that do exclude happiness. Some people may overcome them (as Phillip Carey in Maugham's *Of Human Bondage* overcame the shame of his club-foot). Others may become twisted and perverted by their physical suffering and shame (as was Uhland Herries in Walpole's *The Fortress* or King Richard III in Shakespeare's eponymous play). Physical abnormalities, especially of facial features, also lead to unavoidable (but unwarranted) shame at one's condition (see *TP*, 6.2).

Happiness and mental health   The shame and social unease, fear and timidity, that may follow in the wake of physical defects are psychological sources of unhappiness during phases of a person's life or of their life as a whole. A happy temperament and cheerful spirits are conducive to living a happy life, whereas a melancholy disposition and a pessimistic cast of mind are impediments. These characteristics are largely a matter of natural endowment, and are only partially under personal control and only partially amenable to behavioural therapy. Similarly, the absence of debilitating mental deficiencies that exclude a person from normal human activities is a precondition of leading a happy life. It is commonly said by well-meaning people that mentally defective patients in good care (those afflicted from birth, those suffering one or other of the numerous neural afflictions to which our species is subject, as well as the mental debilities of old age) are happy. This is either false or a misuse of the expression 'happy'. If it were true, then most of mankind, who do not lead happy lives, the mass of whom 'lead lives of quiet desperation' (Thoreau), would naturally wish to be mentally subnormal, would look forward to being demented, or hope to contract Alzheimer's disease. But no one does. Indeed, we all dread such afflictions. What, if anything, is meant is that such patients sometimes display no behavioural signs of misery, discontent, or unhappiness and may often be in a good mood.

We have already noted the role of temperament and character in the attainment of happiness. Those born with a happy temperament and cheerful disposition have a better chance of achieving happiness than those born under Saturn. Similarly, fortitude in the face of misfortune is likely to diminish unhappiness and the ability to look upon the bright side of things is conducive to leading a contented life.

Happiness and need satisfaction
Other preconditions of happiness turn on satisfaction of human needs. We distinguished between absolute needs, socially minimal needs, and relative needs (*HNCF*, 130–7). The satisfaction of absolute needs is necessary for the maintenance of health and fitness, and so is relative to our conception of the optimal functioning of a normal member of our species. They include food and drink, clothing and warmth, shelter and cleanliness, both personal and environmental, and the possibility of safe and peaceful sleep. Socially minimal needs are a function of the current conception of a tolerable life in society. The satisfaction of some of these is a genuine precondition of the possibility of happiness. The satisfaction of others, in a late capitalist age of consumer durables, electronic unendurables, and their acquisition, is not, although members of society are mendaciously persuaded otherwise. Relative needs are dependent on the contingent goals of an agent (what an agent needs in order to engage in some favoured activity or to attain a chosen end). Failure to satisfy the absolute needs of a person is detrimental to his welfare, and normally to his happiness. Their non-satisfaction is likely to contribute to suffering and unhappiness. The satisfaction of socially minimal needs is held to be necessary for the possibility of leading a happy life, but it may merely be believed to be so and *therefore* become so. Their non-satisfaction may be the source of unhappiness due to envy, and to constitute what the agent deems to be relative deprivation. The non-satisfaction of relative needs leads to frustration and distress (deprive a craftsman of his tools and he cannot practise his craft; deprive an artist of his implements, and he cannot fulfil himself in his creative endeavours).

Happiness and social conditions
A person's social conditions are highly relevant to the possibility of happiness. Loving and caring parents who give one a good upbringing that prepares one for life contribute to the child's good – one should choose one's parents with care! A good education enhances life's possibilities and furnishes opportunities of action unavailable to those who lack it, since opportunity is relative to ability. Knowledge endows one with abilities and increases the slender possibility of one's understanding of oneself and others, of one's society and environment. Since discrimination (racial, class, religious, sexual) is a source of fear, hatred, resentment and anxiety, injustice and unfairness, its absence should be included among preconditions of happiness.

Socio-political conditions follow naturally on. It is difficult to lead a tolerable, let alone a contented or happy life, when the basic need for safety and security of person and possessions is not met. The absence of excessive criminality within society, and of warfare between one's society and another, as well as the absence of civil strife, are obvious preconditions of happiness. It is not impossible for anyone to be happy in conditions of extreme danger of warfare (celebrated by Ernst Jünger in his *Storm of Steel* (1920)), but it is unusual and less likely than in peacetime. Next only to security is the need for liberty (hence absence of slavery or serfdom). Servitude diminishes a person's self-respect, and self-respect is a precondition of a happy life or period in one's life. Similarly, as has been argued in Chapters 5 and 6, the absence of humiliation, the recognition of human dignity, and the respect of one's peers are preconditions of happiness.

Happiness can be marred or shattered by a multitude of accidents of fate and by human malice, political upheavals, and economic disasters. We live on the wheel of fortune, and good hap is not in our hands.

5. The epistemology of happiness

Utterances and criteria of happiness   Some first-person utterances of happiness, such as Natasha's at the Rostov ball, are clearly exclamations and so expressions of joy and happiness of the passing hour. They are not evidence-grounded descriptions of one's current state of mind, nor are they immediate descriptions resting on introspective 'observation' of one's 'inner world'. Here there is no space for mistake or even for self-deception (although insincerity and deception are possible). Consequently such exclamations are not truth evaluable after the manner of a third-person description. Rather, in such cases (as with pleasure and enjoyment) truthfulness guarantees truth. The sincere utterance 'I'm so happy', in appropriate circumstances, is a criterion for others to describe the person as being happy or even blissfully happy. But, of course, such an utterance is not the sole criterion for judging another to be happy on a passing occasion. The look on the faces of young lovers, the joyful concentration of a master craftsman on a skilful activity that is evident in his behaviour, or carefree delight in the festive company of friends similarly constitute criteria for judgements of passing happiness. So third-person judgements are true or false, mistake is possible if deception is afoot, and they have defeasible grounds.

Utterances and criteria of unhappiness

Should the same be said of utterances of unhappiness? Before Prince Andrei asked her to dance with him (at the request of Pierre), Natasha was deeply distressed at being neglected, beset with imaginary fears about her inferiority, and very unhappy:

> Natasha felt that she would be left with her mother and Sonya among the minority who lined the walls, not having been invited to dance. She stood with her slender arms hanging by her sides, her scarcely defined bosom rising and falling regularly, and with bated breath and glittering, frightened eyes, gazed straight before her, evidently equally prepared for the height of joy and the depths of misery ... She had but one thought: 'Can it be that no one will come up to me, that I shall not be among the first to dance? Is it possible that not one of all these men will notice me? They don't even seem to see me, or if they do they look as if they are saying, "No, she's not the one I'm after, it's no use looking at her!" No, it can't be', she thought ...
>
> The strains of the polonaise, which had now lasted some little time, began to have a melancholy cadence in Natasha's ears, like some sad reminisce. She wanted to cry.[12]

Such thoughts of unhappiness may be exaggerated, but cannot be mistaken. Their utterance is a criterion of the speaker's unhappiness inasmuch as they are *expressions* of unhappiness. There are, to be sure, other criteria too – facial expression of misery and dejected demeanour. Third-person judgements of someone's unhappiness commonly rest on such criteria, and are defeasible.

Summative utterances

In the case of a relatively prolonged period of time – the joy of a happy honeymoon, or of being isolated with someone loved in the midst of the horrors of war, like Yuri and Lara – the first-person utterance 'We have been so happy here' or 'I have never been so happy as with you these last weeks' is likewise an expression of happiness, although in certain circumstances it may converge on the description 'I was so happy there' or 'Those were the happiest days of my life'. Even such a description does not rest on grounds or evidence, but is a memory utterance. It is susceptible to a certain kind of error, as one's memory may paint the past in

---

[12] Tolstoy, *War and Peace*, II. iii. 16.

rosier colours than it had at the time, as appears to be the case with Karen Blixen's autobiographical memoir *Out of Africa*. However, it is noteworthy that the operators 'It seems' and 'It seems to me' have little if any role here. No one would say 'It seems to me that we have been so happy here' or 'It seems to me that I have never been so happy as with you these last weeks'. By contrast, the comparative mnemonic assertion does allow of such modification, as in 'It seems to me that those were the happiest days of my life'. Similar considerations apply to judgements of unhappiness. The present tense or present perfect tense utterance is an expression of unhappiness. But one's memory may paint one's past blacker than it was. Short periods of grief or misery may have left a deeper imprint on one than the joyful and happy times. So a misjudgement is possible, as is self-deception in cases of motivated error.

When it comes to one's life as a whole, or a substantial period of one's life, a subjective verdict of happiness essentially involves an evaluation and an attitudinal judgement. One must have an overview of one's life and the salient events and experiences in it, of one's activities and commitments, as well as the primary goals one pursued and the extent to which one achieved them. Those goals and engagements must now strike one as worthy and the extent to which one achieved them or was involved in them must be judged to be adequate or satisfactory, even if it falls short of what one may have hoped. Here one must be able to answer the question 'Why do you take yourself to have had a happy life?', as well as 'Why don't you take yourself to have had a happy life?' As we noted, one could not hold that one has had a happy life but have no idea what made it happy or why one takes oneself to have been happy. Here self-deception is quite common. One may insist that one has had a happy life as a defence mechanism, to ward off the pain of recognizing that one has in fact failed in one's life enterprises, or of confronting the sorry fact that one's enterprises were futile and unworthy of the effort one dedicated to them, or were plain misguided (something that the Reverend Edward Casaubon, in *Middlemarch*, cannot face). A happy life, of course, does not mean a life untouched by grief or sorrow – for then it would not be a human life, for it is unavoidable that we lose our parents, and likely that we lose loved relatives and friends. It does not mean immunity to disease or the debilities of old age (should one live that long). The self-regarding virtues of fortitude and self-control aid one through such periods. Such first-person judgements of happiness of life are judgements that on balance things went well, for example, that all things considered one's goals and undertakings were worthwhile and achieved to a satisfactory degree, that one loved and was loved, that one enjoyed genuine friendships, that in general one's relationships with others were good, that one had a reasonable degree of pleasure in one's activities and social relationships, or that one had on the whole fulfilled one's obligations and done some good. Just because self-deception is possible if not common, one's own judgement may not be the final word on the matter and may be overridden by the judgements of others who know one well.[13]

6. Two philosophical traditions

Human beings have hankered after felicity since time immemorial. They have reflected upon its nature long before the dawn of philosophy. From the point of view of a naturalist philosophical anthropology, there are two primary philosophical traditions worth noting. First, the hedonic tradition according to which happiness is to be gained through the pursuit of pleasure (or is identical with pleasure). Second, the Greek eudaimonic tradition according to which living *a worthy and worthwhile* life, doing good and doing well, are the *summum bonum*. In the ancient world, the fount of hedonism was Epicurus and his followers. In the modern world its primary advocates were the utilitarians, especially Jeremy Bentham and John Stuart Mill. Ancient eudaimonism in philosophy stems from Socrates and Plato, and was brilliantly developed into an elaborate doctrine by Aristotle. Stoicism is one branch that ancient eudaimonism subsequently took, a branch that minimized the role of pleasure and welfare in eudaimonia and maximized that of virtue and immunity to misfortune. According to Cicero, the *eudaimon* would retain his equanimity or *ataraxia* even on the rack – totally impervious to his suffering. There is also a powerful religious tradition in Western Christian culture, according to which true happiness is to be achieved only in the afterlife, through the grace of God, and consists in the beatific vision of divine glory. This we shall not discuss.

---

[13] It is also possible for someone to spend a lifetime pursuing a futile goal, or futilely pursuing a goal, without realizing it (e.g. an alchemist). He may be subjectively happy, but his happiness rests on illusion. It is moot whether it should be considered an illusory happiness.

Hedonic tradition   Epicurean hedonism was sketched in the previous chapter. As noted, it was more stoical than sybaritic and conceived of pleasure privatively. Morality was held to be a means to a tranquil life and the pleasures of friendly discourse were highly rated. Modern philosophical hedonism is utilitarian. As we noted, Bentham proclaimed that 'Nature has placed mankind under the governance of two sovereign masters, *pain* and *pleasure*. It is for them alone to point out what we ought to do, as well as determine what we shall do.'[14] With remarkable insouciance he declared that benefit, advantage, pleasure, good, and happiness, for his purposes 'come to the same thing'.[15] The utility of an action, he proclaimed, consists in its contribution to the greatest happiness of the greatest number. John Stuart Mill later wrote:

> The creed which accepts as the foundation of morals *utility*, or *the greatest happiness principle*, holds that actions are right in proportion as they tend to promote happiness, wrong as they tend to produce the reverse of happiness. By 'happiness' is intended pleasure, and the absence of pain; by 'unhappiness', pain, and the privation of pleasure … pleasure and freedom from pain are the only things desirable as ends; and that all desirable things (which are as numerous in the utilitarian as in any other scheme) are desirable either for the pleasure inherent in themselves, or as a means to the promotion of pleasure and the prevention of pain.[16]

Unlike Epicureans, utilitarians hold that it is not pleasure and a life of pleasure that is the *summum bonum* or ultimate good, but rather acting for the sake of the greatest pleasure (happiness) of the greatest number. We dwelt on the flaws of classical utilitarianism in Chapter 8 and will merely note them here. It is mistaken to identify pleasure and happiness, although it is true that a happy life or period of life must involve some pleasure and enjoyment. The philosophical psychology of the utilitarians is woefully inadequate: pleasure is not a sensation. We do not act solely for our own pleasures nor are we egoists of necessity. We commonly act for the sake of interests other than our own, for example of the state, the church, the party, not to mention our family and friends. It is a grave perversion of our concept of

---

[14] Bentham, *An Introduction to the Principle of Morals and Legislation*, ch. 1.2.

[15] Bentham, *An Introduction to the Principle of Morals and Legislation*, ch. 1.2.

[16] John Stuart Mill, *Utilitarianism*, ch. 2.

rationality to suppose that present and future pleasures are the only good reasons for action. A corollary of that confusion is that the importance of backward-looking reasons in our conception of rationality is obliterated by utilitarians. This is an egregious fault: the reason we should keep our word is not because it pays in the currency of future pleasures, and principles of justice and fairness are not valued for hedonic reasons. Utilitarianism is a monistic axiological doctrine, whereas rational social beings, endowed with the powers of sympathy and empathy, cannot but be axiological pluralists.

Eudaimonic tradition The eudaimonic tradition has been less influential than the hedonic one in the early modern and modern eras. It is firmly rooted in the ancient Greek preoccupation with the precariousness of human welfare and prosperity. Solon's views on the matter were mentioned above. This conception was widespread: Pindar (517–438 BC) wrote: 'Briefly the happiness of mortal men waxes, and even then, it collapses when struck down by adversity' (*Pythian*, 8. 92–4). The theme recurs in Sophocles (497–406 BC):

> Sons and daughters of Thebes, behold: this was Oedipus,
> Greatest of men; he held the key to that celebrated riddle;
> Was envied by all his fellow-men for his great prosperity;
> Behold, what a surging tide of ill-fortune swept o'er him.
> Then learn that mortal man must always look to his ending,
> And none can be called happy until the day he passes beyond life,
> free from pain.
>
> (*Oedipus the King*)

It is repeated by Euripides (484–406 BC): 'No mortal is prosperous or happy to the last, for none was ever born to a life free of misery' (*Iphigenia in Aulis*), a motif repeated in *Medea*:

> All things accomplished by mortals are mere shadows.
> Nor am I afraid to say that those who consider themselves philosophers
> And know the art of spinning words, and say that man can live in happiness,
> Well, those men are stupid!
> Wealth may make your life a little better than that of your neighbour
> But no man is ever happy, no one!

If this is indeed the background to the Greek philosophical engagement with the nature of happiness and of a happy life, it is utterly different from the hinterland of the Jewish discussion of human happiness in the Old Testament, and equally different from the much later Christian one. It is also importantly different from Stoicism, which is characteristic of ethical reflection after the collapse of the self-governing Greek city-state (the polis) and coincides with the rise of empires. It determines a quite different approach to ethics.[17]

The primary task of ethics in this context is to teach one how to withstand 'the slings and arrows of outrageous fortune'. The fundamental question underlying the philosophical debates concerning *eudaimonia* is not primarily whether human welfare consists of more or less material or more or less spiritual benefits – although this too is pertinent. Nor is it primarily concerned with the clash between egoism and altruism, which is the obsessive concern of early modern and modern moral philosophy (ever since Hobbes and de Mandeville) – even though this is a concern. Rather it is, above all, how to protect the good of man from the vicissitudes of fortune. We are subject to the whims of the gods and the indifference of fate. The ideal life is a happy one, in which the good of man is achieved and retained until death. But this is highly improbable. We are exposed to the transitoriness of the good, and we are not in control of our fate. So ethics must arm us against reversals of fortune, and teach us to face the inexorable wheel of life. That is achieved by *eudaimonia*, which combines the worthwhile with the worthy, and enables us to retain our worthiness even if we are struck by ill luck.

Vulnerability of the good   We are subject to the whims of the gods. We cannot be or become *eudaimon* at will, since misfortune may deprive us of the good things in life at any time. When fortune smiles upon us and we prosper, we must be content and not crave more. We are prone to hubris, and hubris leads to excess (*koros*). It is manifest in arrogance and avarice, in coveting what is not ours, in showing off our good fortune, power, or wealth. In so doing we go beyond what is permitted to humankind. Nemesis follows swiftly – as the tales of Ixion and of Tantalos teach us.[18] The Greeks had a horror of excess – of flaunting any aspect of the good fortune that the gods

---

[17] See Lauriola, 'From *Eudaimonia* to Happiness'. I am indebted to her insights.

[18] Tantalos was a demigod, favoured by the Olympians to dine with them. However, he shared some ambrosia with mortal men and was condemned to Tartarus, where he stood in a pool of water, with heavily laden fruit trees above his head. When he reached for the fruit to assuage his hunger, the branches rose, and when he tried to drink to quench his thirst, the waters of the pool receded. Ixion was forgiven his terrible sin of

had granted one.[19] That would be unseemly and indecorous, and is wholly beneath the dignity of a virtuous man. Moderation or temperance (*sophrosyne*) was held to be a master virtue that manifests reverence to the gods: hence Pindar: 'It is ever right to mark the measure of all things by one's station' (*Pythian*, 2. 33–4), and 'seek not the life of the immortals, but enjoy to the full the resources within your reach' (*Pythian*, 3. 59–60). Self-restraint and self-control stop us from putting our welfare in jeopardy. It is of great significance that 'Nothing in excess' was one of the two inscriptions inscribed on the entrance to the temple of Apollo at Delphi, the other being 'Know thyself!'

What then were the greatest goods possession of which is constitutive of a good life? Above all was health:

> Health, best of the Blessed Ones to men
> May I dwell with you for the rest of my days
> And may you be kind and stay with me.
> …
> With you blessed Health,
> All things are strong and shine with the converse of the Graces,
> And without you no man is happy.
> (Ariphron of Sikyon, 'Ode to Health')

In an age without medicine or sanitation, health was indeed in the hands of fate. It was not only a precondition of felicity; it was inseparable from the beauty of the youthful human body, which was the object of fervent admiration, celebrated in festive, quasi-religious athletic games and commemorated in monuments and statues. It is in physique and athletic prowess that humanity most closely approximates the gods. But the beauty of youth is ephemeral and inexorably fades with time. Wealth was a good, not for its own sake but for the sake of the things it made possible, including liberality, the graces of social life, and the purchase of beautiful things for the adornment of one's home. But it too is precarious, subject to the vicissitudes of fortune. Conviviality, sex, and drinking parties were among the delights of youth, but they too had to give way to the more sober consolations of age – greater wisdom, more self-control, dignity, and lasting friendships.

---

kin-slaying by Zeus and invited to Olympus, where he lusted after Hera. He was punished by being bound to a wheel of fire, spinning through the firmament.

[19] Niobe boasted of her fourteen children before Letos, mother of Apollo and Artemis, whereupon the two gods slew them all.

It was taken for granted that the esteem of one's fellow citizens in the polis was valuable and one of the benefits of a worthy and happy life.

Virtues as protective of eudaimonia

The virtues were conceived primarily as protective. Courage was a predominant one, not primarily for selfish, self-regarding reasons, but because it was necessary for holding one's place in the shield wall of the hoplites in battle, which was required of every citizen. Of course, courage also strengthens one in facing the hardships of life. A further crucial virtue was *dikaiosyne* – justice, or perhaps more accurately rightness, without which one would be constantly transgressing the laws of the polis and offending one's fellow citizens. But living in harmony with one's fellow citizens is a feature of a worthy and worthwhile life. Moderation or temperance (*sophrosyne*) restrains one from excess in one's treatment of others, granting to each what is his due. *Phronesis* or practical wisdom enables one to judge well what to do in the endlessly changing circumstances of life.

Some such conception of a worthy and worthwhile life was common between the sixth and fourth centuries, and is evident in the writings of poets and playwrights. It is this that provided the material for philosophical reflection, above all by Socrates, Plato, Aristotle, and Zeno of Citium (founder of Stoicism) and their successors. I shall discuss only Aristotle. His account is most comprehensive and is worked out with great subtlety and sophistication. It is a prototype of a naturalist account of a good human life.

Aristotle on eudaimonia

Although *eudaimonia* is commonly translated as 'happiness' and *eudaimon* as 'a happy man', that is, as we have seen, profoundly misleading. It suggests that Aristotle is investigating the same subject as we are – and that is wrong. *Eudaimonia* has no straightforward English translation. It approximates a 'worthwhile and worthy life' or a 'successful and noble life'. So the notion involves a blend of having a good life and leading a good life, of prosperity and virtue. Aristotle did not think that the only ultimate end of any chain of practical reasoning is *eudaimonia*, that it alone justifies all that we intentionally do. On the contrary, he recognized the obvious: we pursue many ultimate ends – for example honour, pleasure, virtue, reason, knowledge, not to mention wealth, fame, and power. But what is distinctive about *eudaimonia* is that it is never pursued for the sake of some other end,[20] whereas

---

[20] To be sure, one may pursue the happiness *of others* for the sake of some other end – happy workers are more productive!

honour, while it may be pursued for its own sake, may also be pursued for the sake of *eudaimonia*. Similarly, pleasure is often pursued for its own sake, as when one goes out on the razzle, but one may also pursue it for the sake of its contribution to *eudaimonia*, for a life without pleasure would be a sad affair. Many of the goals men pursue, such as wealth and power, are not worthy of rational beings. The lives they live are not worthy, and they are not *eudaimon*. Unlike other ultimate ends, *eudaimonia* in itself makes life worth living. The *eudaimon* lacks nothing and in that sense is self-sufficient. Power, pleasure, and acclaim, by contrast, are not self-sufficient and make life neither worth living nor worthy.

It is a tautology that *eudaimonia* makes life worth living. What then is it? Aristotle surveys what is said upon the matter, and then tries to systematize those opinions that pass muster. He resolves the question of the roots and nature of *eudaimonia* by examining the *ergon*, or characteristic activity of man. What is characteristic of man, and of no other creature, is possession of a rational *psuchē* – the power to reason. The characteristic *activity* of man is a life of reason concerned with action. For a power is to be seen in its exercise. Action, guided by reason and based on good desire, is virtue. One possesses virtue if one exercises practical wisdom in one's choices of purpose and of the means to achieve good purposes. In so doing, one must seek and find the mean between extremes. To do so requires possession of virtues, some intellectual and others moral. Aristotle held that, with some exceptions (most prominently, justice), a virtue is the mean between a vice of deficiency and a vice of excess. So, for example, liberality is a mean between meanness and profligacy; courage is a mean between cowardice and rashness, temperance is a mean between indulgence and asceticism, modesty is a mean between shamelessness and bashfulness, and so on. In general, the exercise of any of the virtues is person relative and occasion relative, and is in accordance with practical wisdom (*phronesis*). As von Wright sapiently noted, the path of virtue, unlike the path of duty, is not laid out in advance but has to be decided for each occasion by each moral agent. It is therefore no coincidence that the virtuous deed (the individual act) performed by the virtuous person on a particular occasion is named after the virtue it exemplifies (an act *of courage*, an act *of generosity*, a *just* act), not after the category of acts it instantiates, which may be anything whatever – wiggling a little finger may display extreme courage in appropriate circumstances (the innkeeper's daughter in Alfred Noyes's *The Highwayman*). Aristotle emphasized that activity in accordance with virtue is a source of pleasure: the virtuous person does not act reluctantly, but willingly, and is pleased if his action achieves its purpose.

Aristotle holds that we aim at *eudaimonia* as an ultimate goal. What ends we pursue show how we conceive of *eudaimonia*. Those who lack wisdom will pursue goals that do not result in felicity and which are not worthy, such as power, wealth, or popularity. Those who are wise will lead a life of contemplation in pursuit of knowledge (philosophy), or a combination of a life of knowledge, a life of political service to the community (the polis), and a life of judicious pleasures and enjoyments of the cultivated gentleman.

It might appear that Aristotle's notion of an ideal life, in which a person pursues his happiness in the appropriate way and does so with success, is a remarkably selfish and egocentric ideal. But this is mistaken, especially if one concentrates on his mature work, the *Eudemian Ethics*. The mistake comes from translating *eudaimonia* as 'happiness'. To be sure, a life dedicated to the pursuit of one's own happiness alone would be a selfish one. But the *eudaimon* pursues a *worthy and worthwhile* life. A worthy life requires generosity, kindness, considerateness, compassion, justice, and courage in facing danger in defence of others and in standing up for what is right and condemning what is wrong in the face of the multitude. These virtues are not exercised in selfish behaviour but in altruistic conduct. Their possession is constitutive of the character of the *eudaimon*, and their wise exercise is constitutive of a virtuous life. To be sure, such a person may be struck by misfortunes of indefinitely many kinds that undermine or destroy his welfare. In that case, although he is virtuous, he is not *eudaimon*. For fortune has not granted him an enviable and worthwhile life. Nevertheless, he may well live an admirable and worthy one, and his virtues enabling him to withstand the afflictions and misfortunes of life.

It is noteworthy that the Stoics, Zeno, and Chrysippus were more concerned with inuring oneself against misfortune or ill hap than with preserving and protecting one's welfare and whatever good fortune may have been granted one by the gods. The true *eudaimon*, they held, should be wholly imperturbable to misfortune. This involved a profound transformation in the concept of *eudaimonia*.

7. Happiness and morality

A happy life		We are now in a position to draw our reflections to a conclusion. Someone who is leading or has led a happy life is content with his life and circumstances. He wants for nothing important the absence of which would make him unhappy, that is bored, frustrated, distressed, depressed, miserable, or wretched. He must find fulfilment in his life, in his activities, in his relationships, and in the successful exercise of his talents. He must evaluate his activities and voluntary commitments as being worthwhile and he must be achieving his goals and purposes to a degree that pleases him. With some qualifications in the case of 'lone wolves' and 'rogue elephants', his primary activities must enjoy the respect and appreciation of his peers. He must find a reasonable amount of pleasure and enjoyment in his life – in particular in his close personal relationships and friendships, and in favoured activities. A happy life or period in one's life must be sufficiently robust to withstand the ordinary slings and arrows of fortune – the occasional frustrations and disappointments, periodic failures, and the unavoidable sorrows of life (parental deaths, loss of beloved friends or members of family).

A happy life does not require a sequence of happy phases of life. One may have had an unhappy childhood or youth, yet nevertheless find one's happiness in one's mature years and old age. A happy phase in life need not consist in the predominance of happy episodes. One may be happy during a phase of life as a consequence of one's successful exercise of one's talents and the respect and admiration of others, given that one takes one's primary activities to be worthwhile. The life of a hedonist need not be a happy one. For it is likely that such a person will find much wanting in such a life, such as the responsibilities of love and the selflessness of dedication to another or to a worthy cause, as well as the respect and admiration of one's peers. The perils and evils of the determined pursuit of a life of pleasure are patent in the libertine tradition of John Wilmot, earl of Rochester, in Laclos's *Les Liaisons dangereuses*, in Wilde's *The Picture of Dorian Gray*, in J.-K. Huysmans' decadent *Against Nature*, and in Hogarth's *The Rake's Progress*.

Happiness not the summmum bonum

It is evident that happiness is not the only ultimate goal in life. What is true is that, to the extent that one wants to be happy, one's happiness is always valued for its own sake and not for the sake of some further end. (In this respect happiness resembles *eudaimonia*.) Other ends, such as the happiness of others or the alleviation of the suffering of others, truth and knowledge, service to one's country or to the party and its ideology or to the church and its doctrines, may be pursued for their own sake and as constituent elements in one's happiness. But they may be pursued for the sake of other ends.

These may include personal advancement, honour, power, and fame. There is no master value or *summum bonum* that trumps all. It is not uncommon for people to put the pursuit of their own happiness aside in favour of their duty, as did marriageable girls of royalty and aristocracy who were but pawns in national or international policy and whose duty was to produce offspring for the perpetuation of the dynasty or house. Similarly children may forgo whatever hopes they have of happiness in order to care for an invalided parent, as parents may sacrifice all for the sake of their child or children. Others may abandon any hopes of happiness for the sake of service to their country, or for the sake of justice, or for the sake of dedication to their art or to science. Happiness may explicitly be rejected for the sake of revenge (e.g. Hannibal Barca swore to his father Hamilcar to dedicate his life to the fight against Rome), for the sake of an overwhelming infatuation (e.g. Sanin in Turgenev's *The Torrents of Spring*, who throws away his future happiness to pursue his seductress, Maria Nikolaevna), or for the sake of one's country and one's duty to defend it in war (e.g. Hans Castorp in *The Magic Mountain*).

Happiness, virtue, and morality   What is the relationship between happiness and virtue, or between happiness and morality? The Aristotelian eudaimonic tradition held virtue to be an essential constituent in *eudaimonia*, for *eudaimonia* is activity in accordance with virtue. It is doing well what is worth doing and what we are good at doing, which pleases us and which we enjoy. It is the exercise of good judgement in accordance with reason, which enables us to choose the virtuous mean between the extremes with which we are confronted in our major choices of action and activity. The *eudaimon* must be virtuous, indeed, he must possess all the virtues, since they form a unity and their exercise is determined by practical wisdom. So he must lead a worthy life. And he must have the good fortune to lead a successful and worthwhile life. Evidently, few people are likely to achieve *eudaimonia* – indeed only a select few among the educated upper classes.

Happiness, as we conceive it, comes with no such rare qualifying conditions. It is not at all obvious that *l'homme moyen sensuel* cannot be happy, even if he is not outstandingly virtuous and is certainly not without fault. It is relatively rare for people to live happy lives, partly because the preconditions of happiness are so numerous, and mainly because the sorrows and disappointments of ordinary human life are so many. But being a peasant or a worker, an artisan, tradesman, or banker does not as such exclude the possibility of leading a happy life. One may lead a happy life despite not being particularly courageous or industrious. Certainly wisdom (by contrast with good judgement) or magnificence (by contrast with satisfaction of basic needs) are not required for a man to be happy and to be leading a happy life. What our conception of happiness requires by way of worthiness is primarily *absence of serious vices*, including folly.

Happiness and the wicked A liar or a cheat, a thief or a confidence trickster, a rapist or a murderer, a malicious or malignant person, an envious and jealous one, or someone who is mean and unforgiving is precluded from happiness. As we have seen in Chapters 4 and 5, evil and wickedness breed guilt and remorse, perpetual suspicion and fear, and constant distrust of one's fellow human beings. They normally require concealment of one's past and deception of one's acquaintances and children. These are all inimical to happiness. They are also incompatible with a life appropriate for a human being. There are people who appear to be impervious to their own wickednesses, who carry no inner stigma of evil. They may be said to have no soul, but that is not something they are able to recognize, and the total absence of moral sensibility does not disturb them. They may fervently believe in their ideology and may be triumphant at the success of their murderous projects (e.g. in South West Africa, Anatolia, the USSR, and throughout the Third Reich at the peak of its power). So they may feel fulfilled as far as their evil missions are concerned. But that does not amount to happiness. An evil character, as we have seen, is possessed not by one vice but by many. For the vices too form a hideous complementary unity (Goya's Black Paintings are fiendish counterparts of his youthful rococo tapestry designs of the delights of company in the countryside, and can be conceived as depicting the hidden faces of complementary vices). The vices of the wicked and evil are incompatible with the virtues necessary for love of a spouse, for loving and responsible parenthood, or for genuine friendships.

To be sure, *ex post actu*, evil people may escape justice – as did so many *génocidaires*. However, it is doubtful whether even then a happy life is possible for people who drag long trails of evil behind them, who have manifested the capacity for evil again and again, who can be totally indifferent to the suffering they inflict and were wholly lacking in sympathy for other human beings whom they tortured and tormented. Whether they are capable of the love of another is equally doubtful, although they may well crave love and even demand it.

If they have children, can they reveal the truth of their lives to them? If they have friends, can they divulge their evil deeds? Doubtless they suppress their past and protect themselves by a wall of silence and deception. Their character is rotten and their soul is dead, but they may well be respectable members of their community and may prosper accordingly.

10

The Science of Happiness

1. From eighteenth-century crudity and back again

Flaws in classical utilitarianism Modern utilitarianism has its roots in the eighteenth century, its philosophical blossom in the works of Bentham and the Mills, and its practical fruit in the works of nineteenth-century radical legal and political utilitarian reformers. By the twenty-first century its conceptual flaws had become evident, the theoretical inadequacies of its heir – consequentialism – had become clearer, and the practical applications of monistic axiology in economics had become ever more disastrous. As argued in previous chapters, utilitarianism was informed by a superficial and deeply mistaken philosophy of psychology. It treated being pleased, pleasure, enjoyment, satisfaction, fulfilment, contentment, joy, and happiness as equivalent. But, as we have seen, they are not. In particular, one may be pleased by something or enjoy something without being happy, and one may be happy without concurrently enjoying oneself. Utilitarians held that pleasure, and hence too happiness, are sensations. But that too is mistaken, since, although there are sensations that give one pleasure, neither pleasure nor happiness is a sensation. One can ask someone where he has a pain, but not where he has a pleasure. And to ask someone where he is or feels happy is to ask for a geographical, not a somatic, location. Not only are pleasure and happiness held to be sensations, but they are held to be quantifiable. Bentham, we saw, enumerated seven dimensions of felicity.

The Moral Powers: A Study of Human Nature, First Edition. P. M. S. Hacker.
© 2021 John Wiley & Sons Ltd. Published 2021 by John Wiley & Sons Ltd.

Intensity, duration, and presumably purity can allegedly be summed to give the quantity of pleasure; fecundity, propinquity, certainty, and extent provide further dimensions relevant to the rationality of its pursuit in any particular case. But, as we have argued, it makes no sense to quantify the huge diversity of kinds of pleasure, and comparability of pleasures is very limited. The utilitarians saw clearly that a prospective pleasure is a reason for action, but were quite mistaken to suppose that it is the sole reason a rational or reasonable person can ever have for acting. The account they gave of human motivation was lamentably defective, for they held that we are of necessity egotistic hedonists. But this too is wrong, since we can act for the sake of the pleasure or happiness of others no less than for our own, and we can act for a multitude of other ends, such as truth, justice, the good of our country, or the triumph of the proletariat or of the *Volk*. Any thought to the contrary rested on a superficial conceptual confusion (conflating the logical satisfaction of desire with the object of desire). The conception the utilitarians and their consequentialist successors had of the human person was shallow. Human beings, they held, are in effect mere pleasure or happiness receptacles or desire-satisfying mechanisms. Their lives are better according to whether the receptacle is being filled or their desires satisfied. But there is much more to human felicity than pleasure: there are also autonomy, dignity, and moral worth. There is much more to happiness than satisfying desires: there are also striving after non-hedonic values, fulfilling ideals of life, and integrity. Moreover, desires, apart from natural appetites, are not given but created or induced. They are not value neutral, since some desires are intrinsically wicked or evil, and others are intrinsically good. Consequently, the question of whether a given desire is worthy of being satisfied can always arise. Human beings are not deterministic mechanisms but free agents with the liberty to act or to refrain according to their choice and inclination. Finally, utilitarians and consequentialists were axiological monists, holding all values to be reducible to utility (or desire satisfaction, or fulfilment of preferences), but human beings are axiological pluralists – and that, as we have seen, is no coincidence.[1]

Marginalist economics grew out of utilitarianism

These criticisms, and others too, took a long time to emerge and take reasonably precise form. While the mills of philosophy were

---

[1] These baldly stated criticisms have been supported by arguments elsewhere in this volume and in its predecessors. They are brought upon the carpet again to provide the framework for examining the putative science of happiness.

slowly grinding, the young science of economics was given inspiration by utilitarianism. For quantitative marginalist economics was one of the offshoots of utilitarianism. Utilitarianism inspired micro-economics, the theory of the firm, as well as the theory of international trade. Quantifiable absolute utility, however, was rapidly displaced by ordinal utility. The notion of utility as such, no matter whether absolute or ordinal, proved unmanageable and was replaced by desire satisfaction and subsequently, with the rise of behaviourism, by revealed preferences. Preferences patent in behaviour seemed to possess an objectivity absent in the notion of desire satisfaction. So it must be preferences that are maximized in consumer behaviour.

With the demise of behaviourism and the perceived flaws in economics detached from political economy, a new trend emerged, supported by psychology and neuroscience. Suspicions about the ineliminable subjectivity of individual pleasure, desire satisfaction, and assertions of personal well-being were swept aside in light of the view that neural events and processes could provide the objective scientific backing for a science of felicity.

Science of happiness as a reaction to economic science The idea of a science of happiness appealed to some economists and social theorists who rightly felt increasingly ill at ease about measuring national welfare and progress in terms of growth in gross national product. There was, as we have seen, an insurmountable conflict in utilitarian and consequentialist doctrine between the aggregative principle of 'the greatest happiness' (and hence too the principle of profit maximization) and the distributive principle of 'for the greatest number'. The goal of profit maximization dominated economic science. A variety of doctrines within right-wing economics tried to budget for this (e.g. trickle-down theories), but economists were helpless in the face of, or alternatively unmoved by,[2] the emergence in the late twentieth century of greater inequalities of income than had been seen since the eighteenth century. Moreover, as the twentieth century drew to a close, it became increasingly worrisome that traditional economic theory had no way of sensibly measuring public values such as standards of public behaviour, crime rates and suicide rates, or the variable social effects of rising divorce rates and single-parent families, the value of clean air and clean rivers or beaches, of species conservation and landscape preservation, of the avoidance of global warming, and so on.

---

[2] E.g. Milton Friedman and the Chicago School.

Reversion to happiness maximization

One theoretical reaction to these stresses and strains both in theory and in practice was to revert to eighteenth- and early nineteenth-century utilitarianism in advancing the view that subjective preferences, pleasure, and happiness could be objectively measured after all. Strikingly, Richard Layard observed: 'We also need a revolution in government. Happiness should become the goal of policy, and the growth of national happiness should be measured and analysed as closely as the growth of GNP [gross national product].'[3] Felicia Huppert, emeritus professor of psychology at Cambridge University and director of the Well-Being Institute there has written:

> There is a growing recognition that the goals governments have typically focused on, such as GDP [gross domestic product], are only a means to an end, and that end is happiness. ... It is no longer sufficient to measure economic progress, we must also consider social and even environmental progress, and we must measure subjective well-being.[4]

It is striking that this idea was anticipated in the 1930s by Otto Neurath, writing from a Marxist point of view. He held that 'Maximum profit is the purpose of the individual business in the capitalist economy ... a maximum of happiness, of the enjoyment of life in a community and of utility is the purpose of a socialist economy'.[5]

Neural objectification of subjective happiness

The contemporary aspiration to measure subjective well-being is a reversion to Benthamite speculative utilitarianism with little awareness of the conceptual problems that this endeavour presents. As will become clear, it is supported by empirical psychology and neuropsychology in the form of the doctrine of the necessary possession by normal human beings from infancy onwards of a so-called theory of mind. The hard data for the budding science of happiness are held to be individuals' own statements in response to queries and questionnaires concerning their subjective happiness or well-being. These judgements, seemingly unacceptable to an objective

---

[3] Richard Layard, *Happiness: Lessons from a New Science* (Penguin, London, 2005), 147.

[4] Felicia Huppert, Foreword to Susan David, Ilona Boniwell, and Amanda Conley Ayers (eds), *The Oxford Handbook of Happiness* (Oxford University Press, Oxford, 2013).

[5] Otto Neurath, *Philosophy between Science and Politics*, ed. Nancy Cartwright, Jordi Cat, Lola Fleck, and Thomas E. Uebel (Cambridge University Press, Cambridge, 1996), 29.

empirical science because of their apparent subjectivity, were held to be objectified by the alleged discovery of the 'cortical representations' of pleasure, happiness, and the experience of meaning. For these were hard scientific facts. Kent Berridge and Morten Kringelbach explain that

> Pleasure is mediated by well-developed mesocorticolimbic circuitry and serves adaptive functions … Human neuroimaging studies indicate that surprisingly similar circuitry is activated by quite diverse pleasures, suggesting a common neural currency shared by all. Wanting for reward is generated by a large and distributed brain system. Liking, or pleasure itself, is generated by a smaller set of hedonic hot spots within limbic circuitry. Those hot spots also can be embedded in broader anatomical patterns of valence organization, such as in the keyboard pattern of nucleus accumbens generators for desire versus dread.[6]

The neural science of pleasure, happiness, and even *eudaimonia*,[7] commenced. Like Benthamite utilitarianism, it presupposes the intelligibility of introspective knowledge of the 'inner' in one's own case. It takes it for granted that small children, as they develop a theory of mind, know how things are with themselves prior to being able to apply psychological attributes to others by analogy with their own case. It is therefore unwittingly committed to the intelligibility of a 'private language' in which the meaning of psychological words is given by association or private ostensive definition that links names

---

[6] Kent C. Berridge and Morten L. Kringelbach, 'Pleasure Systems in the Brain', *Neuron* 86 (2015), 646–64, at 646.

[7] See Gary J. Lewis, Ryota Kanai, Geraint Rees, and Timothy Bates, 'Neural Correlates of the 'Good Life': Eudaimonic Well-Being is Associated with Insular Cortex Volume', *SCAN* 9 (2014), 615–18. Eudaimonia was thought of as the conscious and life-long active exercise of intellect and character virtues. Eudaimonia was measured on the six-factor Ryff Scales of Psychological Well-Being. These correlated scales assess autonomy, self-acceptance, purpose in life, environmental mastery, and positive relations with others. This was done by means of questionnaires in which subjects were asked to agree or disagree on a seven-point scale with such questions as 'I tend to be influenced by people with strong opinions', 'I am good at managing the many responsibilities of my daily life', and 'I think it is important to have new experiences that challenge how you think about yourself and the world'. The Ryff scales excluded wisdom, bravery, generosity, and justice. Eudaimonic well-being and its facets were claimed to be heritable. *Eudaimonia* was found to be associated with frontal cortex activity and with grey-matter volume of the right insular cortex. Low *eudaimonia* thus conceived acts as a risk factor for depression and poor physical health. It is interesting to reflect on what Aristotle might have thought of this.

to mental representations.[8] And they presuppose that the psychological attributes of others are inferred from their bare bodily behaviour behind which these psychological attributes are concealed from others. The concealment of the 'inner' from the sight of all save the subject can, it is argued, be transcended by neurological investigations which discover the neural representation of hedonic states and can therefore validate the subject's allegedly introspective reports. I shall elaborate these schematically stated points in section 3 below.

2. How happiness is understood by happiness scientists

Conceptual multiplicity of happiness   Chapter 9 disclosed the complexity of the concept of happiness. It is a concept with multiple centres of variation. In some uses 'happiness' means no more than *being pleased* at or by something and signifies an attitude; the more emphatic use here signifies *being delighted* at or with something. Closely related uses signify *satisfaction* or *gratification*. Further along this dimension lie *fulfilment* and *self-realization*, which are associated with the successful pursuit of life projects. In yet other uses, 'happiness' may signify *pleasure, exhilaration, enjoyment*, or *exultation*. In a different use it indicates a *mood*, such as feeling cheerful, merry, blithe, or even radiant. Moving towards the dispositional end of the psychological spectrum, 'happiness' may signify *absence of discontent*, or, more positively, *being contented, carefree*, and *tranquil*. In one of its most important uses which approximates the venerable notion of the *summum bonum*, it signifies a compound of favourable judgement and reflexive attitude towards a part of or the whole of one's life. To live or have lived a happy life is a gift of the gods, so to speak, and requires gratitude rather than pride. *These centres of variation are not the same.* The logical or conceptual relations that characterize one use differ from those that are constitutive of another. Being happy to hear good news is categorially different from leading a happy life, and having a happy temperament is logically unlike having a happy afternoon walking on the downs. One may wake up feeling happy, that is feeling cheerful – which is a mood;

---

[8] For brief discussion, see P. M. S. Hacker, *Wittgenstein: On Human Nature* (Phoenix Books, London, 1997); for a comprehensive analysis, see P. M. S. Hacker, *Wittgenstein: Meaning and Mind*, vol. 3 of *An Analytic Commentary on the Philosophical Investigations*, Part 1, *Essays*, 2nd edn (Wiley Blackwell, Oxford, 2019), 1–190.

but to be happy to see someone is not to be in a good mood. To be happy to hear the good news is a hedonic intentional attitude, but to have a happy temperament is a non-intentional character trait.

Happiness scientists'
conception of happiness   Having reminded ourselves of these distinctions we may turn to examine what putative happiness-scientists understand by 'happiness' and 'being happy'. Layard, an economist, for example, holds that

> Happiness is feeling good and misery is feeling bad. At every moment we feel somewhere between wonderful and half dead, and that feeling can now be measured by asking people or monitoring their brains. Once this is done, we can go on to explain a person's underlying level of happiness – the quality of life as he experiences it.[9]

Average happiness, he suggests, 'is made up from a whole series of moments. At each moment of waking life we feel more or less happy, just as we experience more or less noise'. Happiness, he supposes, is a continuum: 'happiness begins where unhappiness ends.'[10] Paul Thagard, a philosopher and cognitive scientist, avers that '*Happiness is a conscious emotional experience*. There is something it feels like to feel happy.'[11] Conscious emotional experiences, he explains, have a positive or negative character – happiness is a good feeling. One may feel happy because one enjoyed a joke or because one got an invitation to visit a friend in an exciting city. Such experiences begin and end; for example, 'you start feeling happy when you get an invitation and stop when you get distracted'. Some, like the editors of *The Oxford Handbook of Happiness*, suppose that any logical difficulties can be avoided by

> considering happiness in its broadest sense, treating it as an umbrella concept for notions such as well-being, subjective well-being, psychological well-being, hedonism, eudaimonism, health, flourishing, and so on. In doing this, we hope to achieve a broad conceptualization that is theoretically sound and practically useful.[12]

---

[9] Layard, *Happiness*, 6.

[10] Layard, *Happiness*, 13.

[11] Paul Thagard, *The Brain and the Meaning of Life* (Princeton University Press, Princeton, NJ, 2010), 105 (emphasis original).

[12] Susan David, Ilona Boniwell, and Amanda Conley Ayers, 'Introduction', in David, Boniwell, and Ayers (eds), *The Oxford Handbook of Happiness*, 3.

This merely repeats Bentham's mistakes.

So-called umbrella concepts, if they are anything, are determinables that subsume determinates (like *colour* and *red, orange, yellow,* and so on), not multifocal concepts. But happiness is not a determinable, and pleasure and enjoyment are not some of its determinates. One cannot eliminate logical and categorial differences between different foci of multifocal concepts by declaring the concept an umbrella concept. Determinates of a determinable form a continuum (as, for example, colour forms a continuum from red to violet), but happiness and unhappiness are contraries and do not form a continuum. One may be neither happy nor unhappy, merely contented with one's lot or even just not discontented. It is plainly incorrect to say that 'happiness begins where unhappiness ends'.

It is important to distinguish between being happy and feeling happy. Feeling happy (as Natasha did at the ball, when dancing with Prince Andrei) may be said to be a conscious emotional experience, but to be happy to undertake a task is not an experience at all, nor is having a happy temperament. More importantly, someone who is fortunate enough to lead a happy life is not a person who spends his life feeling happy (let alone euphoric). A happy life will perforce include sorrow and grief (for the death of parents and older relatives, of friends who die before one). It will unavoidably include disappointments and failures that are part and parcel of any normal human life – but a person who has led a happy life will have had the character and fortitude to overcome these.

Reversion to universal hedonistic explanation of motive

Like the utilitarians, happiness scientists take pleasure or happiness (and all their various centres of variation) to provide a single motivating force that explains all human action (or at least all intentional action). 'Happiness', Layard declared, is 'our overall motivational device. We seek to feel good and to avoid pain, not moment by moment but overall … it is impossible to explain human action and human survival except by the desire to achieve good feelings.'[13] We have, Layard asserts, an evaluative faculty that informs us of our situation and directs us to approach what makes us happy and to avoid what does not. From the various possibilities we choose whatever combination of activities will make us feel best. This twenty-first-century science of happiness appears not to be significantly

---

[13] Layard, *Happiness*, 26.

different from eighteenth- and nineteenth-century utilitarian psychology (or indeed from seventeenth-century Hobbesian psychology). We examined in detail the manifold forms of explaining human behaviour in *Human Nature: the Categorial Framework* (7.3–7.4). A reminder is given in list 10.1.

Explanations of action:

- *Explanations in terms of constitutive redescription*: as when we explain the gathering crowd by describing it as a funeral, a demonstration, or a celebration.
- *Explanation by polymorphous redescription*: as when we explain someone's shutting the door as obeying an order, or someone's singing as rehearsing for a recital.
- *Regularity explanations*: as when we explain behaviour as a habit, a custom, a manifestation of a personal disposition, or a character trait.
- *Inclination explanations*: as when we explain behaviour by reference to a preference (with or without a reason) or liking.
- *Explanation by reference to expression or manifestation of emotion*: as when we explain that the tears are tears of joy, of relief, or of grief, or that the trembling is trembling with fear rather than with excitement.
- *Causal explanations*: we explain doings that are not actions, such as stumbling, slipping, or blushing in terms of causes. So too many involuntary or partially voluntary actions are explained causally: he laughed because he was being tickled, cried out because he was hit, sneezed because he had taken snuff.
- *Explanations in terms of agential reasons*: the most common form of explanation of intentional action is specifying the agent's reason for doing what he did. 'Because it is enjoyable/ will give me pleasure' is but one of a multitude of kinds of reason. Reasons are commonly putative justifications. They may be backward or forward looking, and may cite purposes or goals.
- *Motive explanations*: one acts *for* a reason, but *out of* a motive. Motive explanations specify a pattern of reasons.

Explanations of inaction:

- inability – physical, technical (skill), or intellectual;
- prevention, natural or human;
- lack of equipment;
- lack of opportunity;
- ignorance of opportunity;
- forgetting one's intention;
- lack of normative power.

List 10.1   *Diversity of explanations of human behaviour*

The idea that the sole way to explain human action is by reference to the 'desire to achieve good feelings' is misconceived.

3. Psychological and epistemological presuppositions of the science of happiness

Inner/outer picture of the mind   As stated above, the new science of happiness rests four-square on the inner/outer picture of the mind that is characteristic of mind/body dualism and its heir, brain/body dualism, in which all attributes ascribed by dualists to the mind are ascribed instead to the brain (*TP*, 55–6, 364–5). The inner/outer conception of psychological attributes involves the empiricist supposition that each human being has 'privileged access' to his own mental attributes by introspection or inner sense, but has knowledge of the mental attributes of others by inference from their 'bare bodily movements'. That inference is alleged to be analogical: I know when I am in pain, want such-and-such, think thus-and-so, by introspection, since I *perceive* or *apprehend* my pain, my wanting or my thinking. But I cannot perceive the pain, wanting, or thinking of others, since all I can see of them is their 'bare bodily behaviour'. I can know how things are with other people by analogy with my own case. I note that when I am hit, burnt, or cut, I feel pain and cry out, assuage the injured limb, limp, and so on. I observe that when others are hit, burnt, or cut, they too cry out, and so on. So I infer by analogy that they must be experiencing what I experience in similar circumstances. The argument from analogy was advanced by Berkeley in the eighteenth century. In nineteenth-century Britain it was received empiricist doctrine exhibited in the writings of the Mills. On the Continent it received an imprimatur from Kant and was subsequently embraced by von Helmholtz. This conception was assailed in the twentieth century first by successive behaviourists, and in mid-century by Wittgenstein, Ryle, Austin, and Strawson. Its incoherence was definitively demonstrated.[14] How then did this large network of egregious confusions get adopted as a set of empirical truths by psychologists, economists, and cognitive neuroscientists?

Theory-of-mind theory The tale begins with the adoption by cognitive scientists of the idea that all normal human beings, in the first years of life, evolve a so-called theory of mind (*TP*, 380–5). C. D. Frith, one of the pioneers and strongest advocates of the theory-of-mind theory, explained:

> We suspect that people have an everyday theory of mind because they explain and frequently talk about the behaviour of others and themselves in terms of beliefs and desires. Having a theory of mind means that we believe that other people have minds like ours and that we understand the behaviour of these others in terms of the contents of their minds: their knowledge, beliefs and desires. But how can we demonstrate experimentally that people are using their theory of mind to predict the behaviour of others?[15]

The alleged theory of mind takes the general form of the classical argument from analogy. Layard averred that people 'have always been aware of how they feel and have used their introspection to infer how others feel. Since they themselves smile when they are happy, they infer that when others smile, they are happy too. Likewise when they see others frown or see them weep.'[16] But theory-of-mind theory goes beyond the argument from analogy in postulating possession by all normal human beings from an early age of a *theory* concerning the mental attributes of other people, in virtue of which they (or their

---

[14] Wittgenstein, *Philosophical Investigations* [1953], 4th edn (Wiley Blackwell, Oxford, 2009), *passim*; G. Ryle, *Concept of Mind* (Hutchinson, London, 1949), *passim*; J. L. Austin, *'Other Minds', reprinted in Philosophical Papers* (Clarendon Press, Oxford, 1961), 44–84; P. F. Strawson, *Individuals* (Methuen, London, 1959), ch. 3.

[15] C. D. Frith, 'Schizophrenia and Theory of Mind', *Psychological Medicine* 34 (2004), 385.

[16] Layard, *Happiness*, 12.

brain) can draw inferences from observed behaviour of others to the hidden mental attribute that is concealed 'behind it'.

Origins of theory-of-mind theory		It is ironic that theory-of-mind theory began its life among scientists in a 1978 paper by a psychologist, David Premack, and a primatologist, Guy Woodruff, entitled 'Does the Chimpanzee Have a Theory of Mind?'[17] (For more detailed discussion, see Appendix 1.4.) It seems prima facie absurd to ascribe to chimpanzees theories of anything, but Premack and Woodruff were undeterred. Imputing mental states to oneself and to others involves a 'system of inferences' and this is properly viewed as a theory since such states of others 'are not directly observable and the system can be used to make predictions about the behaviour of others'.[18] By 1983 the idea of possessing a theory of mind was extended from chimpanzees to human beings, in particular to small children after the age of four or five by Heinz Wimmer and Josef Perner.[19] Their work bred the so-called false-belief test with a scenario of a pair of dolls (Sally and Anne), a pair of little boxes with a sweet hidden under one of them which Anne moves from one box to another when Sally is called away, and the young child being tested is asked where Sally will look for the sweet when she returns. Autistic children performed very badly on the test, 90 per cent of them saying that Sally will look under the box where the sweet is, rather than under the box where it was when Sally 'left the room'. This test result was alleged to show that autistic children lack a, or have a defective, theory of mind since they cannot ascribe beliefs to others in predicting or understanding their behaviour. The test was widely replicated. It was, however, taken for granted that the four- or five-year-old subjects understood firmly the difference between 'will' and 'should'.[20]

---

[17] *Behavioural and Brain Sciences* 1 (1978), 515–26.

[18] The invocation of a *theory* is no coincidence, since Premack was a pupil of Wilfrid Sellars, who held that ascription of mental attributes to others required a theory. This was an unfortunate legacy of the Vienna Circle, some of whose members also thought that we need a theory in order to infer the existence of material objects from our sense data, since only sense data are given. For further discussion see Appendix 1.

[19] 'Beliefs about Beliefs: Representation and Constraining Function of Wrong Beliefs in Young Children's Understanding of Deception', *Cognition* 13 (1983), 103–28.

[20] This provided the basis for Hanoch Ben-Yami et al. to construct a similar experiment that avoided this question (Hanoch Ben-Yami, Maya Ben-Yami, and Jotham Ben-Yami, 'What Does the So-Called False Belief Task Actually Check?', PsychArchives, https://doi.org/10.23668/psycharchives.2513 (accessed January 15, 2019); see Appendix 1, p. 379 n. 21).

The false-belief test was put to use by Simon Baron-Cohen (a pupil of Uta Frith) in investigations into autism, and was invoked, as we have seen (Chapter 3), in his explanation of the sources of evil. For he held that the root of evil lies in lack of empathy (sympathy) consequent upon defects in the empathy (sympathy) circuitry in the brain.

This, as any reader who has followed the arguments of the previous three volumes of this study will realize, is incoherent. The whole 'inner/outer' conception is radically mistaken. For mental attributes are not hidden from view behind mere bodily movements. What we observe when we engage with our fellow human beings are not mere bodily movements at all, but expressive behaviour, patently intentional action, manifest purposes embedded in the circumstances of human life. Similarly, although we can say what we feel and what we perceive, what we intend and what we want, what we like and whether we are pleased, what we know and what we believe, our ability to do so does not rest on introspection. For introspection is not an inner sense but a reflective disposition that is exercised in our endeavours to understand ourselves, our reasons for thinking, feeling, and acting, our motivations and their roots (*HNCF*, chs 8–10; *IP*, chs 1–3; *TP*, chs 3–4).

First-/third-person asymmetry

In fact, psychological concepts are, for the most part, Janus-faced, involving groundless use in the first-person case. Ascription of psychological predicates to others is rarely inferential at all, but is normally supported by warranting criteria that may be cited when one is asked 'How do you know?' But in the first-person case the question 'How do you know?' (e.g. that you are in pain, want a drink, expect NN to arrive, intend to resign) is typically misplaced. For, in an important sense, I do not *know* and I have no 'privileged access' to anything – rather, I can *say so* and *my word* has a privileged status. I do not *know* that I have a headache, but neither am I *ignorant*.

A consequence of this is that the idea that the child first of all learns how things are with him by processes of introspection, and then constructs a theory of mind to enable him to infer how things are with others, is profoundly mistaken. Moreover the suggestion that children have to engage in something as complicated as constructing a theory of mind before they have adequately mastered their native tongue is preposterous. Constructing theories is a highly sophisticated linguistic activity that presupposes possession of numerous skills that children lack. Both cognitive scientists and neuroscientists suggest that, of course, the theory of mind that every person constructs in childhood is unconscious – it is really the brain that constructs the theory of mind that the child uses. But the idea that the brain, which can neither speak nor understand a single sentence, could construct a theory of mind is even more preposterous. Neither children nor adults have a theory of mind 'deeply buried in the mind/brain' (Chomsky's phrase), for there is no such thing as a 'mind/brain' and no such thing as a brain in possession of a theory. Moreover, what is said to be unconscious is what could be or become conscious. But what would it be for a four-year-old to have a conscious theory of mind? So it would have to be a non-conscious theory – but there is no such thing.

4. Measuring happiness

Basic data for the science of happiness The basic data for the new science is given by people's answers to the question of how happy they are feeling. Kringelbach and Berridge explain: 'The best measure of subjective well-being may simply be to ask people how they hedonically feel just now – again and again – so as to track their hedonic accumulation across daily life.'[21] In the same vein, Eldar, Rutledge, Dolan, and Niv aver that

> Well-being researchers have developed experience sampling techniques that probe participants as to their current subjective state while they go about their daily lives. These techniques, which involve periodically asking participants about their current emotional state and what they are doing, are considered the 'gold standard' to investigating real-world emotion.[22]

There appear to be no qualms about what people understand by the question of whether they are happy at the moment, about whether people are able to pass a reliable judgement on whether they are happy, and about what constraints people may feel when confronted by the question. Apropos the question 'Taking all things together, would you say that you are very happy, quite happy, or not very happy?', Layard remarks that 'In fact most people find it easy to say how good they are feeling and in social surveys such questions get

---

[21] Morten L. Kringelbach and Kent C. Berridge, 'Towards a Functional Neuroanatomy of Pleasure and Happiness', *Trends in Cognitive Science* 13:11 (2009), 479–87, at 480.

[22] Eran Eldar, Robb B. Rutledge, Raymond J. Dolan, and Yael Niv, 'Mood as Representation of Momentum', *Trends in Cognitive Sciences* 20:1 (2016), 15–24, at 16.

very high response rates ... The scarcity of "don't knows" shows that people do know how they feel.'[23] Some researchers, more concerned with continuous happiness and the heritability of a happy temperament, ask: 'Taking the good with the bad, how happy and contented are you on average now, compared with other people?'[24]

Questionable data   One might wonder whether all this research investigates anything more than people's dispositions to answer such questions in a certain way – typically positively. One should worry more about the political correctness of denying that one is happy – it might after all be thought to be un-American or not really British not to be happy. Or it might be thought to be a confession of failure in life, which one would not like to admit to oneself, let alone to others. One does not have to be a Thoreau ('The mass of men lead lives of quiet desperation') to feel suspicious about the results reported by Layard in his 2005 book that in Britain 36 per cent of people hold themselves to be 'very happy', 57 per cent 'quite happy', and a mere 7 per cent 'not happy' (the figures for the USA were 38 per cent 'very happy', 53 per cent 'quite happy', and 9 per cent 'not happy'). Strikingly, Lykken and Tellegen interpreted their results as showing 'that most people are in fact reasonably happy most of the time'. Consequently they were tempted to speculate 'that natural selection tended to favour happy people because they were more likely to mate and raise children and thus to become our ancestors'.[25] Dr Pangloss would have been delighted to hear this mid-American, late twentieth-century speculation, but Voltaire would surely have been sceptical.

Neuroscientific investigation   An attempt to mathematicize research *of happiness*   into well-being or happiness (treated as identical) and to construct a computational and neural model of momentary subjective well-being was made by Rutledge, Skandali, Dayan, and Dolan. One of the foundations on which happiness is assumed to be built, they averred, is the subjective response to rewards. To operationalize their research and to enable them to correlate their behavioural results with neural events, they focused on monetary rewards because these are readily quantifiable. Their subjects were asked to choose successively between options that

---

[23] Layard, *Happiness*, 13.

[24] David Lykken and Auke Tellegen, 'Happiness is a Stochastic Phenomenon', *Psychological Science* 7:3 (1996), 186.

[25] Lykken and Tellegen, 'Happiness is a Stochastic Phenomenon', 186.

yielded a guaranteed monetary outcome and gambles that were riskier but more profitable. Monetary outcomes ranged from losses to gains of up to £3. Every few trials the subjects were asked 'How happy are you at this moment?', which had to be answered by moving a slider. These responses were converted to a 1–100 scale. The reported momentary happiness was found to be a function of expectations and the difference between the predicted reward and the actual reward. Changes in reported momentary happiness in the course of 150 trials were found to be dependent only on the results of relatively recent trials, not on very much earlier ones, for expectations were dependent on recent trials rather than on earlier winnings and losses.[26] The experiment was run in the laboratory on twenty-six subjects and subsequently, using smartphones, on 18,420 participants.

The upshot of this Herculean labour was that one's recent winnings have more effect on one's 'happiness rating' than one's more remote ones; that a subject's expectations and their fulfilment or non-fulfilment play a major role in determining his reported 'happiness' (e.g. a £0 prize decreases 'happiness' if the alternative was winning £2, but increases 'happiness' if the alternative was losing £2). The authors remark that

> In fact in the real world, rewards associated with life decisions are often not realized for a long time (e.g. jobs, marriage) and our results suggest that expectations related to those decisions, both good and bad, do have an impact on happiness.[27]

---

[26] The computational model constructed was:

$$\text{Happiness}(t) = w_0 + w_1 \sum_{j-1}^{t} \gamma^{t-j} \text{CR}_j + w_2 \sum_{j-1}^{t} \gamma^{t-j} \text{EV}_j + w_3 \sum_{j-1}^{t} \gamma^{t-j} \text{RPE}_j$$

where CR is a reward that is certain, EV the expected value of a chosen gamble, and RPE the difference between the expected and actual reward; t is the trial number, $w0$ is a constant term, other weights w capture the influence of different event types, $0 \leq \gamma \leq 1$ is a forgetting factor that makes events in recent trials more influential than those in earlier trials, CR is the certain reward if chosen instead of a gamble on trial j, EV_j is the expected value of a gamble if chosen on trial j, and RPE_j is the difference between the predicted and the actual reward on trial j contingent on choice of the gamble.

[27] Robb B. Rutledge, Nikolina Skandali, Peter Dayan, and Raymond Dolan, 'A Computational and Neural Model of Momentary Subjective Well-Being', *PNAS* 111:33 (2014), 12252–7, at 12256.

The neuroscientific findings of the experiments were that happiness is accompanied by changes in BOLD (blood oxygen level dependent) signals in the striatum argued to reflect phasic dopamine release.

Flaws in the neuroscientific conceptions
　　What is to be made of all this? One worry is obvious: is this an investigation into happiness at all? Would the results have been any different had the subjects been asked 'Are you pleased?'. One suspects not, for it is surely obvious that what is being tested is whether the subjects are pleased or disappointed with the results of their gambles. But being pleased is not one form of being happy in the way in which enjoying oneself is one form of pleasure. A consequent concern is that happiness over a period of time or over a lifetime does not consist of a succession of moments of being pleased. Investigating a succession of moments in which one is pleased with the results of a gamble in the belief that one is investigating happiness is akin to looking for the keys to a car in the night under the lamppost rather than where one dropped them because there is more light under the lamp-post. One can be momentarily pleased at something, but there is no such thing as being momentarily happy any more than there is any such thing as being momentarily grief-stricken or momentarily old-aged. Another puzzlement may arise from the supposition that being happy (i.e. pleased) for a moment excludes being happy (pleased) for much longer than a moment. It would be absurd to suppose that what pleases one are exclusively momentary rewards – in fact that is very rare outside the lives of gamblers. One may be very pleased to have been awarded the Nobel Prize, and one will remain pleased for the rest of one's days. Similarly, one may be exceedingly pleased at the achievements of one's children, and there need be nothing impermanent to that – nor is it a reward of any kind. A further worry is that it is not obvious that the result is an empirical one at all. It is surely a conceptual truth that a person is pleased when he receives more of a benefit than he expected to get or loses less than he expected. It needs no experimentation or mathematical modelling to arrive at this modest truth. Some of the technicalities invented for purposes of mathematicization seem questionable: requiring subjects to move a slider to register how 'happy' they are at a given outcome and then converting this into a 1–100 scale is puzzling. To then say that someone is 20 per cent happier (more pleased) at having won £3 than at having won £2 on the previous test is unintelligible. All it can possibly mean is that he was more pleased (of course) and that he moved a slider up a distance that corresponds to 20 points on an arbitrary scale unknown to

him. Finally, the authors admit that these laboratory and smartphone trials have little to do with the so-called real world. They remark that in the real world rewards associated with life decisions are often not realized for a long time (jobs, marriage), and 'our results suggest that expectations related to those decisions, both good and bad, do have an impact on happiness'. But one did not need their results to know that what one expects of a job or of a marriage will be a partial determinant of one's satisfaction, contentment, or long-term happiness. For, again, it is a conceptual truth that failure to fulfil one's life expectations is a (defeasible) reason for disappointment.

Do neural signals explain why one is pleased?

What of the neural correlations (or so-called representations)? The authors of this investigation declare that 'using functional MRI, we demonstrated that neural signals during task events account for the changes in happiness'. It is unclear what is meant here by 'account for'. To account for something is to explain it. But it is evident that changes in neural signals cannot explain why one was pleased (never mind about happy). What accounts for why one was pleased is that one expected, say, to lose £2 and one won £3. This was a happy (fortunate) surprise, and pleased one. But that is no scientific discovery. The neural signals add nothing to the explanation – at most to description of the conditions of its possibility.

5. Some results of the science of happiness

The absence of conceptual clarity, the failure to draw obvious distinctions in the hedonic and eudaimonic semantic fields, the precipitate employment of mathematical models, the misguided thought that it is the brain that is the real subject of happiness, that feeling happy is actually a brain state (Thagard[28]) is disturbing enough. It may well incline one to doubt whether there really is or indeed can be a science of happiness and meaning, as Kringelbach and Berridge envisage. This suspicion is strengthened by scrutiny of the results of the putative science, when the jargon and pseudo-mathematicization are cast aside.

Results of the science of happiness

Rutledge, Skandali, Dayan, and Dolan declare that the methods of asking people whether they are happy as they go about their daily life 'show

---

[28] 'It becomes plausible to identify emotional feelings with brain states. Your feeling happy is a complex pattern of neural processes' (Thagard, *The Brain and the Meaning of Life*, 107).

that in typical individuals some activities (e.g. conversation and eating) are consistently related to higher happiness ratings'.[29] But it doesn't take an advanced science to show that many people enjoy good conversation and good food. Whether these have anything to do with happiness is dependent on a host of other factors concerning circumstances and temperament. Eldar, Rutledge, Dolan, and Niv assert that 'the main conclusion of the study was that happiness depends not on how well things are going (in terms of cumulative earnings) but whether they are going better than expected'. But this has little to do with happiness, and everything to do with being pleased. Moreover, one does not need such a study to know this. Myers and Diener, we are told, suggested that people who enjoy close personal relationships, who become absorbed in their work and who set themselves achievable goals and move towards them with determination are happier on the whole than people who do not. But this too is no scientific discovery. Nor indeed, is it true without qualification, since it is arguable that the pursuit of evil goals cannot make a man happy since the pursuit of evil ends involves a cluster of deadly vices inimical to felicity (see Chapter 5). Lykken and Tellegen 'are led to conclude that individual differences in human happiness – how one feels at the moment and also how happy one feels on average over time – are primarily a matter of chance'.[30] This is doubtless true, but it could have been learnt from examination of the etymology of the word 'happy' or from the study of Homer and Hesiod. It doesn't need twenty-first-century science to discover that a happy temperament is a gift of the gods and that a happy life is in the hands of Fortuna. Kringelbach and Berridge present as one of the conclusions of their labours the insight that, 'Although it remains unclear how pleasure and happiness are exactly linked, it may be safe to say at least that the pathological lack of pleasure … amounts to a formidable obstacle to happiness'. No one could disagree with that modest claim, but it is surely a conceptual assertion, not an empirical one: one needs no experimental investigations to arrive at it, but only reflection on the relations between the concepts of happiness and of pleasure. Similar doubts can be raised concerning the scientific status of some of their other overviews of the neurobiology of pleasure and happiness:

---

[29] Eldar et al., 'Mood as Representation of Momentum', 16.
[30] Lykken and Tellegen, 'Happiness is a Stochastic Phenomenon', 189.

> Considerable progress has been made in understanding the neurobiology of pleasure in ways that might be relevant [to the contribution of positive 'affect' to happiness] ... One way to conceive of hedonic happiness is as 'liking' without 'wanting', that is a state of pleasure without disruptive desires, a state of contentment. Another possibility is that moderate 'wanting', matched to positive 'liking', facilitates engagement with the world. A little incentive salience may add zest to the perception of life and perhaps even promote the construction of meaning ... Finally, happiness might spring from higher pleasures, positive appraisals of life meaning and social connectedness, all combined and merged by interaction between the brain's default networks and pleasure networks.[31]

Whatever truths are thus stated, they are not contributions of neurobiology. It is true that there are many fortuitous pleasures (things one likes which one comes across by chance and enjoys), but that is quite independent of contentment. One may enjoy a drink or conversation with a friend even when depressed, that is, not content. Moreover, one may be content with one's lot while having numerous desires that one satisfies, indeed especially when one can satisfy one's wish to hear this, see that, or visit here or there. It is trivially true that wanting to visit one's favourite city and doing so 'facilitates one's engagement with the world', at least in so far as one goes sightseeing. Having goals associated with rewards ('incentives') may, if one's efforts are successful, ensure that one is in a good mood. And it is true that if one's valued licit projects are being successfully pursued, one may well find meaning in one's life. But none of this is science, let alone neurobiology. It is rather an admixture of conceptual truth and truism. It is also true that, whatever cortical activity is a precondition for taking pleasure in certain activities, being pleased at certain events, enjoying oneself, being content with one's lot, and leading a happy life must be taking place. But that is no deeper than the fact that one's lungs must be functioning adequately if one is to enjoy the mountain air, that one's digestion must be sound if one is to enjoy the feast, and that if one is in a good mood one is easier to please.

Practical advice given by the science　　Finally, we may turn to the practical advice that has resulted from the new science of happiness. Martin Seligman, proponent of so-called

---

[31] Kringelbach and Berridge, 'Towards a Functional Neuroanatomy of Pleasure and Happiness'. 485.

positive psychology, holds that 50 per cent of one's happiness in life is determined by inherited temperament. All of us, it is claimed, adapt ourselves to circumstances and, after the novelty of a pleasing, delightful, or joyous event or of an enviable and much wanted acquisition wears off, we naturally drift back towards our set norm. So, Seligman avers, one's happiness is a function of one's set norm of happiness (S), circumstances (C), and factors that are under one's control (V). So, to put matters 'formally' (as he does), $H = S + C + V$. But, one can't make homely advice into science by a meaningless formula: a person's happiness is not a sum of happiness, circumstances, and controllable factors, since these are not things that can be quantified and summed. Seligman points out that the level of one's happiness can be raised by circumstances. So, one is advised to live in a wealthy democracy rather than an impoverished dictatorship, to get married, to avoid negative events and emotions, and to acquire a rich social network and a religion. These have a positive effect on people's happiness. By contrast, the accumulation of wealth, education, and moving to a sunnier clime account for only 8–15 per cent of happiness variance.

Laurie Santos, professor of psychology at Yale, gave a course on psychology and the good life that broke all records for attendance. She too made much of our propensity to adapt rapidly to pleasant changes, new acquisitions, new places, promotions, and so on, and to cease to take pleasure in them and revert to our set norm of happiness. Given that this is so, she advised her audience to spend time and money on ephemera, such as enjoyable holidays, where no adaptation is possible and fond memories remain. To maintain high levels of happiness, she suggested one should buy not things but experiences. (Small wonder that advertisers, trying to persuade the public to buy, emphasize *the shopping experience* at Crooks Ltd.) Moreover, one should count one's blessings, and bear in mind one's privileges in possessions and luxuries. To do so, Professor Santos suggested, it helps to deprive oneself, for example, of a hot shower for a day or two.

Social psychology No matter what one thinks of such advice (some of it ludicrous, some wrong, and some Aunt Sally's common sense), it is hardly something to be presented as the fruits of a new science of happiness. One can investigate the social and psychological conditions for human beings to take pleasure in certain things, the heritability of a happy temperament, and the futility of, and harm caused by, many of the pleasures of modernity. Such research is doubtless of paramount importance in the new electronic age with all its temptations (e.g. computer games, smartphones,

Twitter, and Facebook). It is important to know whether children are playing television games at the expense of exercise, whether children's social skills are declining as a result of excessive involvement with Facebook communication, and whether children's computer skills are purchased at the cost of knowledge of literature and history. For such knowledge is needed for a rational educational policy. Sociologists and social psychologists can and should investigate the social causes and consequences of rising divorce rates, of single-parent families, of same-gender-parent families, and so forth. For these can be encouraged or discouraged by financial incentives or disincentives. Well-established statistical correlations are useful for purposes of formation of social policies. But the case for the possibility of a *science* of happiness is non-proven at best. Happiness, as the Greeks knew full well, is in the hands of fortune and good hap. But there can be no science of good luck in life.

A science of misery   By contrast, one *can* investigate the nature of chronic depression or recurrent fits of anxiety, loneliness, and alienation. This is a better field for scientific study than happiness. The concepts involved are sharper. People are clearer about the meaning of expressions such as 'lonely', 'miserable', 'depressed', 'sad', 'frustrated', 'bitter', 'anxious', and 'worried' than they are about the meaning of 'happiness' and the associated hedonic and eudaimonic vocabulary. (Given the patent unclarity of happiness scientists about what happiness is and how to distinguish between happiness, contentment, fulfilment, absence of discontent, enjoyment, pleasure, being pleased, and so on, it is hardly to be expected that unversed members of the public should be any clearer.) People are for the most part more attuned to and attentive to their own suffering than they are to their own felicity. While there is no such thing as being ill with, prostrate with, or incapacitated by being pleased (or by pleasure, enjoyment, or happiness), people may be incapacitated by misery, grief, depression, or anxiety. Consequently their suffering may be a case for treatment. It may be rooted in an explanatory physiological or neurological cause in a way in which pleasure, enjoyment, and happiness by and large are not. For, as we have seen, neurological and psychological preconditions for taking pleasure in something, enjoying something, and leading a happy life are not normally explanatory factors in rendering the pleasure, enjoyment, or happiness perspicuous (exceptions are addictions and other self-harming pleasures, and perversions such as wicked and evil pleasures). The possibility of a science of misery is therefore much more promising than the possibility of a science of happiness. Indeed it already exists in the form of the study of pathological cases of depression, anxiety neuroses, and so forth. This is much more likely to benefit humanity than a putative science of happiness.

One stimulus for the development of a putative science of happiness was dissatisfaction with economists' attempts to evaluate the success of an economic system and the progress of a society by measuring gross national product. What should really be measured, and what Bentham originally thought would be measured by utilitarian calculi, is happiness. It was assumed that a happy society is one in which the greatest number of people are happy for most of the time. But this is a misconception. A happy society is a harmonious one, in which animosities are few, in which good social relations obtain, and in which mutual aid and concern are the norm. A happy society is one in which a sense of community is shared, in which public spaces are the responsibility of all and are loci of community events, in which there is an unproblematic sense of identity that is not purchased by means of xenophobia and hatred, and in which crime rates are low, inequalities of income are not excessive, and power is not for purchase. In a happy society alienation is minimal, and freedom and dignity are respected and preserved. These are political, not psychological, goals and the extent to which they are achieved is not to be measured by asking people how happy they are.

PART IV

Of Meaning and Death

11

The Need for Meaning

1. Meaning

It has nicely been observed that man does not live by welfare alone. One may have enough to live on, but nothing to live for. One may find no point or purpose to one's life and be beset with a sense of the meaninglessness of life. One need not be a Macbeth to think that life is a tale told by an idiot, full of sound and fury, signifying nothing. One may be an adolescent, suffering the *Weltschmerz* characteristic of that difficult phase of life; one may be a young adult, going through the *Sturm und Drang* of early adulthood; one may be in one's middle age, aware that one has more yesterdays behind one than tomorrows before one, disappointed with the course one's life has taken, and disillusioned with mankind; or one may be old, aware of impending death and of the futility of the endeavours of one's life. In such cases, one may be overwhelmed by a sense of the vanity of life in general or of one's own life in particular.

Pleasure, happiness, and meaning The practical hedonist has something to live for, namely pleasure. But it is debatable whether the pursuit of one's own pleasures can render one's life meaningful, just as it is questionable whether a life of pleasure can be a happy one (see Chapters 8 and 9). To be sure, one's pleasures may mean something to one, but an accumulation of pleasures does not make one's life worthy or meaningful. Philosophical hedonists, as we have seen, equate pleasure with happiness, and conceive of the pursuit of pleasure as equivalent to the pursuit of happiness. We have seen reason to doubt this (Chapter 8). It is evident from our investigations of happiness that one cannot find happiness in one's life or in a phase of one's life if one finds no meaning in it. A life devoid of meaning is a life without happiness. But one may find meaning in one's life and in one's activities without being happy. For the meaning one finds in one's life may be given by one's vocation, by the goals and purposes one pursues in the course of the commitments and dedication of a given social role. The pursuit of such goals is compatible with loneliness, sorrow, and unhappiness in one's personal relationships. And, as we shall see below, one may find meaning in one's suffering, as Viktor Frankl did in Nazi concentration camps.[1] In such cases it is the maintenance of one's integrity and consequent self-respect in the face of the horrors of hell that gives meaning to one's existence despite the complete absence of happiness.

Regulative and terminative goals   Goals intrinsic to one's vocation are typically *regulative* – being good at fulfilling one's role (e.g. a good doctor, farmer, teacher, parent); so are goals associated with what one loves doing, for example gardening (and so maintaining a fine garden) or listening to music (and so ever deepening one's understanding of music). The successful pursuit of such goals does not require a consummation or conclusion, although, of course, one's tastes may change and one may lose interest. Other goals have a *terminus* – successfully pursued, they are *achievements*. If not bad, wicked, or evil, it is possible for their achievement to lend meaning to an agent's life (this will be discussed below).

However, to pursue a purpose one has chosen does not necessarily imply that one finds meaning in so doing – it may be no more than an amusement (e.g. collecting stamps or coins). Many of the purposes we pursue are too unimportant in the long run for them to impart meaning to our lives, even though they mean something to one at the time. They lack depth, even though their pursuit may be harmless to others and beneficial for oneself. One might say that they do not touch one's soul.

Meaningful purposes; objective and subjective   For a purpose to lend meaning to our lives it must typically possess a reasonable degree of generality and subsume a multitude of subordinate purposes. A barrister may find meaning in

---

[1] Viktor Frankl, *Man's Search for Meaning* [1946] (Rider Books, London, 2004).

life from the defence of the criminal downtrodden and poor, but not from winning a given argument in court (which may be a triumph). But it is possible for individual acts alone to suffice to give one's life meaning, for example exceptional acts of self-sacrifice or acts of love and compassion (Sydney Carton's sacrifice of his life to save Darnay's). What one finds meaningful in life and what gives one's life meaning is something one finds valuable. Usually it will be something that is valuable, but that is not necessary. Subjective meaningfulness may suffice as long as one's goals are not bad, wicked, or evil. What gives one's life meaning must be something one cares about and that matters to one. It must mean something to one. But not everything one values, not even everything that means something to one, suffices to lend meaning to one's life. It must be serious and not frivolous or trivial. It must transcend selfish and self-centred concerns. Its pursuit must be an expression of one's nature as well as a determinant of one's nature. It must play a role in one's conception of oneself and it must determine in part one's relationship to others and to one's life (see fig. 11.1). So, for example, going on holiday with one's spouse is not normally something that gives meaning to one's life, but it can be. When Plantagenet Palliser sacrifices his appointment as Chancellor of the Exchequer in order to take his unhappy wife, Lady Glencora, on holiday (in Trollope's *Can You Forgive Her?*), that is a profound expression of his nature and the decision has meaning for them both. The life of a contemplative nun, an anchorite such as Julian of Norwich, though it rested on a host of false beliefs, was nevertheless a meaningful, self-transcending life. In such cases, subjective meaning suffices to give meaning to one's life. Of course, not all our goals and purposes are benevolent or value neutral. That evil purposes cannot give meaning to one's life is something that we shall argue below.

G. E. Moore, in his philosophy, and the Bloomsbury Group, in their lives, held that it is above all *experiences*, in particular experiences of love, friendship, and art, that make life meaningful. But that is too narrow. It may also be too selfish a conception – as is evident from the lives of many of the group, whose loves and friendships were often deceitful and callous. Experiences may indeed impart value and meaning to one's life – as the experience of passionate love often does and was extolled by the Romantics as doing. But much else may too: the ideals and achievements of the doers and makers, teachers and thinkers, and the responsibilities of a moral life and of a vocation.

Like pleasure and happiness, goodness and beauty, the meaningfulness one may find in one's life comes in degrees. Certain acts or activities may be of supreme significance to one. Certain phases of one's life may have greater meaning for one, both at the time and later, than others. Accordingly, how meaningful one finds one's life may wax and wane.

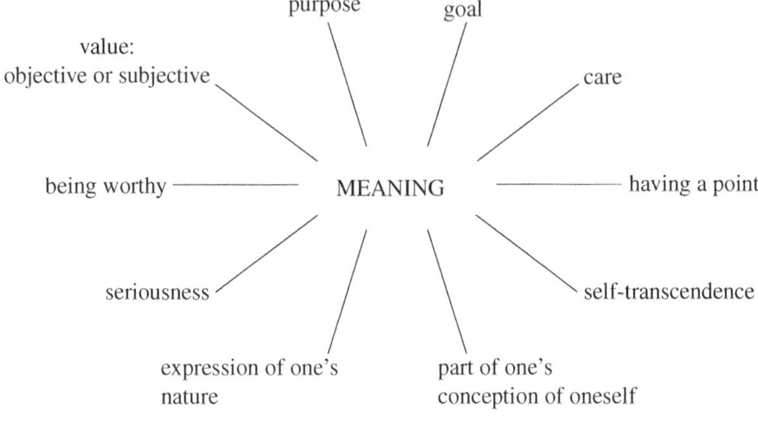

Figure 11.1　*The field of meaning*

Transitive and non-transitive meaning　Before turning to more detailed analysis we must distinguish having *a* meaning and having meaning.[2] Our primary concern will be with the latter. It is non-transitive meaning.[3] As we have noted, one may describe a self-sacrificial deed as lending meaning to one's life, as does Ernie Levy's self-sacrifice in André Schwarz-Bart's *The Last of the Just*.[4] If one is asked 'What does his voluntary death mean?', there is no answer, even though one may explain that his deed means that Ernie's love transcended his desire to live and his fear of the gas chamber and death. The phenomenon of non-transitive meaning is familiar to us from

---

[2] In *HNCF* we found it necessary similarly to distinguish, within the domain of semantics, between a word's having *a meaning* ('red', 'roof', 'rampart') and a word's having *meaning* ('Hey-ho', 'Hello', 'Abracadabra').

[3] The distinction between transitive and non-transitive meaning was introduced by Wittgenstein (see *The Blue and Brown Books*, 178–9).

[4] On a Nazi cattle-truck to Auschwitz Ernie befriends and cares for children whose parents have been murdered or who have died in the torment of the cattle-truck. On arrival at the entrance to these gates of hell, Ernie is allocated to the slave labourers, and the children and Golda, the young woman he loves, to the gas chamber. Ernie insists upon accompanying them, and holds them in his arms talking to them as they all die, endeavouring to give the children a belief in an afterlife – a belief that Ernie himself has lost.

aesthetics: a musical phrase played one way may be quite meaningless; played another way it says something – although one cannot say what it says. Played thus, one may rightly hold, it means something – although there is no saying what it means: it means itself. Just listen carefully!

Subjective meaning and giving meaning to life   We must further distinguish between *something's meaning something to a person*, and something *lending meaning to a person's life*. Many *achievements* may mean something to a person without being of sufficient significance to lend meaning to their life, such as winning in some competitive activity or passing an important examination. One values the achievement, to be sure, but it need not colour one's life and one's assessment of one's life. But some achievements may, for example winning a gold medal at the Olympics or being awarded a Nobel Prize. Some *activities*, such as philately or numismatics may mean much to one, without giving one's life meaning. But many do. One's youthful 'pleasure friendships', as Aristotle called them, certainly mean something to one, but it is one's 'philiac friendships' (see *TP*, 10.2), if one is fortunate enough to have any, that colour one's life, touch one's soul, and contribute to the meaning of one's life – as Montaigne's friendship with de la Boètie did. So too a *relationship* with another person, the love one bears another person, may be so deep that one may avow that without that person life would have no meaning for one. Both in fiction and in fact, in youth and in old age, people have committed suicide on the death of a beloved spouse or have made a suicide pact.

One may find much in life that is no achievement, but nevertheless means something, often a great deal, to one, although it *need not* make one's life meaningful. So, for example, nature means a great deal to lovers of nature, as music means a great deal to lovers of music, or ballet to balletomanes. In general, art in its various forms means a great deal to those with appropriate sensibility and sufficient knowledge to appreciate it. The apprehension of art and nature *may* play a central role in a person's life. Indeed, in the case of a refined aesthete such as Pater, it may endow his life with subjective meaning. But it is not clear that it makes his life meaningful.

Sentimental meaning   One may find individual things that mean much to one because of their associations, or their history. They may have little if any extrinsic value (price), but may be treasured because they belonged to one's parents or grandparents, or because they were given to one by someone one dearly loved, or because they had a role in one's childhood. Such things have *sentimental meaning* and this is commonly perfectly genuine (and not a form of sentimentality).

Illusory meaning　　Because non-transitive meaning is an axiological notion and is essentially tied to value and valuation, it is subject to misjudgement and corruption. Someone may be much taken with kitsch and value it highly (many curators of modern art galleries fall into this category). Such works of art are worthless, although they mean something to their collectors and evidently strike their collectors as meaningful. One might characterize this as *illusory meaning*, for it rests four-square on misjudgement and misguided evaluation. Something similar applies to judgements concerning non-transitive meaning in life and judgements concerning a meaningful life and finding meaning in life. Adolf Eichmann said that he felt he had failed in his mission, since only six million Jews were slaughtered by the Nazis (see below, p. 403). Had he succeeded in massacring all ten million Jews in Europe, he avowed, he would have died a happy man. So it was the murder of Jews that he thought gave his life meaning. This is one form of *illusory meaning* and it is a mark of an evil conscience. Far from exculpating a wicked or evil person, it is an aggravating condition. Nothing that is evil can give meaning to a person's life, for evil is the paradigm of disvalue. To live an evil life is not to live a worthy life. But for someone to live a meaningful life *is* for that life to be worthy. If an evil person conceives of an evil goal as something meaningful (as Eichmann did), that is an illusion – no matter how devotedly he strives for it and how much sacrifice he may make to achieve it. To pursue an evil end one must have not just one vice, but a whole battery of deadly vices. Coupled with cruelty are callousness and indifference to suffering; coupled with humiliation of one's victims is arrogance. Tormenting others is bound up with sadism. Unlimited power over women is linked to rape and sexual torture. And all these, outside the confines of a concentration camp or the Lubyanka, are walled in by lies and deception. In short, the vices form an interconnected family no less than the virtues. When evil is done, a multiplicity of deadly vices is necessarily involved (see Goya's Black Paintings), vices that cannot coexist with virtues such as love, fidelity, honesty, and truthfulness. We shall revert to this below.

However, other forms of illusory meaning, that is meaning that rests on false or misconceived beliefs, *can* impart meaning to a person's life and make it *genuinely* meaningful, as long as it involves neither evil doing nor evil beliefs. For is it not right that the life of a contemplative nun such as Julian of Norwich, or the lives of contemplative Buddhist monks, are meaningful? Their beliefs, though false, are not evil beliefs. It is also clear that their form of life is subjectively meaningful. May it not be genuinely meaningful and not illusory, despite resting on false beliefs? It seems that it may. The reasons it is thus meaningful is that it involves self-transcendence, that it aims to achieve tranquillity of spirit, that it informs their lives and beneficially affects their relationships to others and to the world. (Our observations, as Wittgenstein remarked in another context, are not the groundwork of a theory but the poles of a description. An apparent counter-example is not the refutation of a theory, since we are advancing no theories, but just another kind of example to be accommodated within our descriptions.)

Having meaning
    Non-transitive as opposed to transitive meaning
        Non-transitive meaning in aesthetics as opposed to
           non-transitive meaning in life
Meaning something to a person
    Something's being valued by a person
        Sentimental meaning
Something's contributing to the value a person finds *in* life

A meaningful life
    Objective meaning
        Resting on truth
        Resting on false belief that nevertheless renders a life
           meaningful
    Illusory meaningfulness involving false and evil belief

List 11.1   *Varieties of axiological meaning*

2. The primacy of loss of meaning and the sense of meaninglessness

Numbness to value   It is striking that outside philosophy we do not often speak of the meaning of life or describe someone as leading a meaningful life. We are much more likely to describe someone as having led a worthwhile life or a well-spent life. But it is common to speak of life being meaningless, of lacking a sense of purpose, of someone's no longer being able to see the point and purpose of life in general or of his own life in particular. That suggests that it is meaninglessness that has primacy, that if we wish to clarify what it is for things in life to have meaning, and what it is for a person's life to be meaningful, we must first examine the loss of meaning. This is a common phenomenon in the lives of human beings. It is characteristic of our age, beset as it is with alienation, anomie, and depression. But the sentiment is as old as articulate humanity. It was given its most brilliant poetic expression in the book of Ecclesiastes (written ca. 450–200 BC): 'Vanity of vanities, all is vanity' is the recurrent refrain of the book, interspersed by 'All is vanity and vexation of spirit'. Solomon, a great king in Jerusalem, the fictitious preacher of the book, elaborates many reasons for finding life meaningless that have echoed throughout the ages in philosophy and literature alike. The phenomenon has been felicitously described by Iddo Landau as 'numbness to value'.[5]

This phenomenon, as already noted, characterizes puberty, early adulthood, middle and old age, as well as periods of depression. Its onset may be sudden, usually as a result of a misfortune, or gradual. Hamlet, in fiction, provides a vivid example of the onset of depression and indifference to all value as a result of misfortune: the death of his father and swift remarriage of his mother:

> I have of late – but wherefore I know not – lost all my mirth, foregone all custom of exercises; and indeed it goes so heavily with my disposition that this goodly frame the earth, seems to me a sterile promontory, this most excellent canopy, the air, look you, this brave o'erhanging firmament, this majestical roof fretted with golden fire, why, it appeareth nothing to me but a foul and congested congregation of vapours. What a piece of work is a man, how noble in reason, how infinite in faculties, in form and in moving how express and admirable, in action how like an angel, in apprehension how like a god; the beauty of the world; the paragon of animals; and yet, to me, what is this quintessence of dust? (*Hamlet*, II. 2)

But it is evident, both in fact and in fiction, that numbness to value may arise through questioning what had previously seemed unquestionable.

---

[5] Iddo Landau, *Finding Meaning in an Imperfect World* (Oxford University Press, Oxford, 2017), to which book I am much indebted throughout these pages, as I am to its author for numerous conversations on these themes.

The crisis in John Stuart Mill's early adulthood was precipitated by just that:

> It occurred to me to put the question directly to myself, 'Suppose that all the objects in your life were realized; that all the changes in institutions and opinions which you are looking forward to could be completely effected at this very instant: would this be a great joy and happiness to you?' And an irrepressible self-consciousness distinctly answered 'No!' ... The end had ceased to charm, and how could there ever again be any interest in the means? I seemed to have nothing left to live for. (*Autobiography*, ch. 5)

This sudden transformation led to a nervous breakdown, from which it took many months to recover. Strikingly, what eased the numbness was the Romantic poetry of Coleridge and Wordsworth, which enriched the impoverished utilitarian axiology that had informed Mill's education.

Similar questioning, albeit spread out over a long period of time, is described in detail by Tolstoy, who, having published *War and Peace* and *Anna Karenina* to worldwide acclaim, began to succumb to drastic meaning numbness:

> I was overcome by minutes first of perplexity and then by an arrest of life, as though I didn't know how to live or what to do, and I lost myself and was dejected. But that passed and I continued to live as before. Then those moments of perplexity were repeated more and more often, and always in one and the same form. These arrests of life found their expression in ever the same questions: 'Why? Well, and then?'

> At first I thought that those were simply aimless, inappropriate questions. It seemed to me that that was all well known and that if I ever wanted to busy myself with their solution, it would not cost me much labour – that now I had no time to attend to them, but that if I wanted to I should find the proper answers. But the questions began to repeat themselves more and more often, answers were demanded more and more persistently ...

> My life came to a standstill. I could breathe, eat, drink and sleep, and I couldn't help doing those things, but there was no life, for there were no wishes the fulfilment of which I could consider reasonable. If I desired anything, I knew in advance that whether I satisfied my desire or not, nothing would come of it. Had a fairy come and offered to fulfil my desires, I should not have known what to ask ... I could

> not even wish to know the truth, for I guessed of what it consisted.
> The truth was that life is meaningless.
>
> I felt that what I was standing on had given way, that I had no founda-
> tion to stand on, that that which I lived by no longer existed, and that
> I had nothing to live by.[6] (Tolstoy *A Confession*, III–IV, trans. Aylmer
> Maude)

These are powerful expressions of a significant human condition with
evident philosophical roots. For those who have experienced such
value numbness – and most reflective people will have done – it is all
too easy to understand the sentiments expressed. They are patently
harmful, since the numbing of the sentiments may slowly but surely
lead to the paralysis of the will. That is exceedingly damaging to living a worthwhile life. In so far as the numbness can be assuaged by
sober argument, its intellectual roots must be examined. If they are
infected with philosophical error and misconception, they must be
eradicated. Of course, that does not mean that the numbness to value
will disappear, but it will make it clear that it persists despite the inva-
lidity of the arguments supporting it.

3. The roots of meaninglessness

(i) *Brevity of life and*   The first consideration is the revelation of the
size of the universe   sheer size of the universe and the smallness of
                             the earth and of human beings on it. This was the
shock that the new astronomy of the sixteenth and seventeenth centuries gave to the Christian European world picture. Equally shocking
but platitudinous was the brevity of human life in comparison with
the eternity of the universe. These two limitations notoriously
distressed Pascal:

> When I consider the short duration of my life, swallowed up in an eter-
> nity before and after. The little space I fill engulfed in the infinite
> immensity of spaces whereof I know nothing, and which know nothing
> of me, I am terrified. The eternal silence of these infinite spaces fills me

---

[6] Tolstoy, *A Confession*, trans. Aylmer Maude (Dover, Mineola, NY, 2005), This
prolonged experience of progressive numbness to value and overwhelming sense of
the meaninglessness of life provided the basis for Tolstoy's greatest novella *The Death
of Ivan Ilych*.

with dread. We burn with desire to find solid ground and ultimate sure foundations whereon to build a tower reaching to the infinite, but our whole groundwork cracks and the earth opens with abysses. (*Pensées*, III. 205–6)

This quasi-existentialist reaction to the brevity of life and the infinite spaces of an indifferent universe was to provide the basis for religious existentialism (brilliantly exemplified in the paintings of Edvard Munch) and for the godless existentialism of the twentieth century, as manifest in Sartre's writings and above all in Camus's *The Myth of Sisyphus* and *The Rebel*. The sentiment continues to echo, without the drums, in late twentieth-century analytic philosophy. Simon Blackburn explains that to a witness 'with the whole of space and time in its view, nothing on the human scale will have meaning'.[7]

The idea that the size of the universe should reduce us to despair and rob us of all sense of meaning in our lives is strange. After all, our lives would not be twice as meaningful were we twice the size we are. Similarly, human life expectancy has doubled over the last centuries, but that has not doubled the meaning that human beings find in their lives. Nor would immortality in this vale of tears lend more meaning to our lives – only more tears. What frightens Pascal is not simply the size of the universe or its eternity; it is rather its *indifference* – the eternal spaces of the universe 'know nothing' of him. It is the absence of any affective relationship and intelligible communication between man and the cosmos that terrifies him. But that too is not yet right. The stars at night would be much more frightening than they are at the moment if they were to rearrange themselves into a pattern that reads 'Drink Coca Cola'. Nor would matters be improved if they arranged themselves into the pattern of the words 'Obey my commandments'. For what we ought then to be frightened of would be that we had all gone mad. What Pascal says he is missing is a 'solid ground and ultimate sure foundation whereon to build a tower reaching to the infinite'. But the last time men tried that, they came to dreadful grief. What Pascal is actually missing is the voice and presence of God. That cannot take the form of messages written in the sky.

(ii) *Death of God* It seemed to Ivan Karamazov that if God does not exist, then everything is permitted. The theme of the death of God preoccupied Nietzsche, who did not view it as a reason for despair, but as an opportunity for a transvaluation of values. The silence

---

[7] Simon Blackburn, *Being Good* (Oxford University Press, Oxford, 2001), 79.

of the universe was a pivotal doctrine in Camus's atheist existentialism. But the non-existence of God or of gods does not imply that everything is permitted or that morality has no warrant. All it implies is that if God does not exist then what is prohibited is not prohibited by a divine moral legislator. 'But then where does morality come from?' is a common riposte. The question is deceptive. If 'come from' means 'who ordained it?', then, to be sure, why should it 'come from' anywhere? The 'death of God' does not imply that moral prohibitions on murder, assault, theft, lying, and cheating cannot be warranted in a secular society. A naturalist account of morality, such as the one advanced in the first five chapters of this book, has no need of God to justify moral prohibitions.

(iii) *The atemporality of the meaningful* The absence of a visible God, indeed, the absence of a God, no matter whether visible or invisible, does not deprive what is meaningful of its meaning either. To be sure, the illusion that a godless universe must lack any ethical value is still widespread (since religious faith is now more widespread than in the middle of the twentieth century). Some further aver that if there is no God then all we do lacks ultimate significance. But it is not clear what 'ultimate significance' means. Why should the presence or absence of God or gods affect the value and meaning of the true love between Pierre and Natasha, or of the self-sacrifice of Sydney Carton, or of the creation of sublime beauty by great artists and composers? Does being 'ultimate' signify that the meaning something has lasts for eternity, or will be known by God for all eternity, as it were? But this too seems misguided. Does a profoundly meaningful deed have to be known by others in order to have meaning? If one keeps one's acts of virtue anonymous, does that rob them of meaning or make it impossible for one to view them as giving meaning to one's life? Or is the claim that anything that is, in the requisite sense, meaningful, must remain meaningful for all eternity – which it would do only if God knew of it? But this too seems wrong. Although things that are meaningful are temporal, although meaningful deeds are done at a time and meaningful relations to other human beings obtain for a time, their meaningfulness is no more temporal than the truths of mathematics. Just as '$2 + 2 = 4$' is not true today and false tomorrow, so too the meaningfulness of a human deed is not meaningful today and meaningless tomorrow. Its meaningfulness is not sempiternal but atemporal. Can one say to a person who lovingly and with infinite gentleness helps a suffering dying parent to die: 'What you did was of supreme significance – profoundly meaningful, but it will be meaningless in ten years time'?

(iv) *All will be*   The thought that in time all will be forgotten has
forgotten   vexed many. The pseudo-Solomon wrote:

> For there is no remembrance of the wise than of the fool forever, seeing
> that that which now is in the days to come shall all be forgotten ...
> Therefore I hated life; because the work that is wrought under the sun
> is grievous unto me; for all is vanity and vexation of spirit. (Ecclesiastes
> 2: 16–17)

But this too assumes a misguided perspective. It is true that human
beings crave recognition for their achievements, and great people
wish to leave their mark on history. But the recognition one craves is
not a measure of the value or meaning of one's achievements, but
merely a token of it. And even though a great man's recognition, his
mark on history, will fade and disappear in the sands of time,[8] if his
deeds were truly great and meaningful, that does not detract from
their meaning.

(v) *Ephemerality*   The ephemeral character of human achievements
of all achievement   weighs heavily upon those who cannot achieve the
                       requisite detachment. Again, Solomon articulates
the sentiment:

> I made great works; I built houses; I planted vineyards; I made gardens
> and orchards, and I planted in them all kinds of fruit. I made pools of
> water ... Then I looked on all my works that my hands had wrought,
> and on the labour I had laboured to do, and behold, all was vanity and
> vexation of spirit. (Ecclesiastes 2: 4–6, 11)

But the beauty and wondrous scent of a rose is not rendered less beautiful by its being ephemeral. If it were not ephemeral it would not be
a rose. (Would a magnificent display of fireworks be more magnificent if it could be frozen forever in the night sky?) To create a beautiful garden certainly means something. It involves great dedication
and effort for the sake of the creation of beauty and for the delight of
all who view it. To have created a fine garden may even lend meaning
to someone's life (it surely did to Vita Sackville-West's, who created

---

[8] In Shelley's poem, Ozymandias was King of Kings, and doubtless did mighty
works that lasted long. But they too disappeared, leaving only a broken statue in the
desert: 'Round the decay / Of that colossal Wreck, boundless and bare / The lone and
level sands stretch far away.'

Sissinghurst) – but that the garden will not last for ever does not deprive its creation of significance.

(vi) The tedium of the cycle of life A further ground for despair, cited by Solomon and repeated over the centuries, is the tedium and futility of the cycle of generations – the relentless cycle of life, the seven ages of man from cradle to senility (enumerated by Shakespeare); birth, fornication, and death (as Eliot put it).

> One generation passes away and another generation comes, but the earth abides forever. The sun also rises, and the sun goes down, and hastens to the place whence it rose. The wind blows south, and turns about to the north; it whirls around continually, and returns again according to its circuits. All the rivers run into the sea, yet the sea is not full; to the place whence the rivers come thither they return again. ... What has been is what shall be; and what is done is what shall be done; and there is nothing new under the sun. (Ecclesiastes 1: 4–7, 9)

But this too is misleading. For although the life cycle of all human beings is, barring premature death, the same, it is, for each of us, new. It unfolds for us as the years go by, and the route we trace through the world is unique and unrepeatable, just as we ourselves are unique and unrepeatable. We each experience joys and sorrows, loves and enmities, achievements and disappointments. It matters not at all that others do so too. On the contrary, it is often a solace that one is not alone in one's suffering – which is why ritual and tradition often give us support.

(vii) Evolution and the meaninglessness of nature If seventeenth-century astronomy dealt a weighty blow to the Christian world-view that had dominated European civilization for a thousand years, Darwin's theory of evolution dealt it a mortal blow. It produced a number of destructive arguments and denials of the meaning or significance of human life in general. The spectacle of nature was no longer ordained by God, but was the product of the struggle for life, and the seemingly endless kinds of flora and fauna were not designed once and for all by a benevolent God, but were the product of mutation and savage competition for survival. Nature, as Tennyson wrote, was 'red in tooth and claw':

> So careful of the type she seems,
> So careless of the single life;
> That I, considering everywhere
> Her secret meaning in her deeds,

> And finding that of fifty seeds
> She often brings but one to bear,
> I falter where I firmly trod.
>   (*In Memoriam*, stanza 55)

There are, indeed, times when the spectacle of nature arouses horror and disgust, as was noted by Schopenhauer:

> But the futility and fruitlessness of the struggle of the whole phenomenon are more readily grasped in the simple and easily observable life of animals. The variety and multiplicity of the organizations, the ingenuity of the means by which each is adapted to its element and to its prey, here contrast clearly with the absence of any lasting final aim. Instead of this we see only momentary gratification, fleeting pleasure conditioned by wants, much and long suffering, constant struggle, *bellum omnium*, everything a hunter and everything hunted, pressure, want, need, and anxiety, shrieking, howling; and this goes on in *saecula saeculorum*.[9]

There is no intellectual flaw in this response to nature. Nature lovers have to accept the tension between the beauty and the horror. The horror is not a good reason for the Schopenhauerian despair. The rose is no less beautiful as a consequence of the thorn, and the leopard is no less wonderful for killing a springbok. One must hold in balance the sense of wonder before nature in all its plenitude and the sense of revulsion at the spectacle of *bellum omnium*. The endless killing is indeed meaningless, but that does not derogate from the wonder of nature nor diminish the meaning one may find in its contemplation and observation.

(viii) *Humanity no part of divine plan*   It is, however, not the idea of the meaningless struggle for existence alone that broke the Christian world-view. It was the thought that the creation story of Genesis is a creation myth, that man was not created by God but evolved, like all other forms of life, by endlessly complex and fortuitous processes from primeval slime; that the multiplicity of species were not part of God's design but the result of continuous processes of natural selection and genetic modification. So the world is not an intentional creation and we in it are not part of a divine plan. So our lives have no place in such a plan and cannot derive their meaning from it. So they have, in *this* sense, no ultimate

---

[9] Schopenhauer, *World as Will and Representation*, trans. E. F. J. Payne, vol. 2, 28, ii, 354.

meaning. If the living world is not God's creation, if the story of Genesis is a fairy tale, then living things in general, and animals in particular, were not created for our benefit. And we do not exist for any purpose. Whether or not the earth lies at the centre of the universe, whether or not the sun circles the earth, we are not the centre of attention of a divine creator.

But it is not at all evident why the thought of being the pivot around which the cosmos circles, both literally and figuratively, should provide solace for anyone suffering from numbness to value. Only faith, not reason, can do that. Those who can make the 'leap of faith' are no nearer to understanding the divine plan – that, it seems, is not given to man. Kierkegaard's 'knight of faith' does not grasp any justification for the endless suffering of mankind, but is merely confident that there is one. The agonized cry of the innocent child being tortured continues to ring in the ears of those who have faith, but the plan in which its agony makes sense lies in eternal darkness. This may satisfy Kierkegaard and Alyosha, but not Camus or Ivan. There is no intelligible reason why being a mere mote in the plan of a supreme intelligence should provide any sense of meaning that would otherwise be lacking. Being a pawn in a cosmic game of multidimensional chess is surely not rationally preferable to being a responsible intelligent agent in a universe that lacks meaning in itself, but in which all meaning is given by the sensitivities, activities, creations, and experiences of free beings living in human communities.

(ix) *Interminability of suffering*   A cousin of the previous source of meaninglessness is the horror of the endlessness of human suffering. War is the norm between human societies, peoples, nations, and empires.[10] Political and economic oppression, religious persecution, and suppression of women have been the rule throughout human history, not the exception. Murder, rape, battery, assault, fraud, lying, and cheating are common phenomena in human societies. The sufferings of inherited defects, injury, and disease are ubiquitous, even in today's world. It is all too easy to despair:

> I returned, and saw all the oppression that is done under the sun, and beheld the tears of the oppressed and they have no comforter; and

---

[10] Europe is a continent soaked in blood. Until the advent of the European Union, there had not been seventy years of peace in Europe since the age of the Antonines. It is doubtful whether the other continents are any different.

power belonged to the oppressors, but the oppressed had none to comfort them. So I congratulated the dead, who are already dead, more than the living, who are still alive. But better than both is he who has never been born, who has not seen the evil that is done under the sun. (Ecclesiastes 4: 1–3)

It is true that the spectacle of human suffering is terrible indeed. Some of it is in the hands of nature and ill hap – natural disasters, accidents, diseases, inherited defects of mind and body. Of this one can make no teleological sense. Some of human suffering is the deliberate work of man, and the teleological sense that can be made of it is as the upshot of ill will and evil, or as necessary evil with a warranted goal. For this there can be either no solace or little for those not enmeshed in the non-rational trammels of faith. If one is to find meaning in one's life, then it must be in the face of knowledge of evil and suffering.

(x) Paradox of achievement   A further root of the woes of an Ecclesiastes is the paradox of achievement. If one strives for the achievement of a specific goal, then one's sense of purpose logically cannot survive the achievement. Many of us are familiar with post-examination depression, even if we passed with flying colours. If one conceives of one's life as having meaning only in so far as one is striving to climb the greasy pole of ambition, once this is achieved – so what? The thought was well expressed by Yeats in his poem 'What Then?'

> His chosen comrades thought at school
> He must grow a famous man;
> He thought the same and lived by rule,
> All his twenties crammed with toil;
> *'What then?' sang Plato's ghost. 'What then?'*
>
> Everything he wrote was read,
> After certain years he won
> Sufficient money for his needs,
> Friends that have been friends indeed;
> *'What then?' sang Plato's ghost. 'What then?'*
>
> All his happier dreams came true –
> A small house, wife, daughter, son,
> Grounds where plum and cabbage grew
> Poets and Wits about him drew;
> *'What then?' sang Plato's ghost. 'What then?'*

> 'The work is done', grown old he thought,
> 'According to my boyish plan;
> Let the fools rage, I swerved in naught,
> Something to perfection brought';
> *But louder sang the ghost, 'What then?'*

But this too is misleading. We may doubt the intelligibility of Platonic Ideas, *a fortiori* of contemplating them. But, be that as it may, the argument intimated is wrong. First, although the achievement of a terminating purpose satisfies the desire to achieve it, it does not follow that whatever *psychological satisfaction* one derived from it terminates together with the *logical satisfaction or fulfilment* of the purpose. One may remain proud of one's achievement for the rest of one's days, and its value is not diminished by its having been achieved, even though its pursuit can no longer lend meaning to one's life. Secondly, some ambitions consist in a desire to achieve elevation to a commanding office, and once achieved one may find meaning in the responsible exercise of power. More generally, Yeats averts his eyes from the extent to which goals may be regulative rather than terminative. To the query 'What then?' the answer is 'Then I shall do my very best to be and to continue to be a good writer/artist, friend, husband/wife, father/mother'. And if that too is challenged by 'Plato's ghost', then the response is: 'This is what it is to live a good human life.'

(xi) *The absurd*   A distinctively existentialist source of despair is the thought that the human condition is absurd. Man's craving for meaning is confronted, as Pascal had pointed out, by a silent and indifferent universe. The absurd is the clash between human aspiration to significance and the unreasonable silence of the world. According to Camus, there are three possible responses to this alleged predicament. First, one may commit suicide (like Kirilov in Dostoyevsky's *The Possessed*) – as it were, spitting in the face of the universe in defiance. Secondly, one may adopt a religious response, like Kierkegaard, and make an absurd 'leap of faith'. Thirdly, one may accept the absurdity of the human condition. Therein, Camus averred, lies human freedom. The ability to give one's life meaning is constituted by recognition of the absurd, and by living one's life fully. Sisyphus is the symbol of the existential hero confronting the absurd, which is the rock he pushes daily to the top of the mountain. Sisyphus's triumph comes when he defiantly and scornfully watches the rock crash down the mountain side yet again, and turns to walk downhill to repeat his labour. 'The struggle itself', Camus wrote, 'is enough to fill a man's heart. One must imagine Sisyphus happy.'[11]

This was an intoxicating drink for the youth of the postwar generation of the 1950s, but once the bubbles had disappeared it lost its flavour. It combines deceptive dramatization with bogus heroism. Before jumping to its defence, one must patiently examine whether the term 'absurd' can bear the weight it has to carry in Camus's tale.[12] Something is absurd if it is exceedingly unreasonable, preposterous, ridiculous, silly, ludicrous, or farcical. The human condition – our social, cultural, and individual condition – is none of these. In the first place, the universe is not *indifferent* to our quest for meaning. Only what can be concerned can be indifferent – and there is no such thing as a concerned universe. One might as well rail against one's desk for being indifferent to one's writings. Secondly, human life as such is neither reasonable nor unreasonable, even though we are all unreasonable some of the time, and some of us are unreasonable much of the time. Proposals may be unreasonable, as may plans and projects. They are unreasonable if they are too costly relative to the gain and to alternative proposals, or if they transgress relevant norms of fairness and propriety, or if they are unlikely to achieve their purpose. But human life is not a proposal. It is certainly misguided to represent the human condition as preposterous – preposterous as opposed to what? Perhaps many of us lead meaningless lives, but there is nothing *preposterous* about that – it is only sad. For if they do, that is partly due to adverse circumstances and partly due to misguided but free choices. Finally, the human condition as language-using, self-conscious social animals with limited capacities for good and evil, with highly competitive and killer instincts is anything but ludicrous, farcical, silly, or ridiculous. It is far too sad and full of suffering for laughter – even though much in it is funny, and we are endowed with a sense of humour to appreciate the comical in our lives and to keep a sense of proportion. Nothing in our condition or in our relation to the cosmos *prevents* us from finding meaning or creating meaning in our lives or from living meaningful lives. It is not *given* us (as God is alleged to be); we must forge it in our lives. Finally, Sisyphus is a poor symbol for existential man in a godless

---

[11] Albert Camus, *The Myth of Sisyphus*, trans. Justin O'Brien (Hamish Hamilton, London, 1955).

[12] See Landau, *Finding Meaning in an Imperfect World*. ch. 18.5.

universe, for Sisyphus is not a social creature; he has no one to love and is responsible for no one other than himself. He has neither friends nor family for whom to care. He has no self-transcending task save for courage and tenacity in the face of adversity. His purpose – his task and goal – is indeed meaningless. But it is misguided to suppose that our purposes are uniformly meaningless. On the contrary. We find meaning in love and friendship, in self-sacrifice and dedication, in creativity and service to worthy causes.

(xii) *Mortality*   The final source of the belief that all is vanity is our mortality. The fact that we are doomed to die seems to many people an overwhelming reason for thinking that life is meaning-less, that death reduces us all to nothingness, that all are equal in death:

> The wise have eyes in their heads, whereas the fool walks in darkness: and yet I myself knew that the same fate befalls them all. Then I said in my heart, what happens to the fool, will happen likewise to me. So to what purpose have I made myself more wise? Then I said in my heart this also is vanity. (Ecclesiastes 2: 14–15)

We shall investigate the place of death in life in the next chapter. Here it is necessary only to point out that, although we are all equally mortal, and although when we are dead we are all equally dead, it does not even appear to follow that when we are still alive wisdom is worthless, let alone that no one is wiser or more virtuous than others. Nor is it true of the dead that 'they are all equal now'.[13] They are, to be sure dead, but one was wise and virtuous and another was stupid and vicious; to the first we are still grateful, while the other we hold in contempt. One may have led a worthy and meaningful life, the other a wasted, worthless, and meaningless one.

4. Does life have a meaning?

'The meaning   'What is the meaning of life?' is as misleading a ques-
of life'   tion as 'What is the meaning of a word?', but it is not
                 misleading in the same way. To be sure, in neither case
is there any substantive to answer to the question 'What is the meaning of ...?'. In both cases the wh-pronoun is interrogative rather than relative. In the case of the meaning of a word, there is no object that

---

[13] Thackeray, *Barry Lyndon*, 1.

is the meaning of a word, just as there is there is no object that is the meaning of life. But to the question 'What is the meaning of a word?' one may reply that the meaning of a word is its use, or that its meaning is given by describing or specifying its use or its usage. But no such easy way out is given in the case of the phrase 'the meaning of life'.

Ineffability   Is this because the meaning of life is ineffable? The young Wittgenstein when writing the *Tractatus* evidently thought so:

> The solution of the problem of life is seen in the vanishing of the problem.
>
> (Is not this the reason why those who have found after a long period of doubt that the sense of life became clear to them that they have then been unable to say what constituted that sense?)
>
> There are, indeed, things that cannot be put into words. *The make themselves manifest.* They are what is mystical. (*Tractatus Logico-Philosophicus*, 6. 521–2)

But this seems wrong. One may find meaning in one's life through reciprocated love of another, and one *can* say so. One may have a sense of destiny, and when one finds opportunity to fulfil one's destiny one may well feel that one's life has meaning. One's vocation and sense of vocation may give meaning to one's life. And one may find meaning in a tranquil and happy life cultivating one's garden with a beloved wife and children. There seems nothing ineffable about such cases and such experiences, although it is correct that the kind of meaning in question is non-transitive. What is also true is that characterization of those things that lend meaning to the lives of men and women will not move those who are numb to value, or convince them that not all is vanity and vexation of spirit.

The meaning of life   Our reflections so far suggest that there is no such thing as '*the* meaning of life'. We have noted that one walks on safer ground if one restricts oneself to intransitive meaning. Now we may distinguish the following questions:

- Does life as such have meaning? Does life have a purpose?
- Does the life of animals have meaning or purpose?
- Does human life as such have meaning? Does the existence of our species have meaning? Is there a purpose to the existence of man on the face of the globe?
- Does the life of an individual person have meaning? Does it have a purpose?

It seems evident that once divine cosmic plans are laid aside, then life as such does not have meaning. Nor does it lack meaning, any more than the Pacific Ocean, an eruption of Vesuvius, or an eclipse of the sun have or lack meaning. Similarly, life as such has no purpose. One can ask for the purpose of an artefact – what is it for? what does one do with it? One can ask for the purpose of an action – what intention is it meant to fulfill? what is the point of doing it? One can ask for the purpose of the organ of a living creature or of a part of an artefact or machine – what is its function? But once we have realized that life as such has no role in any cosmic or earthly design, then the question of its purpose is patently illegitimate. One has naturally but misguidedly extended the idiom of purpose beyond its licit constraints. (But that is not incompatible with retaining a sense of the wonder of life in all its multiplicity and fecundity.)

Animal life has no meaning   What then of animal life in particular? The emergence of living forms from primeval chemical concoctions was not in response to any design, but purely fortuitous. The emergence of intelligent life forms very much later was likewise sheer accident – an evolutionary response to the struggle for life in a world of scarce resources. Animal life as such has neither purpose nor meaning. Animals (contrary to the book of Genesis) were not made for us, nor do they exist for a purpose. To be sure, we may grow and breed them for *our* purposes (e.g. for food, as beasts of burden, as pets), but these purposes are not their purposes. Their purposes are manifold, for they pursue goals: to catch prey, to escape predators, to find water, to copulate, to feed their young, and so forth. But the existence of animal life as such has no purpose. Nor can a non-human animal feel its life to be meaningful or meaningless. But we may, and some of us do, find meaning in the plenitude of nature and its beauty.

Rational teleology   The emergence of human beings, their evolution into *Homo loquens*, and their development into cultural beings were wholly haphazard. The process was guided neither by an extrinsic purpose or design, nor by an intrinsic one. What is true is that, once our primitive forebears introduced language and began to engage in discourse, then potentialities of reason and of capacity rationality were released that possess an intrinsic logic of their own, quite different from the mechanisms of evolution. Once the question 'Why?' can be raised, reasons can be given. Once reasons can be given, they can be challenged and debated. Once chains of reasons can be adduced to warrant conclusions, reasoning is in place and rational teleology is afoot. Now both skills and knowledge can be transmitted by word of mouth and explanation. In the fullness of time, reasoning about value will naturally follow. So too will reflection on the meaning of life.

Existence of human species lacks meaning

Does the existence of humanity have meaning? If having meaning consists in being the creation of a purposeful God and being part of a divine plan, then that supposition can be rejected since there is no valid proof for the existence of an omnipotent, omniscient, omnipresent, and benevolent God and no reason either to think that the idea of such a God is coherent or to think the supposition rational. Humanity does not exist for a purpose and the continued existence of the human species has no meaning. However, it is true that the continued existence of humanity matters to us and is a condition for much that we value to have value and for much that we find meaningful to be so. Innumerable multigenerational projects would cease to have meaning were mankind known to be coming to an end in the very near future.[14] All research aimed at benefiting mankind in the long run would become pointless, all long-term constructional processes would cease, and all agricultural projects would cease to matter, as would all parental efforts to ease the lot of their children and grandchildren beyond the immediate present, and so forth.

Meaningfulness of individual lives

Do the lives of individual human beings have meaning? Do their lives have a purpose? We have already seen that some human beings find life purposes of their own – in the pursuit of their destiny, or in their dedication to their vocation. Many human beings find multiple purposes in their lives and at various phases in their lives. They pursue these both simultaneously and successively. Some may be achieved once and for all, others may be abandoned with age, and yet others are regulative and may inform large stretches of a person's life or their whole life. Such purposes mean much to them. In some cases, they give meaning to their lives and make their lives worth living.

5. Finding meaning in human life

Let us try to draw the threads together. Despite appearances to the contrary, the woes of an Ecclesiastes are not supported by good reasons. Not all is vanity and vexation of spirit. Our finitude and our size

---

[14] An idea introduced and explored by Samuel Scheffler in *Death and the Afterlife* (Oxford University Press, Oxford, 2013). See also P. D. James's novel *The Children of Men*.

provide no reason for thinking our lives to be intrinsically insignificant. The 'silence of the cosmos' is neither surprising nor unexpected, and the absence of gods or of one God or another of monotheism does not deprive whatever has meaning of its meaning, let alone render meaningful lives meaningless. If one acts well, justly, and rightly, if one masters one's natural selfishness or self-centredness, the fact that others do not know, or will not know in ten years' time or in a thousand years' time, is irrelevant to the value of one's deed and to its role in lending meaning to one's life.

Alternative pictures of what is transient

We are over-impressed by the transience of life and of everything in it. We find the image of the river of time irresistible – everything is in flux and one cannot step twice into the same river. We further distort matters by supposing that the past no longer exists. Only the present is real, and everything human in it is ephemeral – swept away into the inexistence of the past. But this is only a picture – or a crossing of pictures. We could equally well think differently, adopt a different picture. What is truly transient is not human deeds, but unrealized possibilities. As we go through life we are presented with, or make for ourselves, indefinitely many opportunities for action. Only a minute proportion of these are utilized – the rest are ephemeral, fleeting, mere possible actualities that are never materialized. *These* cease to exist as they pass – they are missed opportunities, most of which will never recur. But the possibilities we actualize in our deeds are inscribed on the adamantine marble of the past. What is done cannot be undone, and what we have done that has meaning can never be taken away from us.[15] This too is a picture, and one that comes less naturally to us than its familiar alternative. But it too can be used. Our deeds are not written on the sands of time, but on the marmoreal irrevocability of the past.

Sources of meaning

Doubtless there are many different ways of classifying those things that commonly serve to prevent a meaningless life – a life that it is not pointless and without worth. The following is but one possible classification among others that aims at maximal generality.

- Acts and activities that transcend one's selfish and self-centred concerns (and so, in one sense of that slippery term, transcend the self). This is above all evident in love and the pursuit of value:

---

[15] The idea is Viktor Frankl's in his *Man's Search for Meaning*.

transcendence of selfishness in loving relationships and in philiac friendships; dedication to justice and to the welfare of others; and so forth.

- Self-realization, that is the development of one's talents in worthy pursuits and worthy social roles (including familial ones).
- Creative labour, ranging from the most mundane to the most sublime. *Homo faber* finds fulfilment in making things, which may be useful or beautiful or both, and may be material (artefacts) or immaterial (music and literature, mathematics, or philosophy). It is important to bear in mind mundane creativity: preparing food for one's family and friends is a creative labour of love, and eating together (as opposed to feeding) involves both recognition of that labour and renewal of familial bonds. Creating a garden is creating a living work of art.
- Cultivation of one's sense of wonder and delight in nature and in art. Human susceptibilities and sensitivities are not equally distributed in this respect. Not all are as blessed as Blake in being able

> To see a World in a Grain of Sand
> And a Heaven in a Wild Flower
> Hold Infinity in the palm of your hand
> And Eternity in an hour.
> ('Auguries of Innocence')

But those who are find that apprehension and appreciation of nature gives much meaning to their lives, and may contribute to a sense of a worthwhile life. Similarly, sensitivity to art is not equitably bestowed, but those who are blessed with it, and who work to develop it, find it life enhancing.

- Development of the distinctive human powers of reason in the quest for understanding. This may lie in the domain of amicable dialogue that cements friendships. It may lie in a life dedicated to the pursuit of knowledge and understanding in any given domain. The list is platitudinous. But there are no mysteries about how to live a meaningful life – it is above all blinkers that stand in one's way, and misfortune that drains one's spirits and weakens one's will. The blinkers are the deadly human vices of selfishness, such as greed, lust, envy, and jealousy, and the equally deadly vices of cruelty, hatred, callousness or indifference to the suffering of others, arrogance, and the unfortunate intellectual flaws of stupidity, ignorance, and lack of understanding, as well as the character failing of laziness (lack of persistence, commitment, and hard work). The misfortunes are lack of parental love, grinding poverty, oppression, insecurity, ignorance, illness, and so forth.

- Finally, meaning can be found in suffering, in the manner in which a person meets overwhelming adversity without flinching, in retention of inner autonomy. Suffering can crush a human being, rob one of one's will and of one's self-respect, deprive one of one's very individuality and one's sense of one's own humanity. Viktor Frankl, in *Man's Search for Meaning*, dwelt on this. Everything can be taken from man but one thing, he wrote, the last of human freedoms: to choose one's attitude to what befalls one and to maintain one's humanity. This chilling thought had already been expressed by Dostoyevsky: 'There is only one thing that I dread: not to be worthy of my sufferings.' Human beings, uniquely, can turn suffering into an achievement.

On relativity and constancy

It might be objected that what human beings find meaningful in their lives is wholly culturally relative. We may find that love and compassion, pursuit of knowledge, and understanding are among the things that lend meaning to our lives, but in other times and other cultures such things would be dismissed as effete and foolish. The ideals of life and the conception of a worthy life that was to be found among warrior peoples in times gone by, such as Homeric heroes, Vikings, the hordes of Genghis Khan, and the Aztecs were profoundly different. This must be conceded. Some such societies were in themselves evil – as is patent in the blood-intoxicated arts of the Aztecs and in what we know of their deeds. Genghis Khan, in a recorded address to his troops, is alleged to have outlined the ideal life of a Mongol warrior: to slash off the heads of their enemies in triumphant battle, to seize and rape the wives and daughters of their defeated foes, and to sell their children into slavery.[16] But these are the misconceived evil ideals of one of the greatest *génocidaires* in human history. There are, as was argued in

---

[16] Mycetes, in Marlowe's *Tamburlaine*, expresses the same sentiment:

> That I may view these milk-white steeds of mine,
> All loden with the heads of killed men,
> And from their knees, even to their hoofes below,
> Besmer'd with blood – that makes a dainty show.
>
> (I. ii)

Chapter 5, limits to intelligible relativism. Not anything goes. The warrior ethos of Homeric heroes or Viking raiders presents a more primitive stage of human moral understanding and of the understanding of human nature. Even such savage societies have some grasp of the manner in which friendship and loyalty can lend meaning to life, and love between husband and wife, between parent and child, can ennoble life and make it meaningful. But there is much that they did not understand and, in the case of the Homeric ethos, that it would take centuries for anyone to understand. Mankind did not automatically realize that one should respect and protect the stranger in one's land. Nor did the thought of the House of Hillel that one should not do unto others what one would not have them do unto oneself come naturally. It was not immediately obvious that equality before the law is a moral condition for obedience to the law. That genuine tolerance[17] of religious beliefs is a requirement of rationality and of morality had to await the writings of Sebastian Castellio in the sixteenth century. That freedom of thought and talk is morally required by creatures who can reason and think is a late arrival on the human stage, and is still not everywhere realized. So too is the idea of the formal respect owed to rational beings (see Chapter 5). In the short history of civilization, moral insights and an understanding of human nature and of the moral potentialities of mankind occur rarely, and take root with difficulty. There are advances and regresses. The third to the fifth centuries were regressive as Christian fanaticism destroyed the culture of Rome; so were the Dark Ages which followed the fall of Rome, and so were the sixteenth and seventeenth centuries, in which Europe tore itself to pieces in the atrocities of wars of religion. So too was the short twentieth century (1914–89) in which mankind surpassed itself in evil doing. Its sequel looks ominous. For we are destroying Yggdrasil, the Tree of Life that upholds the world, and Ragnarök, the burning of the world, will follow as night follows day.

---

[17] As opposed to the pseudo-tolerance granted by Christianity (e.g. Aquinas) and Islam to the Jews.

12

The Place of Death
in Human Life

1. What is death?

Eschatology Throughout much of human history most people conceived of death as a transitional event. The physical organism was what died – the soul, the person, the locus of consciousness, survived the death of the body. In what form and in what place it survived was subject to widespread disagreement both within and across cultures and societies (see fig. 12.1). Most commonly, the soul was held to be transposed to another domain of existence. This was often thought to be a place of joy and happiness in reward for a virtuous life, or of suffering and punishment for wrongs done on earth. On some eschatological views, resurrection will occur only at 'the end of days'. For some of them, this will involve the perfection of this world, in which all suffering and evil will cease, while others envisaged a future heavenly world. The fate of the soul was likewise disputed: on some views it would be resurrected together with the body; on others it would enjoy a purely spiritual existence. Such beliefs, characteristic of Zoroastrianism and its offshoots, and of different branches and phases of Judaism, Christianity, and Islam, have an obvious incentivizing social function. However, the afterlife was also conceived to be a dark, dank, cold world and the form in which the soul was held to continue to exist was that of a twittering ghost (in Sheol according to the ancient Hebrews, in Hades according to the Homeric Greeks). The

The Moral Powers: A Study of Human Nature, First Edition. P. M. S. Hacker.
© 2021 John Wiley & Sons Ltd. Published 2021 by John Wiley & Sons Ltd.

social function of such belief is opaque. Other societies and cultures (e.g. Buddhism and various forms of Hinduism) cleaved to a doctrine of transmigration of souls, which could be reborn not only in human but also in animal form in an indefinitely long cycle of reincarnation. According to Buddhists, only continual moral improvement could lead to liberation from the interminable recurrence of life and *dukkha,*[1] which life entails, and to the attainment of selfless nirvana. As should be evident from the previous three volumes of this essay on human nature, these eschatological doctrines, which have and have had great appeal to most of humanity, are open to conceptual challenge.

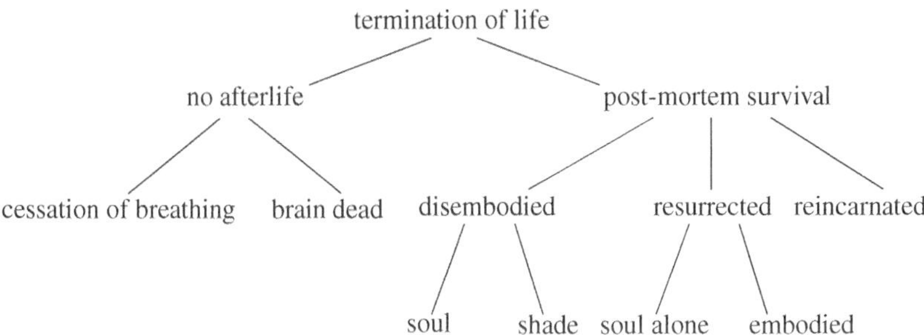

Figure 12.1 *Different conceptions of death and its aftermath*

Secular conceptions of death　　An alternative, secular, conception of death is as the permanent cessation of all life-sustaining biological functions. Accordingly, the death of the physical organism *is* the death of the person or human being (and of the soul and mind properly conceived as foci of distinctive capacities). As a result of medical advances in the last half century, it has become increasingly problematic to identify death in this way. For it is possible to keep a human being alive for years by means of life-support machinery, long after the permanent cessation of all *consciousness-sustaining* biological functions. For a human being may be brain-dead, but continue to breathe and metabolize nutriments, and so on. This has led to the identification of death with the *permanent cessation of consciousness.* However, the identification of such a terminus is medically problematic, for recovery from profound long-lasting

---

[1] Suffering, the ephemeral character of life, and vulnerability or dependency on chance.

coma is not unknown. Some neuroscientists claimed to have discovered that fMRI scans of patients who appear to be in a permanent vegetative state show brain activity when they are exposed to a highly specific auditory stimulus instructing them to imagine doing something.[2] This brain activity (in effect BOLD (blood-oxygenation-level-dependent) signals in specific parts of the brain) was held to be characteristic of thought and imagination. If so, then at any rate some cases of apparently permanent vegetative state might actually be limiting instances of locked-in syndrome, in which consciousness persists during paralysis of all volitional activity save for an eyelid. However, doubt has been cast on the validity of the fMRI scans and of the associated reasoning.[3] Such borderline cases will not be investigated here.

Human awareness of mortality However death be conceived, human beings are the only creatures that are aware of their mortality. All other animals live in the perpetual present, although their temporal world, so to speak, is not punctiform. The conscious impact of the immediate past (e.g. the terror of being chased) lasts for a short while, not in thought or *ex post actu* reflection but in distress, and anticipation of the immediate future guides their reactions to perceived danger and their pursuit of appetite satisfaction. But not being language users, they have no concepts of past, present, or future. There is nothing that they can do that would count as *thinking about* what happened yesterday or yesteryear, or as thinking about tomorrow or the years to come. This being so, they can have no *concept* of dying or of death, and so too no *conception* of their own mortality. So animals have, and can have, no fear of death, although they can, adequately for their needs, differentiate between a dead animal and a living one. They cannot, however, contemplate their own death with either fear or equanimity, since they cannot contemplate their death at all.[4] Of course, they fear predators and pitfalls, rightly associating them with danger and harm. They struggle desperately to preserve their lives. This is sometimes

---

[2] Adrian M. Owen and Martin R. Coleman, 'Functional Neuro-imaging of the Vegetative State', *Nature Reviews* 9 (2008), 235–43; Adrian M. Owen, Martin R. Coleman, Melanie Boly, Matthew H. Davis, Steven Laureys, and John D. Pickkard, 'Detecting Awareness in the Vegetative State', *Science* 313 (2006), 1402.

[3] P. Nachev and P. M. S. Hacker, 'Covert Cognition in the Persistent Vegetative State', *Progress in Neurobiology* 91 (2010), 68–76.

[4] Although elephants react to elephant skeletons with patent distress. Many animals we raise for slaughter show distress and fear at the smell of blood. This does not mean that they fear death or are aware of their mortality.

misleadingly characterized as 'predatory death anxiety'. Such instinctive fear is necessary for the survival of any kind of developed sentient organism, and has been selected for throughout evolutionary history. Without an intuitive 'fight or flight' instinct, no species could survive in the face of predators. That was, sadly, proven in the case of flightless birds living on islands without predators, such as the elephant bird in Madagascar, the dodo in Mauritius, and the various kinds of moa in New Zealand. These birds were rapidly rendered extinct as soon as man and rats were introduced. Our own natural fear of dying and of death (thanatophobia) is an aspect of our animal nature. But with us, *Homo loquens*, it is suffused with cognitive powers characteristic of language users. No other animal can long for death, view death as a release from suffering, or reflect on the meaning of death.

Our unique awareness of our own mortality is not innate but acquired in experience. Were the child not witness to the deaths of other animals and of family members or acquaintances, or were the child not told by its parents of the inevitability of death, it would never occur to it that it is doomed to die.

Sanctity of life   Life sometimes appears to be the most valuable thing of all. Many would exchange everything they possess in order to preserve their lives – but many would sacrifice their lives for their spouse or lover; for their children; for their country, faith, or ideology; or in pursuit of adventure. To sacrifice one's life for another is often conceived to be the greatest sacrifice of all – but sometimes continuing to live and to bear the duties and obligations of one's vocation or role may be a far greater sacrifice than death. We commonly speak of *the sanctity of life*, although it is far from clear what we mean by this ready-made phrase. In contemporary Western cultures suicide is commonly held to violate the sanctity of life, is prohibited by religious institutions, and, for the most part, is frowned upon by legal institutions, even though it is no longer illegal. Most Western jurisdictions penalize any attempt to assist someone in committing suicide. The death penalty is often thought to be the most severe punishment of all, far worse than life imprisonment. Life is often said (misleadingly) to be all we have, and its loss is said (equally misleadingly) to be the greatest loss we can sustain. So life is conceived by some to be infinitely valuable. Nevertheless, there are those who hold that life is a curse, that suffering predominates and is irremediable. Accordingly death is held to be a welcome escape from the evils of living.

2. An afterlife

Different forms of afterlife	Attitudes towards death vary according to the acceptance or rejection of belief in an afterlife. Such beliefs characterize monotheist religions, although it is noteworthy that early Judaism prior to the Babylonian exile (587 BC) appears to have advanced few eschatological doctrines. Mention of Sheol is rare and obscure in the Hebrew Bible and appears to be largely a residue of pagan beliefs. Dualism of body and soul, survival of the soul after death, doctrines of the resurrection of the flesh, and beliefs in a heaven and a hell as adopted in Western monotheism appear to have originated in Zoroastrian doctrines pervasive in Babylon, whence they spread throughout the Middle East. According to Flavius Josephus, among the Jews in the late Second Temple period, survival after death was rejected by the educated aristocratic school of Sadducees. Survival of the soul, in some form or other, was accepted by some Pharisees, and was embraced by the Essenes. The theological grounds for belief in an afterlife, or in a resurrection of the body at 'the end of days' are unclear, since these subjects are hardly discussed in the Old Testament, save for obscure intimations in the apocalyptical books of Ezekiel, Second or Deutero-Isaiah, and Daniel. Eschatology plays a far greater role in the New Testament. Purported knowledge of the immortality of the soul, of an afterlife, of the second coming, and of resurrection was obtained by means of *revelation* granted to the prophets, apostles, Jesus, Paul and his followers, and evangelists, and were duly embedded in the New Testament and numerous apocryphal and Gnostic writings.

In the *Apology*, the *Phaedo* and the *Meno*, Socrates and Plato's Socrates intimate or embrace the idea of the survival of the soul after death. Unlike the Bible, with its reliance on revelation, Plato advances *arguments* in favour of immortality. First, the soul is the principle of life, and as such must be immortal. For the principle of life cannot die. Second, the doctrine of anamnesis, defended in the *Meno*, allegedly makes plausible the existence of the soul prior to birth and the onset of worldly experience. Thirdly, the soul is the form of life and, like all Platonic forms, is indestructible. Finally, body and soul are separable – the soul being indivisible and incorporeal and therefore indestructible (since destruction is decomposition into constituent elements) and so immortal. This argument for immortality of the soul or mind was to be revived in the early modern era by Descartes. Our sole concern here is with the bare idea of the post-mortem survival of the soul.

The philosophical worry is not so much about the truth of this claim as about its intelligibility.

The soul or mind is supposed to be an immaterial substance with a persistent identity through time, that is a subject of thought and experience with a spatial and temporal location, although not a space *occupier* since it does not consist of matter. It is this that is supposed to be immortal. The intelligibility of this idea can be challenged. First, the soul thus conceived is not an intelligible subject of experience or thought. For any substance concept to be intelligible, there must be an associated principle of individuation that will determine what counts as an individual substance of that kind and what counts as the same substance again. The principle of individuation for human beings is the identity of the relatively persistent living, physical, spatio-temporal organism. But there is no principle of individuation for souls or minds conceived as disembodied beings. There is no way to distinguish a single such being thinking such-and-such thoughts and having such-and-such experiences and a thousand such beings all thinking the very same thoughts and having the very same experiences. Secondly, a similar difficulty arises with regard to diachronic identity. There is no way to determine whether a soul at one time is the same being as a soul at an earlier or later time. For all we can tell (as Kant pointed out), perhaps each soul persists for only a moment before passing all its thoughts and experiences on to a fresh soul, just as a billiard ball passes on its momentum to the next billiard ball in a series. Thirdly, the imagined soul substance is supposed to be the bearer of thoughts and experiences. But in virtue of what can we ascribe a given thought or experience to such a subject? Suppose a mind or soul is said to see a tree in the garden. There is no such thing as experiences that are no one's experiences. So what makes the visual experience of the tree the experience of a given soul? One might suppose that it is the relation of *possessing* that attaches the experience to the soul. But *to have* a thought or experience is not to own or possess anything any more than having a train to catch, having a birthday next week, or having a headache. It is not a *relation* of any kind although possession is the form of representation we deploy in our descriptions (for more detail, see *IP*, 265–6). The subject of thought or experience is the human being who manifests it in behaviour, in speech, action, or behavioural reaction, or who would do so were appropriate conditions met (e.g. if asked). But disembodied souls or minds do not behave, that is, *there is no such thing* as a disembodied spirit behaving. So there can be no such subjects of experience.

Fourthly, the imagined mind or soul is disembodied but is supposed to see or hear things in its environment. So it must have a spatio-temporal location. But since it is immaterial, how can one distinguish one soul seeing the tree in the garden from a thousand souls all seeing the tree in the garden from exactly the same point in space at the same time? Moreover, how can a being that has no eyes see? What is its cone of vision? Can it *look* in a given direction? How would it do so without a head? So, the idea of an immortal disembodied soul or mind is not at all promising.

Idea of physical resurrection incoherent
     If disembodiment fares badly, does the resurrection of the body on the day of judgement or at the end of days fare better? No. It suffers from similar defects. First, what would make the resurrected body *my* body? It cannot be the same matter as that which constituted my corpse when I died, since that has long since turned to dust, been eaten by worms, or been otherwise destroyed. It cannot be that it looks like me, or looks as I did when I died. For, in principle, presumably, there might be 25 or 250 similar bodies all 'resurrected' – which one would be me? And if there is no telling, then there is no telling whether any given single one is me or not. Secondly, what makes the soul embodied in this resurrected body me? If I ceased to exist on death, and am resurrected, duly embodied, at the end of days, what ensures that it is me as opposed to a facsimile? For again, there is no criterion of identity that can ensure that it is me as opposed to PMSH Mark II (which can come in as many copies as the gods please). If it be replied that what is resurrected is not a conjoined mind (or soul) and body, but a single substance – a human being – this too is of no avail. For the same question arises here too. What makes this 'resurrected' human being me? That it both looks like me, purports to remember everything I remember, and shares all my character traits? But that could be true of 250 such beings.

So, post-mortem survival does not appear to be a very good bet.

Agnosticism
     Others have held an agnostic position on the past and future of the soul. The Venerable Bede, in his *Ecclesiastical History of the English People*, describes an agnostic thane as asserting:

> When compared with the stretch of time unknown to us, O king, the present life of men on earth is like the flight of a single sparrow through the hall where, in winter, you sit with your captains and ministers. Entering at one door and leaving by another, while it is inside it is

untouched by the wintry storm; but this brief interval of calm is over in a moment, and it returns to the winter whence it came, vanishing from your sight. Man's life is similar; and of what follows it, or what went before, we are utterly ignorant. (2. 13)

The image is beautiful, but the agnosticism unnecessary. For we do know that nothing came before and nothing follows.

Incoherence of idea of post-mortem verification Moderns such as Moritz Schlick (leader of the Vienna Circle) have accepted agnosticism about immortality itself, but have claimed that the survival of the soul after death is an empirical proposition that *will be* verified or falsified *after death*.[5] To be sure, if there is no life after death, we shall not know that post-mortem survival has been falsified. Disregarding all the above considerations for the sake of argument, if there *is* life after death, shall we be able to verify it? Before the matter can be discussed, Schlick would have to give us some idea of what is to count as existing after death – in what form is such post-mortem existence being envisaged and is it coherent? Secondly, what is to count as verification? After all, can I verify that I exist here and now? Certainly not by repeating to myself 'Cogito ergo sum'. Others may verify that I exist, but I cannot. For anything I might do presupposes what it is meant to prove. To be sure, the mistake is to suppose that whether we survive after death is an empirical question at all. It is not and we do not.

It should be borne in mind that arguments against the intelligibility of the immortality of the soul are not arguments against possession of a secular soul distinct from the mind (see Chapter 5). For to speak thus of the soul is a form of representation of moral powers and sensibilities.

3. The valuelessness of life

Mortality renders life valueless Many of us will have experienced the feeling that life is worthless, that all is vanity, or even that death will be a welcome escape from misery. Such feelings are

---

[5] Moritz Schlick, 'On the Meaning of Life' (1927), in *Philosophical Papers*, vol. 2, ed. Henk L. Mulder and Barbara F. B. van der Velde-Schlick, trans. Peter Heath (Reidel, Dordrecht, 1979), 112–29. Empirical evidence commonly adduced for immortality includes séance experiences (popular in the interwar years), apparent memories of a previous existence that appear to be confirmed, and near-death-experiences.

characteristic of the growing pains of youth, of the *Weltschmerz* of early manhood and womanhood, of the disappointments of middle age, and of disillusionment with humanity in old age. Many writers have given expression to the thought that all is vanity. The imaginary King Solomon of Ecclesiastes argued that life is valueless precisely *because* we are mortal and doomed to die, and all our labours and achievements are fated to be forgotten. It is death itself that makes life worthless:

> Then said I in my heart, As it happeneth to the fool, so it happeneth even to me; and why was I then more wise? Then said I in my heart, that this also is vanity. For there is no remembrance of the wise more than of the fool for ever: seeing that which now is in the days to come shall all be forgotten. And how dieth the wise man? As the fool. Therefore I hated life; because the work that is wrought under the sun is grievous unto me: for all is vanity and vexation of spirit. (Ecclesiastes 2: 15–17)[6]

Suffering renders life valueless

Other writers in the ancient world held life to be worthless, but not because of death, rather because suffering vastly outweighs joy, because any happiness achieved is fragile and is likely to be destroyed. Far from death rendering

---

[6] Similar kingly despair was expressed by James Shirley (1596–1666) in *The Contention of Ajax and Ulysses*:

> The glories of our blood and state
>   Are shadows, not substantial things;
> There is no armour against Fate,
>   Death lays his icy hand on kings;
>   Sceptre and crown
>   Must tumble down,
> And in the dust be equal made
> With the poor crooked scythe and spade.
>   …
> The garlands wither on your brow
>   Then boast no more your mighty deeds;
> Upon Death's purple altar now
>   See where the victor-victim bleeds;
>   Your heads must come,
>   To the cold tomb;

But, unlike the Preacher in Jerusalem, Shirley adds:

> Only the actions of the just
> Smell sweet, and blossom in the dust.

life meaningless, death is to be welcomed inasmuch as it brings the
sufferings of life to an end. Such sentiments were commonly expressed
in the harsh world of ancient Greece, both in the archaic period and in
classical Greece. Theognis, in the sixth century, wrote in despair:

> Not to be born – never to see the sun –
> No worldly blessing is a greater one!
> And the next best is speedily to die,
> And lapt beneath a load of earth to lie.

Similar sentiment was expressed by Sophocles in *Oedipus at Colonus*
at the end of the fifth century (written in 406 BC, first performed
posthumously in 401 BC):

> Not to be born at all
> Is best, far best that can befall;
> Next best, when born, with least delay
> To trace the backward way.
> For when youth passes with its giddy train,
> Troubles on troubles follow, toils on toils
> Pain, pain, for ever pain;
> And none escapes life's coils.

Much the same idea informs the legend of Cleobis and Biton (see
above p. 261) in which death in the prime of one's life is the best of
gifts the gods can bestow.

The most pessimistic and life condemning among modern philosophers is Schopenhauer. He advanced a variety of arguments in *The
World as Will and Representation* (1818). First, life is a constant striving after goals. If they are not achieved, disappointment ensues; if they
are achieved, they lose their charm immediately as our insatiable will
moves on to new objects after which to hanker. This is true of some
possession-involving goals and some achievement goals, but not all (a
collector's delight in his collection need not diminish with time, nor
need an athlete's pride in an Olympic gold medal). It is not true of goals
with no terminus, such as living a certain kind of valued life (of a
scholar, doctor, or gardener) or being a certain kind of virtuous person
(see Marcus Aurelius's *Meditations*). Secondly, old age is full of infir-
mity and illness. This is true for some, but not for all. Even when true
it does not negate the value found in the rest of a person's life, nor need
it, save *in extremis*, deprive the decrepit of the forms of goodness available to the aged – the joy in children and grandchildren, the pleasures

of friendship, the delight taken in nature or art. Thirdly, the apparent (subjective) duration of pain is greater than that of pleasure and enjoyment. This is true in some cases, but not in others. Even when it is true, it does not imply that life is worthless, let alone meaningless. Fourthly, relief at the cessation of pain does not last long, but disappointment at the loss of pleasure may endure. Indeed it may be so deeply felt that one wishes one had never had the pleasure in the first place. Again, this may be true in some cases (e.g. Turgenev's protagonist Sanin in *The Torrents of Spring*, who ruins his life for the sake of his passion for Maria Nikolaevna). But some joys may be lasting, many passing pleasures may be replaced by others, and some goals need not be pleasure involving at all, yet give meaning to one's life. Fifth, we feel pain, but not painlessness; care, but not freedom from care; fear, but not its absence. The great blessings of life are health, youth, and freedom: we are not conscious of them while we have them, but only after we have lost them. So what seem positive values in life are in fact merely privative. The only reality is the reality of suffering. This seems at best a partial truth. Even if it were true, it would not follow that imperfect health reduces life to misery, or that middle age is bereft of joys or meaning. Not all forms of loss of the freedoms of youth are felt to be irksome. One may still be free to pursue one's goals and projects and to cultivate cherished, loving relationships, and if fortunate one may find fulfilment and happiness in so doing. Finally, Schopenhauer observes, boredom is a persistent form of human suffering. This may be true, but not everyone is susceptible to boredom and few are bored all the time. One may have projects and hobbies, and find joy and meaning in love and friendship. In short, Schopenhauer exaggerated the miseries of life and underestimated the sources of pleasure and enjoyment, of meaning, fulfilment, and self-transcendence.

4. The value of life

Two forms of life affirmation Some philosophers and poets, endowed with a happier temperament or more fortunate in their lives than Ecclesiastes, Theognis, or Sophocles find life – human life – to be a blessing. They laud the joys and pleasures of life, the contemplation of the beauties of nature, the happiness of reciprocated love and of loving friendship. A variety of conceptions and attitudes is evident (see list 12.1). Some may glory in the beauty of life and the wonder of human experience of joy and happiness, as well as grief and sorrow:

<table>
<tr><td>

Ihn schau ich mit wachsendem
Entzücken.
Von Sturz zu Sturzen wälzt er
jetzt in tausend,
Dann aber tausend Strömen
sich ergießend,
Hoch in die Lüfte Schaum an
Schäume sausend,
Allein wie herrlich, diesem
Sturm ersprießend,
Wölbt sich des bunten Bogens
Wechseldauer,
Bald rein gezeichnet, bald in
Luft zerfließend,
Umher verbreitend duftig-
kühle Schauer!
Der spiegelt ab das menschliche
Bestreben.
Ihm sinne nach, und du
begreifst genauer:
Am farbigen Abglanz haben
wir das Leben.

</td><td>

I watch with ever growing rapture
The waterfall that surges down the
crag,
From fall to fall it plunges, pouring
headlong,
Splitting into a thousand streams,
Roaring and foaming high into
the air.
How splendidly the many-coloured
rainbow
Arches high in ever-changing
constancy
Out of the raging torrent brightly
painted,
With mist-cool showers scattering
all around it.
In it is all mankind's endeavour
mirrored.
Reflect on it and you will grasp the
truth:
In that bright splendour we have
life itself

</td></tr>
</table>

(Goethe, Faust, II. 4715–27)

Others may find life awful, and the suffering it involves unbearable – but nevertheless hold it to be good in itself. Dylan Thomas, in a Nietzschean spirit, affirmed life with all its suffering:

> And death shall have no dominion.
> Under the windings of the sea
> They lying long shall not die windily;
> Twisting on racks when sinews give way,
> Strapped to a wheel, yet they shall not break;
> Faith in their hands shall snap in two,
> And the unicorn evils run them through;
> Split all ends up they shan't crack;
> And death shall have no dominion.[7]

---

[7] Dylan Thomas, 'And Death Shall Have No Dominion'. The title and refrain are a reference to St Paul's Epistle to the Romans, 6: 9: 'Knowing that Christ being raised from the dead dieth no more; death hath no more dominion over him.' Thomas puts the same image to non-Christian use.

Life is valuable because of the beauty of nature and the wonder of love.
Life is worthless *because* of its suffering.
Life is valuable *despite* its suffering.
Life is worthless because of our mortality.
Life is valuable only on condition of our mortality.
Life is valuable because its joys and fulfilments outweigh its suffering.
Life is valuable irrespective of the values or sufferings it involves.
Life is valuable not in itself but as a precondition of finding or creating meaning.

List 12.1   *A variety of options*

What leads other poets and philosophers to complete despair, Thomas celebrated in the face of death. Despite all horrors, life in itself is a triumph.

Immortality would deprive life of meaning   Nietzsche himself held that the very idea of immortality robs life of all meaning. In his view the finality of death, the mortality of man, is a condition of finding meaning in life:

> If one shifts the centre of gravity of life *out* of life into the 'Beyond' – into *nothingness* – one has deprived life as such of its centre of gravity. The great lie of personal immortality destroys all rationality, all naturalness of instinct – all that is salutary, all that is life-enhancing, all that holds a guarantee of the future in the instincts henceforth excites mistrust. *So to live that there is no longer any *meaning* in living: *that* now becomes the 'meaning' of life.[8]

Assuming that death entails the permanent cessation of consciousness, if death is held to be bad, it cannot be so because of what it is like. For, unlike an afterlife in Hades or in heaven, it has no hedonic or anti-hedonic properties. Rather, the evil of death, if it is an evil (natural or human), must consist in what it deprives one of. And what it deprives one of is life – not mere biological life, to be sure, but life as experience – conscious experience. Here we must distinguish the

---

[8] Nietzsche, *The Anti-Christ* (1895), trans. R. J. Hollingdale (Penguin, Harmondsworth, 1968), §43.

unqualified yea-sayers such as Goethe, who praise human life because of the beauty and wonder it contains, the dark yea-sayers like Dylan Thomas who affirm life despite its horrors, and those such as Thomas Nagel, who affirm life as such irrespective of what it is like.[9]

Third form of life affirmation The most 'formidable benefits' of human life, Nagel averred, are, at their most general, such items as perception, desire, activity, and thought. These, he held, are good irrespective of their good or bad 'content' – they are common to both the joys and the miseries of life. But they are partly constitutive of human life and it is they that make life worth living irrespective of its happiness or suffering, its achievements and its failures. Strikingly, Nagel contended that the value of experience is cumulative: the more the better, so the longer one lives the more value there is to one's life – a view that stands in striking contrast to Elina Makropulos's in Karel Čapek's *The Makropulos Case*, which will be discussed below.

It is not evident what Nagel has in mind.[10] Individual *instances* of perceiving something, wanting something, or doing or thinking something are not constitutive of human life. They may be agreeable, indifferent, or dreadful, but it is not evident that they are valuable (enjoyable, useful, beneficial, good of their kind) independently of what they are experiences of. Assuming them to be conceived to be 'conscious experiences', they cease during sleep. So is sleep, which 'knits up the ravell'd sleeve of care' not an element in what is of value in life? Or is what Nagel holds to be valuable the possession of the *powers* of perception, conation, action, and thought? Certainly it is part of the nature of man to have these powers. Are they good for man, that is, are they beneficial? The question is curious. Being able to see can't be said to be good for one; rather, being blind is deleterious and detrimental to one's welfare. It is possible for any normal human being to want indefinitely many kinds of things and things of a kind. But is that beneficial, or enjoyable, or useful? We have the power to do a multitude of things and kinds of thing. We would certainly be much worse off if we were deprived (e.g. through paralysis) of our abilities to act. But merely having these powers does not endow

---

[9] Thomas Nagel, '*Death*', *reprinted in his Mortal Questions* (Cambridge University Press, Cambridge, 1979).

[10] His position is illuminatingly criticized by Joseph Raz in his *Value, Respect and Attachment* (Cambridge University Press, Cambridge, 2001), ch. 3, and by Mikel Burley, 'Immortality and Meaning: Reflections of the Makropulos Debate', *Philosophy* 84 (2009), 529–47, to both of which I am indebted.

our lives with meaning or make them good lives. Possessing the distinctive powers of mankind does not make people happy, although losing them is a source of much misery. Rather, possession of the normal powers of humanity enables us to engage in the distinctive activities of mankind and to enjoy the distinctive hedonic passions. It is only in this sense that it can be said to be beneficial. And, of course, they also make it possible for us to undergo the distinctive sufferings under which we labour – which are normally anything but beneficial. Certainly possession of our natural powers does not make our lives meaningful – that, as we have seen, depends on our temperament, on good fortune, and on how we deploy our powers. They do not make life worth living.

Is being alive a good? All of us, Nagel averred, are fortunate to have been born, although, of course, it is not a misfortune to anyone not to have been born. On this view, life – our living – as such is good. However, it is not clear what this slogan means. After all, life (in the sense here relevant) is a variety of existence – in von Wright's sense of 'variety': it is the form of existence of organisms, by contrast with the forms of existence of inanimate material objects, of stuffs, of numbers, of values, or of sensations. Excluding for the moment the idea of continuing to exist after biological death, then death terminates life – bringing the existence of living beings to an end. So is the claim that life is valuable to a human being tantamount to the assertion that existence is a good for man? It is far from clear what *that* means. No doubt the abstract nouns are deceptive. So presumably what is meant is that existing – continued existence – is good, that is, being alive is good. To be sure, the utterance 'It is good to be alive!' is familiar. But it is not an utterance anyone would make when in great pain, when profoundly depressed, or when undergoing terrible suffering and persecution. It is an expression of joy typically made in the prime of health, at moments of exultation. But it gives no support to the general claim that being alive as such is good. Who is it good for? Not for someone suffering the pains of damnation. It is, after all, not *beneficial*. It is not conducive to *good health*. It cannot be said to be *useful*. One is not *better off* existing than not existing. The most that can be said is that this utterance can perhaps be construed as a Nietzschean exclamation of affirmation. Living is then conceived to be inherently good. But religious arguments apart, there is no reason that speaks for this conception.

A more plausible claim is the obvious one that, in so far as there is anything good about being alive or existing, it is simply as a *precondition* of what is valuable. It is no less of a precondition of what is bad, wicked, and evil, of what is harmful and painful. In itself, life – the mere fact of living – is neither good nor bad, neither beneficial nor harmful, neither healthy nor unhealthy, neither utile nor useless. To be sure, most people, most of the time, want to continue living. But that is not usually because they take the view that living in itself is good no matter how much suffering it involves. Rather it is because they have ongoing projects they wish to complete, or because they still find much happiness and joy in personal relationships, or because they hope for better times and have an inextinguishable zest for life, or because, like Raskolnikov, they wish to atone for the wrong they have done. Or, commonly, because they fear death.

5. Living for ever

Reasons for wanting to live longer Death is inevitable. It may occur at any phase in life. Death in old age is generally held to be less deplorable than death in one's prime. The old have had the full span of normal human life, have had whatever opportunities life has presented, and, if fortunate, had the chance to create opportunities for themselves by their own endeavour. They may have used those opportunities well or poorly; they may have seized them or let them slip by. They may have been lucky or hapless. Is the desire to prolong life in old age reasonable? It may be so, if there are important family obligations to discharge (e.g. to care for orphaned grandchildren), or if one has important projects upon the completion of which one has set one's heart, or if one retains a zest for life and finds pleasure in action and experience. These provide reasons for wanting to continue living.

Premature death, however, is commonly deemed to be tragic, but not because merely being alive is a good of one kind or another, let alone good in itself. The desire to live longer, the desire not to die, is fully intelligible and rational in the case of those about to be cut off in their prime, with commitments and endeavours incomplete – as in the case of a recently married spouse (Catherine Barkley in Hemingway's *A Farewell to Arms*), a parent with young children (David Copperfield's mother), or a war leader who dies at the peak of his powers before giving his people complete victory (e.g. Judas Maccabaeus, Giovanni dalle Bande Nere). The desire to live longer is fully intelligible and rational for youth – for those who have not experienced much of life, who don't want to die because they have not yet lived – they may not have had the opportunity to find themselves, to seek fulfilment in life, to experience deep love, to learn what is and what is not meaningful. Chidiock Tichborne, aged twenty-three and condemned to be hanged, drawn, and quartered for his part in the Babington Plot wrote the following lines the night before his execution in 1586:

> My prime of youth is but a frost of cares,
> My feast of joy is but a dish of pain,
> My crop of corn is but a field of tares,
> And all my good is but vain hope of gain.
> The day is gone and yet I saw no sun,
> And now I live, and now my life is done.
>
> The spring is past, and yet it hath not sprung,
> The fruit is dead, and yet the leaves are green,
> My youth is gone, and yet I am but young,
> I saw the world, and yet I was not seen,
> My thread is cut, and yet it was not spun,
> And now I live, and now my life is done.

Fear of death/fear of dying

The desire to prolong life is not the same as fear of death, and fear of death is not the same as fear of dying. A multitude of considerations may warrant the desire to prolong life within the framework of the normal span of human existence. But the desire to prolong life does not imply fear of death – great warriors, such as Alexander the Great or Charles XII of Sweden, were certainly not afraid of death, but wanted to continue their life's work. Nor does fear of dying, which is primarily a matter of fear of the suffering involved in departing life, imply fear of death. One may fear death without fearing the pains of dying, as Dr Johnson did. And one may fear the pains of dying without fearing death – as do those who seek assisted suicide.

Prolonging life indefinitely

If one believes, as Nagel does, that life in itself is good come what may, and also that the good that life furnishes is cumulative, then one may find the prospect of a *greatly* prolonged life very attractive. Indeed, one may even fantasize about immortality and think that living forever, not in a mysterious and unimaginable afterlife, but in this familiar life on earth, is desirable. If life as such is good, then it may seem that prolonging life on earth would be a good for man. And if prolonging life is good, then eternal life would seem to be best of all.

Artificial prolongation of an individual's life was dramatically depicted in Karel Čapek's play *The Makropulos Case* (and Janáček's slightly different eponymous opera). In the play an opera singer Elina Makropulos is depicted as having drunk an elixir of life 300 years earlier at the court of Rudolph II, and as having lived through the ensuing centuries increasingly blasé, joyless, weary of life, and incapable any longer of forming genuine human attachments and commitments. The story provided the theme for a prolonged philosophical debate initiated by Bernard Williams.[11] Like Elina in the drama, Williams held that such prolongation of life would exhaust one's fundamental or 'categorical' desires, the pursuit of which gives meaning to one's life. Life would therefore become intolerably boring and tedious. Others hold that fresh categorical desires would always arise, giving point and purpose to one's continued existence and ensuring a sense of fulfilment. They deny the exhaustion of goals and purposes and look forward to spending hundreds or even thousands of years engaged in studying, say, mathematics, and then thousands of years in perfecting their skills as a pianist, and so on indefinitely. Yet others hold that the opportunity cost of the things we value is a condition of their having value to us at all. A given goal or objective is valued precisely because of what has to be forgone in order to pursue it. But if one has an infinite amount of time, then nothing has to be forgone for the sake of a given goal, but only postponed. The only sacrifice made in choosing, it is argued, is in the temporal ordering of objectives pursued.

The difference between the greatly extended longevity of the Makropulos case and earthly immortality seems irrelevant for our reflections on mortality and the place of death in human life. What is important is the idea that if death is, as it is commonly held to be, an evil, then it follows that life is a good and that indefinite prolongation of life would be desirable. As Williams pointed out, the fantasy must presuppose that such immortality ensures a continued sense of one's own identity – otherwise the tale is pointless. The moot question is whether such fantasies of immortal life on earth make sense.

Assuming for a moment that they do, it is wrong to suppose that everything we value would lose all value because there would be no opportunity cost for anything save what is linked to temporal ordering.[12] For scarcities would not cease just because one became immortal or because everyone became immortal. Unique desiderata would still retain their rarity value and would still be destructible. Human

---

[11] Bernard Williams, 'The Makropulos Case: Reflections on the Tedium of Immortality', in *Problems of the Self* (Cambridge University Press, Cambridge, 1973), 82–100.

[12] As pointed out by Raz, *Value, Respect and Attachment*.

relations would still involve reciprocity: one would be able to marry or have love affairs or enjoy deep friendship with an infinite number of people in the infinite time available to one, but, of course, only if they reciprocated one's feelings. But they may not do so, no matter what the ordering of one's advances. Nor is the ordering of desiderata irrelevant to welfare or happiness. Whether one chooses to become a farmer first and a biological scientist later, or vice versa, will make a great deal of difference to one's 'career'. One can only have one's first experience of a given source of pleasure, delight, or joy once – so it matters greatly with whom one has one's first love affair. And so on.

Solitary immortality   We must distinguish between one person alone being granted such immortality, while the rest of mankind is unchanged, and all of mankind being granted immortality. Let us assume first that it is oneself alone who is granted eternal life (as it is Elina Makropulos alone who is given an elixir of life). Ageing has to be excluded,[13] so presumably one is granted immortality at a certain biological age, which never changes. How are we to envisage health? Is the immortal never ill? Is he or she never injured? If he loses a limb, does it grow back? Can the immortal be killed or will mortal injury be miraculously cured (as in the film of *Pandora and the Flying Dutchman* (1951))? How will indefinitely long life affect one's faculties? Presumably eyesight will not deteriorate with the passing of time. But what happens if an eye were lost in an accident? would one's memories extend indefinitely, or infinitely? If not, one's sense of identity would be jeopardized. If they do, is one to imagine memories going back thousands if not millions of years? Is that intelligible? While *we* share our memories with those we love, if we were immortal, after a mere human lifetime there would be no one with whom we could share memories of childhood and youth, of joint endeavour and adventures. But a shared life and

---

[13] Immortality coupled with ageing is wittily depicted by Swift in *Gulliver's Travels*, book 3, section 10. In the kingdom of Luggnagg, Struldbruggs are occasionally born. They are immortal. Gulliver, before he meets them, fantasizes about all the things he would do if he could live for ever: he would first amass riches for 200 years, then dedicate himself to the study of the arts and sciences, then become a historian, so that he would be 'a living treasury of knowledge and wisdom' and the oracle of the nation. He then discovers that, as the Struldbruggs age, they have all the follies and physical infirmities of old men or women, and are opinionative, peevish, covetous, morose, vain, talkative, incapable of friendship, envious, and racked by impotent desire. On encountering them, Gulliver concludes: 'My keen appetite for perpetuity of life was much abated. I grew heartily ashamed of the pleasing visions I had formed; and thought no tyrant could invent a death, into which I would not run with pleasure from such a life.'

experience makes up much of what we value in our lives as human beings. With us mortals, there is a psychological limit upon the loves and philiac relationships we can sustain – we cannot enjoy deep, cherished friendships with many people, and we cannot enjoy loving marriages or intimate partnerships with many either. If one were to live for ever, one would outlive all such relationships. Would one be able to form fresh philiac relationships as one's closest friends die? Could one grieve the loss of generation after generation of bosom friends? Is one to envisage an infinite number of marriages with successive people one passionately loves? Could one sustain the grief of the loss of a beloved spouse infinitely many times, and with the same depth of feeling? It is natural for most human beings to wish to have children. Does our immortal crave children? Are we to envisage him having infinitely many children from infinitely many spouses? Could a human being live through the death of them all, mourn them all, and still enjoy a fully human emotional life? Is it intelligible for human will-power and fortitude to last for ever? Would one never weary of the struggle?

The more one presses the fantasy, the less human the subject becomes. If we were to envisage creatures superficially like us who are *all* immortal, the problems multiply. The Malthusian consequences are patent. So is one to envisage these immortals restricting themselves to two children in their late twenties or early thirties in their very first marriage or love affair? Do their procreative instincts cease thereafter despite their continued copulative ones? Or do their copulative instincts disappear? Do their paternal and maternal attitudes persist through eternity or do they progressively fade away as their two children arrive at their thousandth birthday? If they enjoy an infinite number of love affairs, are these limited by strict birth control? Do those who become creative artists produce an infinite number of paintings, write an infinite number of novels, compose an infinite number of pieces of music? All with the freshness and novelty of their 'youth'? It is surely evident that we have very little idea what we are talking about. The only model we have of such beings are the gods of Olympus. Their main amusement seems to have been to interfere deleteriously in the lives of human beings. They certainly do not seem to offer us a picture of an enviable form of existence.

6. Thanatophobia – the fear of death

Mortality intrinsic to human nature   On reflection, then, there seems little attractive about immortality, and indeed little human about it. The more one tries to conceive it, the less clear it becomes and the more alien it is to creatures such as ourselves. It is striking that monotheistic beliefs in an afterlife are extraordinarily chary in their descriptions of endless life in heaven, as chary as they are generous in their ghoulish descriptions of hell. Even after reading Dante, we are little clearer about the daily routine in paradise. Indeed, it is not even clear whether there is one. (Do they sleep? Arise refreshed and ready for breakfast? Have a shower before getting dressed? Relieve themselves? Brush their teeth after breakfast? We know that they entertain themselves by watching the sufferings of the damned.) Mortality is an *intrinsic* feature of human nature. It forms an essential part of the framework of our thought, feeling, and action – something that, once learnt, we unreflectively take for granted no less than our bipedality and bilateral symmetry.

Fear of the suffering of dying

We distinguished between fear of dying and fear of death. It is not true that those whom the gods love die young. But it is surely true that those whom the gods favour, other things being equal, die quickly and painlessly. Unfortunately, those whom they favour thus are relatively few today. Most people die slow, humiliating, and sometimes painful deaths, for not all pain can be eliminated by opioids. Modern medicine, coupled with modern irrational attitudes towards death, ensure that dying is as prolonged as possible. Fear of dying in the manner most people now die is eminently reasonable – as reasonable as fear of pain, of incapacitation and dementia, of physical dependence on others for the most elemental bodily functions, and of the humiliation that this involves. It is, however, noteworthy that it needs but a reorientation of thought, a reorientation taken by Martin Eden in Jack London's eponymous novel, to view the pains of dying as the last indignity and suffering that life can inflict upon one. Fear of dying, thus conceived, is fear of the final blow life can deal one.[14]

---

[14] At the end of the novel Martin Eden, disappointed and disillusioned, commits suicide by drowning himself: 'Down, down, he swam till his arms and leg grew tired and hardly moved. He knew that he was deep. The pressure on his ear-drums was a pain, and there was a buzzing in his head. His endurance was faltering, but he compelled his arms and legs to drive him deeper until his will snapped and the air drove from his lungs in a great explosive rush. The bubbles rubbed and bounded like tiny balloons against his cheeks and eyes as they took their upward flight. Then came pain and strangulation. This hurt was not death, was the thought that oscillates through his reeling consciousness. Death did not hurt. It was life, the pangs of life, this awful, suffocating feeling; it was the last blow life could deal him' (Jack London, *Martin Eden*, ch. 46).

As noted above, fear of dying is distinct from fear of death, although in practice the two may be intermingled. Fear of death, for those who believe in an afterlife is fear of hell-fire or of some other form of horrible post-mortem existence (e.g. as shades in Hades, as described by Achilles in Homer's *Odyssey*). Fear of death, for those who do not believe in an afterlife, is fear of no longer existing (see table 12.1).

Fear of an afterlife that is worse than earthly life	Achilles in Hades: 'Glorious Odysseus: don't try to reconcile me to my dying. I'd rather serve as another man's labourer, as a poor peasant without land, and be alive on earth, than be the lord of all the lifeless dead' (Homer, *Odyssey*)
Fear of the unknown	Shakespeare: 'Who would fardels bear, / To grunt and sweat under a weary life, / But that the dread of something after death, / The undiscovered country from whose bourn / No traveller returns, puzzles the will, / And makes us rather bear those ills we have / Than fly to others we know not of? (*Hamlet*, III. 1)
Fear of painful dying	Tolstoy: '"Why have you done all this? Why have you brought me here? Why, why do you torment me so terribly? ..." He did not expect an answer and wept that there was not and could not be an answer. The pain rose again, but he did not stir, did not call out. He kept on saying to himself: "Well, go on, beat me! But what for? What have I done to You? What for?"' (*The Death of Ivan Ilyich*).
Fear of non-existence	Larkin (see below)

Table 12.1   *Forms of thanatophobia*

Epicurus endeavoured to eliminate thanatophobia by arguing that 'death is nothing to us', since death implies cessation of all conscious experience. But what has no conscious experience is nothing to us. So there is no reason for fearing death.

Fear of non-existence   This fails to console those who simply fear non-existence, such as Philip Larkin, who thought the argument specious. For, he wrote, the mind blanks at the thought of

> ...the total emptiness for ever,
> The sure extinction that we travel to
> And shall be lost in always. Not to be here,
> Not to be anywhere,
> And soon; nothing more terrible, nothing more true.
>
> This is a special way of being afraid
> No trick dispels. Religion used to try,
> That vast moth-eaten musical brocade
> Created to pretend we never die,
> And specious stuff that says *No rational being*
> *Can fear a thing it will not feel*, not seeing
> That is what we fear – no sight, no sound,
> No touch or taste or smell, nothing to think with,
> Nothing to love or link with,
> The anaesthetic from which none comes round.
>
> ('Aubade')

The poem is striking and powerful (even more so in its entirety). But its reasoning is nevertheless confused. For after death we no longer exist. Larkin confuses the seventh age of man: 'second childishness and mere oblivion, sans teeth, sans eyes, sans taste, sans everything'[15] with death. When we are dead we shall not be blind, but rather neither blind nor sighted. We shall not be deaf, but neither deaf nor able to hear. We shall not be *deprived* of touch, taste, or smell. We shall not *lack* people to love or befriend – we shall not exist.

Sleep and death　　A common attempt to give quietus to thanatophobia is as old as tombstones, namely the comparison of death to sleep and rest. It is a conceit that we still maintain in our ready-made 'RIP'. It was articulated by Socrates in the *Apology*. After his condemnation by the jury, Socrates addressed them, pointing out that death is either a migration of the soul to another place, or it is annihilation. If it is annihilation it is absence of all consciousness, just like a dreamless sleep. If so,

> death must be a marvelous gain. I suppose that if anyone were told to pick out the night on which he slept so soundly as not even to dream, and then to compare it with all other nights and days of his life, and then were told to say, after due consideration, how many better and happier days and nights than this he had spent in the course of his life – well, I think that the Great King himself, to say nothing of any private person, would find these days and nights easy to count in

---

[15] Shakespeare, *As You Like It*, II. 7.

comparison with the rest. If death is like this, then I call it gain, because the whole of time, if you look at it in this way, can be regarded as no more than one single night. (*Apology*, 40d–e)

It is debatable whether this is not subtle Socratic irony, since dreamless sleep is not a gain relative to the life of enquiry and virtue that Socrates advocated. But it may well be nothing to fear. This solace for thanatophobes has been endlessly repeated through the ages.[16] Nevertheless, the analogy is a poor one, as pointed out by Edward Greenwood in his poignant poem 'A Meditation':

> We wake from sleep, but do not wake from death
> Though death has often been conceived as sleep,
> But, when we sleep, there is the sound of breath
> In death it is just silence that we keep.
>
> And the soul's journey was just make believe,
> As children animate a wooden toy,
> And no amount of grief in those who grieve
>
> Will resurrection ever turn to joy.
> This is a truth few are inclined to speak,
> A thought most minds are glad to set aside,
> The power of contemplation is too weak
>
> To grasp the nothingness of those who've died.
> In memory we seem to hear their voices,
> And see them live again in some past scene,
> And for a moment then our heart rejoices
>
> Over what now can't be, but once has been.
> Until we realize we can't embrace
> Those we have lost and whom we loved the best,
> For what is nowhere now in time or space
> Can neither be in motion, nor at rest.

Sleep is an interruption, death a termination. Sleep may be good or poor, depending on how peaceful or disturbed and restless it is, but death (as opposed to dying) is neither. Good sleep refreshes one, for

---

[16] Thus John Donne in his 'Holy Sonnets':

> Death be not proud, though some have called thee
> Mighty and dreadful, for, thou art not so …
> From rest and sleep, which but thy pictures be,
> Much pleasure, then from thee, much more must flow.

one wakens well rested and invigorated – but there is no waking from death. A good sleep is one of the pleasures of life, experienced as one falls comfortably asleep and as one wakens invigorated. But death is no pleasure. We feel sleepy, feel the need for sleep – we do not *feel* 'deathy', and do not feel a need for death as opposed to a desire or longing for it. One may long for sleep and one may long for death, but the former longing is an appetite – it is recurrent, is satiated when satisfied, and after a while recurs. The longing for death is no appetite, and to die is not to satisfy an appetite. The criteria for sleeping are that one breathes and is behaviourally responsive to sensory stimuli even if one does not awaken. The dead neither breathe nor respond behaviourally to such stimuli. Despite our tombstones and platitudes, the dead are not at peace, nor do they rest. What is true is that the deceased's suffering is over.

Lucretius: no harm to one after death   Lucretius, Epicurus' poetic follower, elaborated further reasons for not fearing death, given that death is cessation of existence. His arguments have been very influential, both in antiquity (Cicero, Seneca, Plutarch) and after the rediscovery of his work in the Renaissance (Montaigne, Dryden, Hume, Schopenhauer). Lucretius distinguished more clearly than Epicurus between two different arguments. The first is that there is nothing to fear in death, since what we fear are evils that may befall us. Evil can befall only a subject of experience. But, after our death, we shall no longer exist. So no evil can befall us in death. This has been criticized on the grounds that, just as one can be harmed in one's life by events of which one is ignorant (e.g. false and demeaning rumours circulating about one, one's spouse's infidelity), so too one can be harmed in death. Because of false rumours in life, one may be debarred from a wide range of possibilities (of advancement, of access to desirable social circles, etc.). Is that not a harm done to one, even if one remains ignorant of it? So one may be harmed, even though one neither suffers nor is cognizant of the harm. Is it not so with post-mortem harm? No. It is in one's interest to have those possibilities of action available, whether one knows of them or not, for *were* one to wish to avail oneself of them, one *would be* debarred. So it makes sense to speak of being harmed. Not so in death. For there are no opportunities for action for the dead, and they cannot be harmed by the foreclosing of opportunities. To be sure, after one's death one's great works (e.g. of architecture) may be destroyed, one's reputation may be besmirched, one's corpse may be maltreated. We may acknowledge the sadness, horror, or tragedy of such events without acknowledging that the dead person is harmed. After all, he does not *suffer* – it is his reputation that suffers; it is his corpse that gets maltreated, not him; it is his interests that are damaged, not his person. Of course, his family may be harmed by this and suffer greatly.

Lucretius: prenatal and post-mortem non-existence   The second, more renowned Lucretian argument concerns the alleged symmetry between prenatal and post-mortem events. We are not distressed that we have been unable to experience events prior to our birth: there is nothing *horrible* about not having witnessed Socrates discoursing, Mozart playing, or Callas singing. These occurrences took place without us. But that fact does not strike *fear* in our hearts. We shall not witness events after our death. These future post-mortem events are on a par with antenatal events. There is nothing *fearful* about them, nor any reason for gloom and lamentation.

Lucretius exaggerated the symmetry between prenatal and post-mortem events. First, we commonly have a multitude of interests in the future that are not matched by interests in the past. We have an interest in the fortunes of those we love, especially in the fate of our children. We have an interest in the fate of the projects for which we laboured and that we valued. We have a concern for the future of the community and country to which we belong, as well as a concern for the future of humanity and of civilization, of nature and the natural world.[17] Secondly, the past is written in marble, but the future is open. We wish to live to see how things with which we are currently familiar will eventuate. We naturally find it regrettable that someone died without knowing the turns of immediate history – that so many heroic figures (e.g. Sophie Scholl and the members of the White Rose) died without knowing that their cause would triumph, or, much more modestly, that many composers, artists, and writers died without knowing that their work would receive its due recognition (e.g. Bizet, van Gogh, Gerard Manley Hopkins). Thirdly, Lucretius does not do justice to the extent to which those who die premature deaths may rationally desire to live longer.

Solace of death   When all is said and done, death is often tragic for those who die prematurely. It is a reason for sorrow and even despair for those who die without completing a task they

---

[17] Just how much the future of mankind matters to us is well brought out by Samuel Scheffler in his *Death and the Afterlife*. The value of much that we do and of much that we cherish is dependent on the presupposition of the continued existence of mankind.

had undertaken, completion of which meant much to them. It is a source of regret for those who still take pleasure and find meaning in life. It is, however, no evil for those whose suffering is intolerable, who are overwhelmingly weary of life, and who have lived out their lives. At some stage in the natural life cycle of mankind, one's physical powers decline, one's mental powers begin to wither, one's role is over, and it is time to leave the stage. Then death should be nothing to fear.

> From too much love of living,
>   From hope and fear set free,
> We thank with brief thanksgiving
>   Whatever gods may be
> That no life lives for ever;
> That dead men rise up never;
> That even the weariest river
>   Winds somewhere safe to sea.
>   (Swinburne, 'Proserpine')

Appendix 1:
On Animal Beliefs and Animal Morality

1. Animal morality

The appearance of morality It is argued by some ethologists and primatologists that chimpanzees and bonobos (and perhaps other animals too) have a morality.[1] It is well known that they are able to respond differentially to the emotions of others. They often display sympathy and emotional contagion, as well as consolatory behaviour in response to another's suffering after a fight. They appear to have something akin to the moral emotion of resentment, but not of indignation. They have limited cooperative abilities that are exhibited in performing joint tasks which presuppose coordinated action if either of the partners is to benefit. They manifest reciprocity for favours done. They form cooperative bonds to mutual advantage, and often behave in ways that are aimed to benefit others. They have regular forms of conflict resolution and reconciliation. And they display group animosity towards outsiders. In order for all this to be possible, it is generally argued, chimpanzees must be able to have beliefs and thoughts, to recognize the mental attributes of their fellow chimpanzees, and to ascribe thoughts, beliefs, and desires to them; they must also be capable of reasoning and acting for reasons; and

---

[1] This appendix adds to and draws together numerous fragmentary observations on animal powers in this and the previous three volumes.

they must be self-conscious. It is correct that these psychological powers and susceptibilities are indeed necessary for a creature to have a morality, a moral conscience, and knowledge of good and evil. What is debatable is whether chimpanzees, bonobos, or any other non-human animal has these or has them to an adequate degree.

The contribution of evolution All the claims about the powers and pronenesses of chimpanzees are advanced in defence of a form of naturalism that nicely accords with the arguments of Chapter 1. On this view, the roots of human morality are the product of evolutionary processes already visible among social animals with established social hierarchies. Far from social animals being innately self-interested, with no concern for the welfare of other individual animals in their social group, let alone of the group as a whole, evolution has evidently selected for cooperation.[2] It is patent that attentive, caring, and protective female mammals whose young mature very slowly are likely to be more successful in breeding than neglectful ones (irrespective of whether the animal is a social creature or not). It is obvious that aggressive social animals can survive in groups only if patterns of conflict resolution are instinctive (as they patently are among, say, wolves). Since such social animals need social bonding if the group is to survive, it is evident that evolution will select for forms of reciprocity: positive to encourage cooperation and negative to discourage deviance and free riding. If a group is to survive in conditions of limited resources, its members must be capable of cooperating to ward off aggressors. It is not surprising that ferocious chimpanzee intergroup hostility also serves to enhance in-group solidarity.

Morality is not the upshot of rational self-interest What all this shows is that morality is not the product of social contracts between self-interested individuals. Moreover, the model of contractual agreements between such creatures is not adequate to justify the moral practices of creatures who are not essentially self-interested. So one style of rendering morality intelligible to ourselves, namely social contract theories such as Hobbes's, is profoundly

---

[2] It is doubtful whether it makes sense to claim that non-social animals are self-interested in a full-blown sense of the term. For not only do they have to raise their young, but, much more importantly, because the very notion of self-interest presupposes the ability to be motivated by long-term goals, to evaluate possible outcomes of actions, and to estimate relative probabilities. Such abilities are the prerogative of language users. Other animals, social or solitary, often pursue their interest, but are not, in this extended sense, even self-interested.

misleading. So too are its modern cousins, which attempt to derive fundamental moral and political principles from the requirements of self-interested rational human beings living together in society.[3] Nor are moral principles and moral virtues to be explained in game-theoretic terms as the mutually most advantageous compromises between rational self-interested parties. Human beings are not such creatures, and everything we know of our nature, of the nature of our closest animal relative the chimpanzee, and of evolution speaks against the presuppositions of such enterprises. For it cannot be argued that, if moral and political principles can be derived from the selfish nature of rational self-interested creatures living together for mutual advantage, they will also be suitable to render intelligible the moralities of creatures such as human beings, who are not wholly self-interested or selfish, who are endowed with limited sympathy, who possess potentialities for friendship and love, who are susceptible to religious belief and political ideology, and who have an innate proneness to tribalism and group solidarity.

Animal social behaviour and morality categorially different

That there is a large degree of behavioural and psychological continuity between the rest of the animal kingdom and human beings is not rationally disputable. That a wide range of the characteristic features presupposed for the possibility of moral value, moral virtue, a moral conscience, and knowledge of good and evil is to be found in non-human animals is well established. That human beings, with rational powers of thought and reasoning, recognizing a variety of forms of goodness on the one hand, and a variety of virtues on the other, evolved gradually over a very long period of time from humanoid apes cannot reasonably be denied. The development of a moral being was doubtless gradual rather than precipitous,[4] at least as gradual as the emergence of a language with logical connectives and quantifiers. Nevertheless, the logical differences between a pre-moral, non-language-using animal and a rational, moral being (i.e. an animal with knowledge of good and evil, with the ability to feel remorse for acknowledged wrong-doing, and bearing responsibility for what it does or omits doing) are categorial. Here in the conceptual

---

[3] As David Gauthier tried to do in *Practical Reasoning* (Clarendon Press, Oxford, 1963) and *Morals by Agreement* (Oxford University Press, Oxford, 1986).

[4] There is no such thing as a morality gene that might spontaneously begin to generate animals with moral powers and sensibilities.

domain, I shall argue (in opposition to Frans de Waal[5]), we have a transition from quantity to quality. What makes for such a categorial difference is a language sufficiently evolved to endow the speaker with abilities of intellect and will, as well as the susceptibility to novel or extended passions that flow from this.

2. Animal thinking, animal thoughts, and animal memory

Animal thinking   The limits of possible thought are the limits of the possible expression of thought – assuming no internal or external constraints (*IP*, 393–7). What this means is that the thinkings that it makes sense to ascribe to a being are limited to those the creature can (but need not) express in its behaviour by speech or action (both expressive and intentional). This is a conceptual truth, not an empirical one. It describes an internal relation between possibilities, not between actualities. It is a grammatical proposition: an expression of a rule for the use of the word 'thought' and its cognates that is elucidated by a description of what counts as thinking (*IP*, ch. 10). In particular, no psychological ('inner') events that may take place in a non-language-using creature can constitute thinking anything. An animal that cannot speak cannot say anything to itself and mean what it says. Even if it can conjure up a mental image (which is moot),[6] no mental image by itself is tantamount to a thought nor is having a mental image tantamount by itself to thinking (*IP*, 406–16). Does this mean that animals cannot think? No; that would be precipitate. Animals can know certain things to be so, how to do certain things, where certain things are, and how to look for them. For such knowledge, in the case of an animal, is manifest in the exercise of appropriate non-linguistic abilities. If a dog sees a cat in the garden, it knows a cat is in the garden (sight being a cognitive faculty) and may

---

[5] Frans de Waal has, over the past few decades, provided systematic observations of chimpanzee and bonobo behaviour to back up Darwin's great insights, which rested largely on anecdotal evidence. While accepting de Waal's naturalism, I shall disagree with him on important conceptual issues and on his interpretations of some of his experimental results.

[6] It is moot precisely because what makes a mental image of something the image of what it is consists in what the agent does and says. But nothing in the behavioural repertoire of a non-language-using animal could constitute criteria for its having or having had a mental image of something.

chase it accordingly. If it hears its master's footsteps outside, it may rush to the door wagging its tail, knowing its master is about to come in (hearing too is a cognitive faculty). A dog may know how to open a door with its paw, or how to find its way home. It may know where it buried a bone, and where its food bowl is kept. In each case, the abilities constitutive of knowing are exhibited in doing. Thinking or believing are default positions when knowledge fails. If a dog chases a cat up a tree, and then stands below the tree barking excitedly despite the fact that the cat has long since jumped to another tree and escaped, we may describe the dog as thinking the cat is in the tree. For knowing something to be so is factive – it entails that things are so, whereas thinking or believing things to be so are not.

The limits of animal thought   In the absence of mastery of a language, the horizon of animal thought, that is of what an animal can intelligibly be said to think, is very close. Lacking devices for temporal reference, animals cannot intelligibly be said to think anything involving reference to yesterday or yesteryear, or to tomorrow and the years to come. For there is nothing in its behavioural repertoire that could manifest its so thinking and that could therefore count as a criterion for its so thinking. Lacking mastery of the use of quantifiers, animals cannot be described as thinking anything general. For nothing in their behavioural repertoire could manifest generalization. Indeed, animals cannot intelligibly be said to think thoughts at all. For thoughts (Fregean propositions) are bearers of truth and falsehood. Obviously a non-language-using animal cannot be said to think something to be true, since it lacks the concept of truth. Moreover, the kinds of things that are true are assertions, declarations, statements, rumours, suspicions, hunches, and guesses, in short *sayables*, that an animal can neither utter nor understand. An animal can be said to think or believe something to be so, but not to think or believe something (some thought, proposition, or other sayable) to be true (or false). A dog may *think* or *believe* that the cat is now up the tree at the foot of which it is barking, but it cannot be said to *have the thought* that the cat is in the tree. Such a thought cannot be said *to occur* to the dog, or *to cross its mind*. It makes no sense to suppose that an animal can *have a thought* that it *registers* for future use, that it can *call to mind* for reconsideration, and that it can *reflect on* and *revise*. These are prerogatives of language users. The idea is given powerful poetic expression in Edwin Muir's poem 'The Animals':

> They do not live in the world,
> Are not in time and space.
> From birth to death hurled
> No word do they have, not one
> To plant a foot upon
> Were never in any place.
>
> For with names the world was called
> Out of empty air,
> With names was built and walled,
> Line and circle and square,
> Dust and emerald,
> Snatched from deceiving death
> By the articulating breath.
>
> But these have never trod
> Twice the familiar track,
> Never never turned back
> Into the memoried day.
> All is new and near
> In the unchanging Here
> Of the fifth great day of God,
> That shall remain the same,
> Never shall pass away.
> On the sixth day we came.

Animal thinking redescribes but doesn't explain   Saying that an animal is acting thus-and-so because it thinks things to be so is not to explain the animal's teleological behaviour in terms of its reasons. To say of the dog that it is barking up a tree because it thinks that there is a cat there is not to say that the dog's reason for barking up the tree is that it thinks this. For it makes no sense to describe the dog as reflecting on a reason antecedently to its action, as weighing this reason against reasons for desisting, let alone as acknowledging this as its reason for doing what it did *ex post actu*. An animal may recognize alternatives, and opt for one rather than another. But it makes no sense to suppose non-language-using creatures to deliberate, weigh reasons and relative advantages, and come to a conclusion on the basis of reasoning. For that they would need a wide range of powers that it makes no sense to ascribe to non-language-users. There is a great deal that a creature must be able to do and to learn before it can be said to deliberate or weigh reasons.

Animal teleology falls short of rationality It would be equally mistaken to suppose that because animals engage in purposive behaviour – do such-and-such in order to attain a certain goal – their reason for doing such-and-such is therefore to attain that goal. That is true of language users but not of animals that lack a language. It is true of language users because they can do and comprehend a multitude of other things. For a language user can view its goal as warranting its action, as providing a justification or excuse for doing what it does in pursuit of that goal. But an animal cannot bear in mind the goal of attaining such-and-such for further or future consideration; it cannot be said to know what reason it had had for pursuing its goal after the event. It cannot take responsibility for its action as done for that purpose. Nor can it regret having acted in pursuit of that goal, wish it had not done so, or resolve never to do so again, although if it has a mishap it may well never do so again.

Reasoning Conceding that animals can think something to be so is not to concede that animals can think, that is reason or engage in reasoning. To reason is to deduce consequences from assumptions, infer explanations from data, derive conclusions from evidence. To come to a conclusion on the basis of reasons presupposes the ability to weigh different considerations for and against something's being so, to judge that this proposition is more likely to be true than that one because ..., or that this course of action is better than that one, because .... A creature can be said to think practically, to reason, only if it can reflect that things' being so warrants acting thus-and-so. It must be capable of conceiving of one proposition or of a set of propositions as *warranting* a conclusion, or *realize* that a given array of propositions *justify* a certain conclusion. A creature can be said to act *for a reason* only if it *can reason*, can think. Lacking a language, animals cannot be said to do things for reasons or to act on reasons, although, of course, they have purposes in doing what they do; they act for the sake of a goal; and they display intelligence and adaptability in their pursuit of their goals.

3. Counter-arguments and their rebuttal

Animal communication These negative claims are highly controversial and have been controverted. They have been challenged in a multitude of ways. It has been argued that animal communication is sufficiently sophisticated to count as primitive language. And if animals can master a primitive language, then surely they can be said to have primitive or 'proto'-concepts? Similarly, animals can surely know and believe things – they can know *that* things are so, and believe *that* things are so. So they have propositional knowledge. But surely propositional knowledge requires concept possession and hence the ability to reason. Intelligent animals such as apes, corvidae, and dolphins can solve quite complicated problems and puzzles. Surely that means that they can think and reason, draw inferences from premises or assumptions, and come to conclusions on the basis of evidence. And that is proved by the story of Chrysippus's dog.[7] Or so it seems.

There is no doubt that many kinds of animals employ quite complex arrays of calls, cries, and signs in their natural environment. There is, however, no evidence to suggest that such calls are tantamount to speech or that the sounds emitted amount to uses of a language.[8] Reiteration of calls until a response is achieved is not evidence for linguistic utterances – newborn chicks' frenetic squeaking until they are given food is not a request. There is no evidence for understanding or not understanding, and none for corrections of misunderstanding or for explanations of meaning. Nor is there evidence for grammatical form and structure, or for licit and illicit combinatorial possibilities.

Animal concepts/ proto-concepts
The moot question is whether the animal systems of communication with which we are familiar are sufficiently robust to warrant ascribing mastery of concepts. Those who attribute such powers to animals are prone to

---

[7] The dog was chasing a rabbit by its scent. Coming to trifurcation of the path, it lost the scent and sniffed the first two paths, whereupon without any further sniffing it raced down the third path. This is alleged to show reasoning by disjunction elimination (cf. Hans-Johann Glock, 'Animal Minds', in Brian P. McLaughlin (ed.), *Philosophy: Mind* (Macmillan, London, 2017), 97) or Bayesian probabilistic reasoning (Michael Rescorla, 'Chrysippus's Dog as a Case Study in Non-linguistic Cognition', in Robert W. Lurz (ed.), *The Philosophy of Animal Minds* (Cambridge University Press, Cambridge, 2009)). As pointed out (*IP*, 101 n.), Sir Kenelm Digby solved this problem four centuries ago in his *Two Treatises* (Paris, 1644). Hunting dogs sniff the ground when they lose the scent. As soon as they regain the scent, they lift their heads and race off in pursuit of their prey, since the scent is so strong that they need not sniff the ground to keep it (for Digby's words, see *IP*, 102; I owe the reference to Hanoch Ben-Yami). But, even if we disregard this, a creature cannot be said to engage in reasoning unless it manifests this ability on many different occasions, in a diversity of contexts (not just sniffing paths), and in a variety of inference patterns – not just a disjunctive syllogism. (A child that can move a chess king from one square to an adjacent one has not mastered a rule of chess.)

[8] There is evidence to suggest that bonobos in captivity can master the primitive use of some 400 icons, given very intense training. Whether these impressive skills are tantamount to mastery of the relevant concepts is not altogether clear. If they are, then to be sure, we may ascribe concept possession and rudimentary mastery of language to these bonobos.

do so on the grounds that animals have discriminatory abilities and can differentiate between something that possesses a given property and something that does not. Or, a little more demandingly, that animals classify things as having or lacking some property in a deliberate and considered manner, and correct mistakes. These considerations do not warrant ascribing concept possession to non-language-using animals. Nor do they warrant the weaker position of ascribing to them 'proto-concepts'. For, in so far as that is a stipulation, it is an unnecessary and potentially confusing one. It is unnecessary because everything we are warranted in saying about animals' recognitional and discriminatory powers can be said without this innovation. It is potentially confusing because it seems, by its very terminology, to give additional support to the claims that animals can have thoughts and beliefs, possess 'propositional knowledge', and can reason from premises to conclusion. But, whatever apparent support there is for these theses, nothing further is added by claiming that prelinguistic discriminatory powers are evidence for possession of proto-concepts. The most that can be said is that some animals possess abilities that are, for humans, the preconditions of concept possession.

Concepts and concept possession   Concepts, as has been argued (*IP*, 128–9, 379–87) can be said to be abstractions from the use of a word in a given language. That assertion, though true, does not go deep – but it suffices to emphasize the usefulness for certain purposes of detaching one's arguments from a given natural language in order to address features common to many different languages and to reflections by different thinkers who speak different languages. The question 'What is a concept?' is as little useful as the question 'What is a number?' – both induce misleading answers, since the nominals invite the introduction of unhelpful nominata (e.g. 'an abstract object', 'a class of classes'). The pivotal question is not 'What is a concept?' but rather 'What is it to possess a concept?'. To possess a concept is to have mastered the use of a word or phrase that expresses that concept, so it is a complex ability. One must be able to employ the word in the performance of acts of speech and to explain in rudimentary ways what one means by what one said.[9] One must be able to respond intelligently to others who employ the word in speech-acts in a language one knows. For some concepts, recognitional abilities

---

[9] I deliberately disregard here the case of people who had full mastery of a language and then suffered from crippling paralysis, as in cases of locked-in syndrome. In this context that is irrelevant (just as the existence of three-legged dogs is irrelevant to the truth of the statement that it is of the nature of dogs to be four-legged).

are necessary conditions for their possession, for example concepts of perceptual qualities (colour, sound, taste, smell, and tactile quality concepts). For others they are not, for example concepts of distance and speed, of temperature or barometric pressure. But for no concept is a mere recognitional ability sufficient for the ascription of concept possession. For, in order to possess, for example, the concept of red, one must not only be able to differentiate red things from things that are not red, but one must also know that red is a colour (so if something is red, then it has a colour), that red is darker than pink (so that if one thing is red and another thing is pink then the first is darker in colour than the second). One must grasp that red can be ascribed to visibilia, but not to sounds, tastes, smells, or tactile qualities. One must understand that if something is red, then it is either extended or it is a dot or flash of light. And so on. In short, one must grasp the grammar of a word that expresses the concept in question and know its place in the web of words (or conceptual scheme). Of course, abilities can be possessed to a greater or lesser degree – so mastery of the use of a concept word may be greater or lesser. How much of the grammar one has to have mastered in order to be said to possess the relevant concept will doubtless vary from case to case and context to context (all of us know what 'a metre' or 'a kilogram' means, but few of us know the whole complex social practice in which its use is embedded (the role of the standard kilogram, for example[10])). But a mere recognitional ability dissociated from grasp of any other concepts in the conceptual field in which the putative concept is located is not sufficient for the possession of any concept. Here one may indeed say that 'light dawns gradually over the whole'.

Identifying, classifying, considering

It might be thought that all that needs to be added to a recognitional ability for it to be tantamount to concept possession is that the animal be able to identify things with a given property and differentiate them from things that lack it, and to sort and classify them accordingly. But one must not confuse sorting with classifying. Classifying is indeed the exercise of a concept subsumption ability, but sorting is not. Does an animal qualify as classifying items if it can not only sort them but also *correct misallocations*? If all the latter amounts to is

---

[10] The standard metre is now obsolete, having been replaced by the distance that light travels in a vacuum in 1/299792458ths of a second. To be sure, this is no more familiar to the vast majority of speakers who correctly use the expression 'a metre' than the role of the standard metre rod in Paris, and later in Sèvres, was. But the standard kilogram soldiers on, even though few know about it.

that when, for example, the animal sorts red and blue pieces of wool into piles and notices a blue piece of wool in the red pile, it removes it and puts it into the blue pile, then that does not betoken possession of the concepts of red and blue. All it shows is a recognitional ability, an associated ability to sort, and an ability to spot items that do not belong to the pile in which they are located. The ideas that animals submitted to such tests recognize their own mistakes or realize that they have made a mistake have no warrant. Could one not hold that the animal not only sorts items, but classifies them as such-and-such, as opposed to classifying them as so-and-so? For then one might argue that reallocation betokens the animal's realization that it has made a mistake inasmuch as it has classified an F as a G? Indeed, may one not hold that such an animal judges *a* to be an F or judges it to be a G? That in turn implies that the animal's sorting involves consideration, deliberation, and choice in judging between different apprehended options.

One cannot licitly help oneself to the concepts of consideration, deliberation, and apprehension of options so quickly. Of course, a chimpanzee may engage in problem-solving, scratch its head (or its armpit), grimace, exhibit frustration, and manifest pleasure at achieving a reward for solving a problem. But it takes much more than that to consider, deliberate, and examine the available options. What determines whether it makes sense to say of a creature that it is examining available options, considering, or deliberating is not merely what it does in the circumstances (scratches its head, pulls a face, looks at the sky, shuts its eyes) but what it can do and does in other circumstances. Behaviour is a criterion only in appropriate surroundings. Behaviour that counts as a criterion for examining options, considering, or deliberating in our case does not count as a criterion in the case of a non-language-using animal. The background is missing.

A creature that can consider whether *a* is an F must also be able to check whether *a* is an F and weigh reasons for and against its being an F. In order for an animal to deliberate about what to do, or what option to choose, it must be able to consider possibilities and to examine their implications or outcomes in advance of acting. None of this is possible – makes sense – for an animal that has not mastered a language. I shall say more on this matter below when I consider the intelligibility of non-language-using animals reasoning from premises to conclusions and drawing conclusions from evidence.

Animals and propositional knowledge

It has been argued that animals not only know how to do things, when to do things, and what to do in certain circumstances, but that they also know or believe that things are thus-and-so.[11] To know how to open a box is to know that one must lift its lid. Similarly, animals can not only perceive things (e.g. boxes, balls, other animals) and events (e.g. an explosion), but also perceive *that*. For they can perceive *that* things are so (e.g. that the bone is on the table, that the bone is in its bowl) and, since seeing is believing, animals can believe that things are as it perceives them to be. For only on that supposition can one explain that the dog does not take the bone when it is on the table (for which it gets punished), but that it does take it when it is in its bowl (which is licit). For, if it merely perceived the bone, the table, and the bowl its behaviour could not be explained. So it seems that one must conclude that an animal can be said to perceive *that* things are so, to *perceive facts*,[12] and to *perceive information*. This is an important result, since it is often argued that animals cannot have propositional knowledge ('knowledge that') on the grounds that propositional knowledge presupposes concept possession, and animals are held not to possess concepts.

This argument is wrong. There is no such thing as perceiving facts, although one may learn many facts by perceiving whatever one perceives. Facts are not located in space or time, but what we see, hear, smell or taste is. Our talk of facts is, for the most part, merely a way of presenting knowables. Similarly, one cannot perceive information, only acquire information from what one perceives (and from what one is told, when someone imparts information to one). It is not necessary to describe the dog as seeing the fact that the bone is on the table and as seeing the information that the bone is in its bowl in order to explain why it did not eat the bone on the table but did eat it from its bowl. To be sure, the dog saw the bone on the table, and later saw the bone in its bowl. But surely, when the dog sees the bone in its bowl, it knows that the bone is in its bowl? Again, one does not have to describe things thus. There is nothing amiss in describing the dog as knowing the bone is in the bowl – the that-clause is unnecessary. That does not mean that it is improper, only that in some philosophical contexts it can be misleading and is thus best avoided. But, one may respond, surely when asked what the dog knows or believes, the answer is that it knows (or believes) that things are so. Yes, of course. For this is the canonical form of the answer to this kind of wh-question,

---

[11] This is argued by Glock, 'Animal Minds'.

[12] See Fred Dretske, 'Seeing, Believing, and Knowing', in Robert Schwartz (ed.), *Perception* (Blackwell, Oxford, 2004), 268–86.

as it is to the demand for a report of speech: 'What did he say?' 'He said that things are so', even though what he said was 'Things are so'. Reporting belief in the form of reported speech is perfectly innocuous as long as one does not take it, in philosophical contexts, to imply that non-language-using animals possess 'propositional knowledge' (i.e. know a proposition to be true) or that they have 'propositional attitudes' (i.e. attitudes towards propositions).[13] In trying to unravel the subtle differences between animal and human cognitive and cogitative abilities, one is well advised to grant that animals can intelligibly be said to know, to think, and to believe things to be so, but to be wary of granting that animals possess 'propositional knowledge' or 'knowledge that', or have thoughts and beliefs. For these nominals are misleading, and introducing them (quite unnecessarily) in the context of an examination of the cognitive and cogitative powers of animals is a first step on a very slippery slope down which one need not go. The slippery slope is patent when it is argued that (i) since the dog saw that the bone is on the table, it must possess the concepts of a bone, a table, and of one thing being on another; or (ii) since the dog thinks, believes, or knows that the bone is on the table, or later that it is in the bowl, thinking thus requires no concepts. That may be a tempting idea, but it is altogether mistaken, as becomes evident when one reflects what such ascriptions of concept possession and cognitive powers can possibly amount to in the case of a dog. There is nothing it can do that it could not do without concept possession and such cognitive powers, and there is nothing in the dog's behaviour that requires explanation in these terms.

Animal memory Just as animal knowledge is constrained by the forms of knowledge-exhibiting behaviour available to it, and hence is limited by the lack of a language, so too there are constraints on animal memory. An animal can intelligibly be said to remember what it can express or manifest in its behaviour. A dog may

---

[13] That knowing, believing, hoping, suspecting, expecting that *p* are *not* attitudes towards propositions was argued in detail in *IP*, 168–70. To be sure, there are such things as attitudes towards propositional-like items, such as approving, endorsing, dismissing, ridiculing, being amused by utterances, declarations, statements, and announcements. What is striking to note is that these genuine propositional attitudes *do not admit that-clauses* (see Bede Rundle, *Mind in Action* (Clarendon Press, Oxford, 1997), 53). One can take up an attitude towards a statement, declaration, or announcement, but not towards *that things are so*. I cannot ridicule that the government is strong and stable, or endorse that someone is qualified but only the announcement that the government is strong and stable, or the declaration that someone is qualified.

remember where it buried a bone, but not that it buried it there yesterday. An elephant's recognitional powers are excellent and long-lasting. If a keeper inflicted pain on the elephant, and then left the zoo and returned ten years later, the elephant may recognize him straight-away and fling a lump of dung angrily at him. Does it remember that the keeper hurt it many years ago? No, it merely associates the keeper with being hurt. It makes no sense to claim that the elephant remembers that the keeper inflicted pain on it ten years ago, or even just a long time ago. For what could constitute the remembering? Certainly not a memory experience (see *IP*, ch. 9). For, in so far as there is any such thing as a memory experience of a past event, it presupposes memory and cannot explain it. For what could give a current experience reference to the past event which it is said to be of? But the sight of the cruel keeper may arouse in the elephant a fear of being hurt, since the keeper has hurt it in the past. An animal can remember how to do all manner of things that it has learnt to do in the past. But it makes no sense to say that it remembers learning those things in the past. These logical or conceptual constraints on animal knowledge and memory have ramifying consequences for ethologists and their descriptions and explanations of animal behaviour. That is evident in the following remark:

> Of all existing examples of reciprocal altruism in non-human animals, the exchange of food for grooming in chimpanzees appears to be the most cognitively advanced. Our data strongly suggest a memory-based mechanism … Apart from memory of past events, we need to postulate that the memory of a received service, such as grooming, triggered a positive attitude toward the individual who offered this service, a psychological mechanism known in humans as 'gratitude'.[14]

There can be no memory of past events in a non-language-using animal, only present associations consequent upon past events. It makes no sense to say that a chimpanzee remembers being groomed by another chimp last week, yesterday, or merely in the past. Rather, it may associate a chimpanzee with being groomed by it. For *that* memory is something that can be expressed in the chimpanzee's behaviour, for example in displays of affection, and in presenting itself to the other chimp in postures that invite grooming. We do not have gratitude here, although

---

[14] Frans de Waal, *Primates and Philosophers: How Morality Evolved* (Princeton University Press, Princeton, 2006), 44.

we do have its primal source and seed, which will sprout only in linguistic water. Similarly questionable reasoning is patent in an explanation of the behaviour of scrub jays which, having pilfered another's cache in the past, will then secretly recache their own food if they notice another jay watching them when they originally cached it. For the explanation advanced was that scrub jays 'relate information about their previous experience as a pilferer to the possibility of future stealing by another individual, and modify their recovery strategy appropriately'.[15] Here too the behaviour of the scrub jay is over-intellectualized. For there is nothing in a scrub jay's behavioural repertoire that would warrant ascribing to it current knowledge of previous experience as opposed to current associations caused by previous experience. Non-language-using animals cannot be said to possess information, but only wh-knowledge, which can be exhibited in behaviour. Similarly, the supposition that a scrub jay might entertain the possibility of another conspecific stealing its cache makes no sense. But it may well have learnt to associate another scrub jay's watching it cache food, with unease regarding the locus of the food.

Animal reasoning We can now turn to animal reasoning. We have disposed of Chrysippus's dog above (n. 7): not only does it not give an example of animal reasoning, but it shows the absence of any reasoning at all in the example it advances. Putting aside the question of whether it makes sense to ascribe concept possession to an animal that has not mastered a language, we now need to examine the question of whether it make sense to say of such an animal that it can reason from premises to conclusion or from evidence to what it supports. To reason is to deduce consequences from assumptions, infer explanations from data, or derive conclusions from evidence. A creature can be said to think, to reason, only if it can conceive of one proposition or of a set of propositions as *warranting* a conclusion. To think or reason is not literally to engage in an activity of 'making transitions' in thought from premises to conclusion, or of 'moving' in thought from one proposition to another that follows from it. These are but metaphors for apprehending a formal *because* (as an Aristotelian might put it). It is to apprehend a normative nexus, not a causal one. It is to see such-and-such premises as *justifying* such-and-such a conclusion. To do that, of course, is to

---

[15] Nathan J. Emery, Joanna M. Dally, and Nicola S. Clayton, 'Western Scrub-Jays (Aphelocoma californica) Use Cognitive Strategies to Protect their Caches from Thieving Conspecifics', *Animal Cognition* 7 (2004), 37–43.

apprehend the premise or premises as *reasons* for the conclusion. A creature can be said to act *for a reason* only if it *can reason*, can think. Non-language-using animals cannot be said to do things for reasons or to act on reasons, although, of course, they have purposes in what they do and they display intelligence and adaptability in their pursuit of goals (*HNCF*, ch. 7). For a chimpanzee to manifest pleasure at having found a solution to some practical problem it confronted does not show that any reasoning was involved, only that its goal was attained.

Responsibility, answerability   This is why non-language-using animals are not responsible for what they do. To be answerable for one's deeds, as rational human beings are, is to be able to answer the question 'Why did you do it?'. To answer that question is to give one's reason for doing what one did (or to offer an excuse, e.g. 'I was drunk'). To give one's reason for doing what one did is, other things being equal, to take responsibility for what one did *as done for that reason*. Animals have two-way powers and can do what they want, act for the sake of a goal (e.g. to satisfy their appetites or curiosity) but, as Aristotelians understood full well, they have no will. In that sense they may be said to be wantons, that is in bondage to their wants.[16]

Swift and slow neural pathways for decisions   Some psychologists, ethologists, and neuro-scientists do not try to elevate chimpanzees to the heights of reason, but to derogate human beings to the depths of non-rational spontaneity.[17] So, for example, it is now commonly argued among neuroscientists that there are two separate neural pathways involved in decision-making, one swift and 'too rapid to be mediated by cognition and self-reflection often assumed by moral philosophers', the other slow enough to enable intention formation and reasoned action.[18] The claim originates in the work of Robert Zajonc,[19] who argued that 'affect and cognition are

---

[16] I do not mean that they lack second-order desires (as was suggested by Harry Frankfurt). I doubt whether there is any such thing as a second-order desire as opposed to wishing that one had a desire or wishing that one wanted something. To be sure, this too is something we humans can do, but other animals cannot.

[17] Jonathan Haidt, 'The Emotional Dog and its Rational Tail', *Psychological Review* 108 (2001), 814–34, is an example of such an argument. He holds that human beings do not act for moral motives but out of social intuition. The moral reasons given *ex post actu* are, he suggests, mere rationalizations.

[18] De Waal, *Primates and Philosophers*, 169.

[19] In particular his paper 'Feeling and Thinking: Preferences Need no Inferences', *American Psychologist*, 35 (1980), 151–75.

under the control of separate and partially independent systems that can influence each other in a variety of ways, and that both constitute independent sources of affect', and suggested, in the spirit of William James, that 'affect often precedes cognition'. The idea of separate neural pathways in the brain controlling rapid reaction as opposed to slower thoughtful response was advanced by the research of Joshua Greene and his colleagues.[20] De Waal suggests that

> We may therefore be less intentionally altruistic than we like to think. While we are capable of intentional altruism, we should be open to the possibility that much of the time we arrive at such behaviour through rapid-fire psychological processes similar to those of the chimpanzee reaching out to comfort another or sharing food with a beggar. Our vaunted rationality is partly illusory.[21]

This conclusion could be drawn only as a consequence of not understanding what it is to act for a reason, to do something intentionally, and to be responsible for what one does. What neural pathways are involved is irrelevant to the rationality or reasonableness of what one does in such contexts. Moreover, transmissions along neural pathways are not psychological processes. Affect does not precede cognition (*TP*, 97–103), since one's knowledge in such cases is not an antecedent state but an ability triggered by what one perceives (*IP*, ch. 4). This needs clarification.

Most of what we do intentionally is not preceded by antecedent planning or intention formation. One is offered a drink at a party and reaches out to take a glass – no reflection is involved: one simply picks it up. If asked why one did so, one may answer that one likes a pre-prandial drink or that one is partial to a glass of champagne. One leaves a room and turns the lights off as one leaves – no thought of doing so crossed one's mind: one simply flipped the switch. One's purpose was indeed to switch the light off. One switched it off intentionally – not accidentally or by mistake, and not automatically either. For one can say why one flipped the switch (one might admit that one did it automatically if one had been told to leave the light on). One will take responsibility for switching off the light, and if there is someone in the

---

[20] Joshua Greene, 'Emotion and Cognition in Moral Judgement: Evidence from Neuro-imaging', in Jean-Pierre Changeux, Antonio R. Damasio, Wolf Singer, and Yves Christen (eds), *Neurobiology of Human Values* (Springer, Berlin, 2005), 57–66.
[21] De Waal, *Primates and Philosophers*, 169.

room whom one did not notice, one will apologize for doing so. Similarly, if one is walking down the street with a friend, and one's friend slips and is about to fall into the path of a car, one will, as quick as a flash, reach out a restraining hand and save her from disaster. (Affect does not 'precede cognition', since one can see the impending danger.) One did exactly what was morally called for. Moreover, one did it intentionally, not accidentally. If asked after the event why one caught one's friend's arm, one would reply, 'I did it to save her from falling in front of the car.' (Indeed, one might rightly be congratulated for one's quick thinking, even though no thought crossed one's mind. To do something without reflection or deliberation does not imply doing it mindlessly.)

To this one may reply: if all this is correct, then surely it is correct in the case of the chimpanzee too? For it too may act for the benefit of its companion, even though it does not say so after the action. But there is a crucial difference: a human being *can* explain and justify what he did *ex post actu*, but a chimpanzee cannot. And, precisely because we can answer the question 'Why did you do that?', we are answerable for what we do. And because we can answer that challenge we are also normally morally responsible for what we do intentionally, even though no antecedent intention was formed.

4. Animal knowledge of other animals' minds

Theory-of-mind theory   As noted in Chapter 10, David Premack (a psychologist who studied philosophy under Wilfred Sellars and Herbert Feigl) and Guy Woodruff (a primatologist) published an unfortunately seminal paper in 1978 entitled 'Does the Chimpanzee Have a Theory of Mind?' The logical positivist legacy was evident in the supposition that ascribing psychological attributes to others requires the formation of a theory. Premack and Woodruff declared that 'an individual has a theory of mind if he imputes mental states to himself and others. A system of inferences of this kind is properly viewed as a theory because such states are not directly observable and the system can be used to make predictions about the behaviour of others.' To be sure, the idea of imputing was stretched to include merely responding differentially to the manifest mood and temper of their conspecifics, since animals cannot ascribe attributes to a subject. But they can respond differentially to their conspecifics being angry or affectionate, looking at something or watching it, or wanting and trying to get something. This was wrongly taken to amount to recognition. What Premack and Woodruff effectively argued was that, in order for one chimpanzee to recognize that another chimpanzee is angry, friendly, or frightened, it must be able to impute such mental attributes to itself, and it must have a theory of mind. It is only because chimpanzees have a theory of mind that they can engage in *mind-reading*. Mind-reading is necessary because anger, feelings of friendliness, and fear are not observable to others, who can perceive only 'mere' behaviour. So a chimpanzee, in virtue of having a theory of mind, can identify the mental states of other chimpanzees, which it infers from the observed behaviour. By 1983 this misconceived idea was extended from chimpanzees to human beings.[22] Attributing a theory of mind to children after the age of four or five, and explaining their failures in the false-belief test[23] was influenced by Jean Piaget's so called genetic epistemology (or 'cognitive ontogeny'). Piaget claimed that children prior to the age of four or five are unable to understand conservation of quantity (in particular the sameness of the quantity of water poured from a narrow tall glass into a wide but shallow glass).[24] The comparable idea of a theory of mind and of mind-reading caught on and came to play a major role among cognitive scientists, psychologists, and cognitive neuroscientists (*TP*, 380–1). It is now a standard part of the explanatory apparatus of professional psychologists.

---

[22] Heinz Wimmer and Josef Perner, 'Beliefs about Beliefs: Representation and Constraining Function of Wrong Beliefs in Young Children's Understanding of Deception', *Cognition* 13 (1983), 103–28.

[23] A child, let us say Jill, is presented with the following scenario: a pair of dolls Sally and Anne are on the table together with two small boxes and a sweet. Anne puts the sweet under box A. Sally is called to help Mummy in the kitchen. While she is away, Anne removes the sweet from beneath box A and puts it under box B. When Sally returns, Jill is asked where Sally will look for the sweet. If Jill replies 'under box B', this allegedly shows that Jill does not yet have a theory of mind according to which she can impute beliefs to Sally, namely that Sally will believe that the sweet is under box A, where it was before Sally left the room. The account has been challenged by the Ben-Yamis' slightly different experiment, since the received experiment takes it for granted that the child fully understands the difference between 'will' and 'should' (Ben-Yami et al., 'What Does the So-Called False Belief Task Actually Check?').

[24] Piaget's claim concerning children's ability to grasp conservation of quantity was challenged by Peter Bryant. As noted, the false-belief test is currently under challenge. In both cases a linguistic misunderstanding (of the use of 'more' for quantities of liquid, and the use of 'will' and 'should' in the false-belief test) was interpreted by psychologists as a cognitive limitation.

Qualms about possessing   To the untutored this may indeed seem
a theory-of-mind   bizarre. First, it is unintelligible to ascribe
theory building to non-language-users. Neither
chimpanzees nor small children could conceivably construct a theory
of any kind. For a theory, in its most minimal sense, is a conjecture or
hypothesis. In a weightier sense it involves a universal generalization
and a set of initial conditions that conjunctively warrant an inference
(there are other forms a theory may take). But it is at the very least
unclear what it would mean for a non-language-using animal to conjecture or hypothesize anything, let alone construct a hypthetico-
deductive theory. Secondly, it is equally misguided to suppose that
adult human beings have a theory of mind. To be able to use the
psychological vocabulary is no more to have mastered a theory of
mind than mastering the use of the vocabulary of furniture is mastering a theory of furniture. This egregious idea was reinforced by the
habit (encouraged by American philosophers such as Sellars and
Quine) of speaking of our ordinary third-person psychological statements as constituting a 'folk psychology' and of our psychological
vocabulary as a theoretical vocabulary.[25] If that is nonsense, then
holding that chimpanzees construct a theory of chimpanzee minds is
nonsense on stilts. Thirdly, Premack and Woodruff, Sellars and Feigl,
all assume that it is intelligible for a creature to know or think of its
own mental attributes or mental states antecedently to grasping the
grounds for the ascription of mental attributes and mental states to
others. But, as we have seen again and again in our discussions in this
and the previous three volumes, that is unintelligible – it would require
the possibility of a 'private language'. In the case of the chimpanzee
things are even more dire, since there is no such thing as a chimpanzee
ascribing a mental state *to itself* or *thinking of itself* that it sees the
alpha male, that it is frightened, that it is inclined to run away, or that
it wants to have some of the food in the cage. Finally, it is misguided
to suppose that psychological attributes are concealed behind bodily
behaviour. That is to succumb to the pernicious 'inner/outer' picture
of dualism. But there is nothing concealed when an animal writhes on
the floor squealing in acute pain; nothing is hidden when an animal

---

[25] For detailed criticism, see P. M. S. Hacker, 'Eliminative Materialism', in Severin
Schroeder (ed.), *Wittgenstein and Contemporary Philosophy of Mind* (Palgrave Macmillan,
Basingstoke, 2001), and 'Passing by the Naturalistic Turn: On Quine's Cul-de-Sac', in
Wittgenstein: Comparisons and Context (Oxford University Press, Oxford, 2013),
187–208.

roars in rage with hackles raised, teeth displayed, and face contorted; and nothing is unobservable about an animal's fear when it grovels whimpering with terror in the face of threat or danger. For one chimpanzee to see that another chimpanzee is threatening it, it needs no theory any more than if someone were to pull a knife on us we would need a theory to interpret the behaviour as a threat.

Self-consciousness and mirror recognition   Many primatologists claim that animals are no less self-conscious creatures than we are, and that they have a 'sense of self'.[26] This far-fetched claim rests on animals' ability to recognize their reflection in a mirror (something that chimpanzees, gorillas, orang-utans, rhesus monkeys, Asian elephants, and bottle-nosed dolphins can do). This allegation rests on the thought that, since they can recognize themselves in a mirror, they can recognize their selves in a mirror. Nothing could be more mistaken. These animals can recognize their faces in a mirror (also their hands and legs, and in the case of male vervet monkeys their bright blue scrotum, in which they take much more interest than in their face). They can be said to be recognizing themselves, but they cannot be said to be recognizing their selves. For the self is not an object that might be recognized by perception or by anything else. The ability to recognize one's reflection in a mirror does not mean that one knows that one is a self (among other things, because one isn't – one is a human being, a chimpanzee, gorilla, elephant, or dolphin). Recognizing one's face in a mirror does not imply that one has a sense of self (or even of 'selfhood') any more than recognizing one's foot in a mirror implies that one has a sense of feet (let alone of 'foothood'). 'Self-consciousness' has a variety of meanings, but the ability to recognize one's face in a mirror is not one of them.[27] To be self-conscious is not to be conscious of a self and to have the power of self-consciousness is not to have a recognitional ability, but a cogitative one. Someone is self-conscious in the relevant sense if they have the power and inclination to reflect critically on their motives and attitudes, on their past deeds and thoughts, on their relationships with others, and so forth. This has absolutely nothing to do with the ability to recognize reflections in a mirror. Moreover, it is patently beyond the powers of a chimpanzee.

---

[26] See e.g. Gordon Gallup, 'Chimpanzees: Self-recognition', *Science* 167 (1970), 86–7, who pioneered this research.

[27] For explanation of the varieties of self-consciousness, see *IP*, ch. 1.8.

Belief evaluation   It has been suggested that many animals are capable of consciously evaluating their own beliefs. In an experiment in the 1990s a dolphin was trained to distinguish auditory pitches of different frequencies and rewarded for a correct response (questionably described as 'judging correctly'). It was also trained to signal 'opt out', and then received a lesser reward. As the experimenter moved the higher and lower pitches closer, making the task more difficult, the dolphin would sometimes, after hesitation, choose to opt out. 'I'm not sure of what I'm hearing', it seemed to be saying. Macaque monkeys, it was claimed, behave similarly and 'spontaneously generalized the opt out behaviour to memory tasks'. 'I'm not sure I'm remembering right', they seemed to be thinking. 'Since the macaques appeared to reflect for several seconds before deciding to opt out, there is good reason to believe that their judgements were arrived at consciously.' Thomas Nagel replied to this: 'I am open to the possibility that some nonhuman animals evaluate their own beliefs and motives. It is an empirical question.'[28] But that is precisely what it is not. Being unsure whether one is remembering correctly presupposes possession of the concept of memory, hence too of the concepts of past, present, and future, as well as mastery of the use of the first-person pronoun in thought – and if in thought, then in speech. But only language users possess these abilities. Hesitation before action shows the animal to be unsure, but not unsure in judgement, rather unsure in action, that is it hesitates – which is where we started!

Animal morality   Finally, we should examine the claims for animal, and in particular chimpanzee, morality. No one can dispute that chimpanzees have powers of sympathy in that they manifest emotional contagion, can identify and recognize individual members of their tribe, and can bond with a select few (forming groups). They sometimes share their food with another chimp, and show their favourable disposition towards another by grooming him or her. They exhibit consolatory behaviour towards another who has come out worse in a fight with a third chimp. They engage in cooperative activities. However, none of this shows that they have a morality or a moral conscience, or that they follow moral norms or principles. In recent years de Waal and others have produced evidence that seems to

---

[28] Jim Holt, 'Known Unknowns', *New York Review of Books* (21 March 2019), https://www.nybooks.com/articles/2019/04/18/rational-animals-known-unknowns (accessed 17 January 2020); exchange of letters between Thomas Nagel and Holt, *New York Review of Books* 66:7 (18 April 2019).

suggest that chimpanzees have a sense of fairness and justice. Very crudely, the evidence is given by a set of experiments done on pairs of chimpanzees or capuchin monkeys in adjacent cages. When they perform a desired sequence of actions, they are rewarded by a piece of cucumber (a little-valued edible) or by a grape (a highly valued edible). Each can see what the other gets for the same 'work'. Rewarding the two animals differentially for performance of the same task resulted in the one who got the lesser reward (a slice of cucumber rather than a grape) throwing it angrily at the experimenter. This was taken to show that the chimp felt it was unfair to be given a lesser reward than the other chimp. This is curious, since it seems to show no more than that the chimp wanted a grape too. De Waal noted that, when one of the monkeys was given a grape and the other was not, no such angry behaviour ensued. It was only manifest when the monkeys received 'money tokens' for task performance, which were then exchanged for rewards of food. Then, if there was differential reward, the 'underprivileged' monkey became angry. But there is no reason to suppose that the anger manifested a sense of justice, as opposed to disappointed expectations (perhaps a primitive form of 'food-envy'). De Waal concedes that what would perhaps approximate to showing a sense of justice and fairness would be if the favoured chimpanzee were to share its tasty grape with the chimp who received only a miserable slice of cucumber. But even this would show no more than rudiments of fairness: a willingness to share. For this to constitute a sense of justice, there would have to be some kind of behaviour that expressed the principle of fairness that fortuitous goods should be divided equally unless there is a reason that militates against equity, or that equal labour requires equal reward. But there is none. What is actually displayed, de Waal suggests,[29] is egocentric inequity aversion in cooperative circumstances, equitable reward being more likely to maintain cooperation. But this falls far short of principles of justice or fairness that require allocentric inequity aversion.

Darwin, who, to be sure, was no more a Darwinist than Marx was a Marxist, may have the last word: 'A moral being is one who is capable of comparing his past and future actions or motives, and of approving or disapproving of them. We have no reason to suppose that any of the lower animals have this capacity.'[30]

---

[29] See Sarah F. Brosnan and Frans B. M. de Waal, 'Evolution of Responses to (Un)fairness', *Science* 346:6207 (2014), 1251776.

[30] Darwin, *The Descent of Man* (Murray, London, 1871), vol. 1, ch. 3, p. 88.

5. Animal emotions

It should be obvious that the only emotions that can intelligibly be attributed to animals are emotions the ascription of which is not logically dependent upon the animal's mastery of a language. So too it should be evident that the only intentional objects of animal emotions are those that do not presuppose concept possession on behalf of the animal. For the behavioural repertoire of animals is limited to non-linguistic behaviour. The forms animal emotions may take are accordingly curtailed.

Animal fear and anger   Given that almost all animals are subject to threat and danger from predators or competitors, concepts of fear have a ready application in all cases where fear-expressive behaviour in the face of actual or expected threat is exhibited. Lack of fear can sensibly be ascribed only to an animal that can exhibit fear. Mere avoidance behaviour in the absence of any visible somatic expression provides a warrant for fear attribution only in a very thin sense (if at all). For avoidance behaviour may be exhibited by primitive animals without any manifestation of fear. Animals that do manifest fear may do so without any avoidance behaviour – as when they fight against the odds with a predator threatening their young or intruding on their territory. In general, it makes sense to say of such animals that they experience anxiety, trepidation, fear, panic, and terror. However, animals cannot be said to fear death, since they have neither a concept nor a conception of death (see Chapter 12). What they fear is the danger of a predator, and being caught by it.

Given that the quest for food is so arduous in nature (especially, though not only, for predators), given territorial instincts, given competition for mates, and given the need to protect offspring, displays of preparation for attack or aggressive defence give ample space for concepts of anger to get a grip. For animals may bare their teeth, adopt postures preparatory for attack, growl or hiss, raise their hair, move their tail in distinctive ways, and so forth. So it is fully comprehensible that this kind of behaviour in such contexts provides criteria for ascribing to such animals irritation, annoyance, anger, and rage, as well as feelings of hostility.

Grief or distress?   Where offspring are parent dependent, forms of bonding and preferential treatment are obviously selected for, both between parents and between parents and offspring. The extent to which these engender emotions, such as love and affection, or sorrow and grief (as opposed to mere passing distress) in the event of death is, in the case of many kinds of animal, unclear. It may well be that a maternal chimpanzee will not allow the corpse of its off-spring to be taken from her for a day or two, and that she displays evident signs of distress. But, once she has abandoned her dead baby, is her apathetic behaviour a criterion of persistent grief and sorrow? Can she be said to know that her baby has died? Can she be said to remember that it has died? Without having a tensed language or some other linguistic means of referring to the past, what could such knowledge amount to? However, the persistence of apathetic behaviour and loss of appetite can be fully explained by the elevated gluticorticoid levels that persist for about a month after the death of a near relative.[31]

Love or affection? [26] That bonded animals can exhibit mutual affection is undeniable. They take evident pleasure in each other's company, care for, and often take care of each other. Some such kinds of animals mate for life, and may display deep distress at the loss of their partner. Does this warrant the ascription of love to them? Love, despite its evident biological roots, is a conceptually far more complex emotion than anything available to non-language-users (*TP*, ch. 10). It involves having reasons for behaving in certain ways towards the creature loved; it provides motives for action; it commonly implies not only desire, but a desire to be desired; it is bound up with a desire to please, a desire for one's love to be recognized, and a recognized need to be cared for. This complex battery of feelings requires correspondingly complex thoughts and reflections. The affections and bonding of animals provide the soil from which love may grow, but only after being fertilized by mastery of a language. Amicable relations between animals rest on shared gratifying experiences and on smell, taste (licking), and sight, but not on recollections of shared experiences.

Cognitive emotions What then of the more cognitive and doxastic emotions? We have no hesitation in saying that a dog expects its master, when, on hearing his footsteps outside, it wags its tail, jumps to its feet, and barks excitedly. But, as Wittgenstein noted, it makes no sense to say that the dog now expects its master to return home tomorrow.[32] For there is nothing in its behavioural repertoire that could manifest its expecting this here and now. To do so, the dog would have to have mastered a language with means of temporal reference. Similarly, a cat or dog may expect to be fed when

---

[31] Gluticorticoidal hormones such as cortisol are strongly correlated with stress.
[32] Wittgenstein, *Philosophical Investigations*, §650.

it sees or hears its master opening a bag of pet food. It may want to be fed. But it cannot hope to be fed, for to hope that something will happen requires one to know or believe that it is neither impossible nor certain that it will. But it makes no sense to ascribe such modal knowledge or beliefs to a non-language-user. Animals can evidently display interest and curiosity, and may react to their explorations and discoveries with surprise (which, according to our classificatory scheme, is not an emotion but an agitation) – but not with wonder, let alone awe.

Human emotions   Let us now turn to distinctively human emotions. Although our most primitive emotional responses are rooted in our animal nature (fear, anger, affection), the objects of emotion and the kinds of emotion that are possible for a language-using creature far outstrip anything attributable to other animals. There is nothing surprising about this, for our emotions, unlike those of other animals, are woven on a doxastic warp with a dense cogitative weft, and run through with threads of judgement and evaluation. Many emotions presuppose concept possession. Only a language user can feel regret, guilt, or remorse – for such emotions presuppose possession of the concept of the past coupled with the memory of having performed a certain act in a certain context, and a negative evaluative judgement on what was done. In the case of feeling regret, knowledge of alternative courses of action that were available is presupposed, as well as knowledge of the opportunity cost of what one did. In the case of guilt and remorse, knowledge or belief that what one has done was intentional (reckless or negligent) wrong-doing is presupposed. In the case of remorse, the agent must know or believe that he has wronged another, wish that he hadn't, and desire to make good the wrong done. No doubt the seeds of such emotions are latent in animal behaviour – in the submissive behaviour of animals to an alpha male or female in a family, pack, or tribe when the animal has deviated from a regularity in the order of allocation of benefits, for example in feeding or grooming. But such emotions themselves are not patent in animal experience. There is a long journey from the cringing dog that has made a mess on the carpet to Raskolnikov.

Only a language user can feel pride or humility, for these are, for the most part, emotions of self-evaluation and judgement. Of course, we speak of the proud bearing of a race horse, or the proud look of a lion. But here we are speaking of no more than the evident health and vitality of a beautiful animal in the prime of life. It is one thing to be endowed with admirable natural qualities, quite another to know that one is so endowed, to know that one's qualities surpass those of others, and to take pleasure in the fact that they do.

> Some glory in their birth, some in their skill,
> Some in their wealth, some in their bodies' force,
> Some in their garments, though new-fangled ill;
> Some in their hawks and hounds, some in their horse;
> And every humour hath his adjunct pleasure,
> Wherein it finds a joy above the rest.
>
> (Shakespeare, Sonnet 91)

And if pride cannot lie within the emotional horizon of animals, humility cannot either. For just as reasonable pride consists in knowing and taking pleasure in one's merits, reasonable humility consists in knowing one's limitations.[33] Both consist in knowing one's place and the place of one's merits in the order of one's society and one's times, and sometimes also in the order of history and of human possibility. A social animal may know its place in the hierarchy of the group, knowledge that consists in its cringing before its superiors and in its deferred eating at the common prey.

Rudiments of human emotion in animals

The rudiments of such emotions as envy, jealousy, and humiliation (after defeat) can be found in the lives of animals, but not the full-blown feelings or the refined vices. An animal that slinks away after a lost fight, looking abject, is showing the rudiments of feeling humiliated, but only the rudiments: it cannot be distressed at what others will think of it; its self-respect cannot literally be damaged, since there is no such thing in animals, but only its faint rudiments; and it cannot resolve to get its revenge. One animal may patently want some good it perceives another animal to have – in particular food or mating precedence. To display this much in behaviour is perfectly possible for many animals. But that falls far short of the emotion of envy, let alone of its cousin *Schadenfreude*. For envy commonly involves not only wanting what another has, but also wanting that the other should not have it and believing that one's standing in the eyes of others, if not also in one's own eyes, is diminished by the other having the envied good – thus Saul's envy of David:

And the women answered one another as they played, and said, Saul has slain his thousands, and David his ten thousands. And Saul was very wroth and the saying displeased him; and he said, They have ascribed unto David ten thousands, and to me they have ascribed but

---

[33] See Arnold Isenberg, 'Natural Pride and Natural Shame', repr. in Amélie Rorty (ed.), *Explaining Emotions* (University of California Press, Berkeley, 1980), 355–84.

thousands: and what can he have more but this kingdom? And Saul eyed David from that day and forward. (1 Samuel 18: 7–9)

The cognitive and cogitative components of envy do not lie within the compass of anything a non-language-user might be said to know and think. It is, however, striking that something akin to spite is exhibited by the vervet monkey, which will occasionally destroy the cache of food of another monkey without eating or stealing any of it. Similarly, the roots of sexual jealousy can be found among the beasts, but not the flowering branches and poisonous fruits that we find among mankind. An alpha male will indeed guard his females. Evolutionary theory has an explanation for the selection of such behavioural tendencies. If another male attempts to copulate with one of the females, the alpha male may be willing to fight ferociously to prevent the insemination of any member of his harem. One might well see in this behaviour the roots of the human emotion of jealousy. But jealousy is a far more complex emotion, and its manifestations in human behaviour far more variegated and thought laden. The history of someone's jealousy is typically both long and complex, involving a multiplicity of other inter-woven emotions. It may involve a sense of vengeful outrage (Browning's 'The Last Duchess'), murderous anger at the unfaithful partner (Othello, Pozdnyshev in *The Kreutzer Sonata*), resentment and hurt at the actual or imaginary infidelity (Trevelyan in *He Knew He was Right*). It may take the form of a desire for vengeance against the person who has, or is believed to have, won the sexual or emotional favours of the partner (the attempted murder of Cassio in Othello). The infidelity of the part-ner may breed insecurity, anxiety, and fear of permanent loss (Swann). Jealousy may ramify into feelings of sexual or emotional inadequacy, loss of self-esteem, and feelings of helplessness.

Aesthetic emotions   Finally, it should be equally obvious that animals can have no aesthetic emotions. Of course, many of them have the sensibility required for an appreciation of beauty. Some animals are responsive to colours. They have colour preferences that commonly play a crucial role in sexual selection. Strong colours in birds indicate good health and absence of parasites, and that provides an obvious evolutionary explanation of why females are attracted to males with such plumage. Some creatures seem to take pleasure in the arrangements of coloured objects – as bower birds apparently do in their endeavour to attain a mate. Many animals have auditory powers that are greater than ours. They may be responsive to rhythmic thumps, and show a preference for certain sound patterns. But the fact that cows produce more milk to the sound of Beethoven as opposed to the sounds of Beatles does not show musical discrimination. It is not given to animals (nor to all humans) to feel awe in the face of the sublime – in art or in a landscape, or to feel reverence before great works of art. Aesthetic understanding and appreciation lies beyond their meagre powers, for that demands not only sensibility, but also intellect – an awareness, as Kant argued, of universality in the particular, and of objectivity in the subjective appreciation of works of art.

Appendix 2:
Diabology: Satan, Lucifer, and the Devil in Western Thought

Human responsibility for evil-doing The endeavour to explain both natural and human evil is as old as serious reflection. There are two contradictory movements of thought. The simplest and sanest is to conceive of ourselves as having by nature the potentiality for both good and evil. We are free beings with the power to do or to refrain from doing as we please, and we possess knowledge of good and evil. So there is no difficulty of principle in identifying the sources of human wickednesses and evil. They lie in us. This was the conception of the authors of the Old Testament and the dominant view of later rabbinical Judaism (and was later embraced by Pelagians (see below)). Loving obedience to the commandments of God ensures a life of righteousness. But most human beings will, from time to time, succumb to the immoral impulses of *yetser ha'ra* (the evil inclination in man): 'For there is not a righteous man upon earth that doeth good and sinneth not' (Ecclesiastes 7: 20). But *we* are responsible for our wicked or evil deeds, not a devil or evil spirit that possesses us or with whom we might have made a pact (like Theophilos of Cilicia[1]).

---

[1] The story of Theophilos is a sixth-century legend of a Cilician monk who, enraged at being stripped of his offices and dignities by a new bishop, approached a Jewish magician, who took him to the Devil, with whom Theophilos made a pact. When the Devil came to take his soul, Theophilos repented and appealed to the Virgin Mary to

'Satan' in the Hebrew Bible　　　To be sure, the word *satan* occurs in the Hebrew Bible. However, it is not the name of an evil spirit, akin to 'Angra Mainyu' and later 'Ahriman' in Zoroastrianism, or 'Satan', 'The Devil', or 'Lucifer' in Christian doctrine. In fact, the word has been extensively mistranslated and misunderstood in the Septuagint, Vulgate, and the King James translation of the Old Testament. The Hebrew word *satan* can occur as a common noun, taking a definite article, or as a verb. As a common noun it means opposer, adversary, antagonist, accuser, or challenger, signifying a function rather than a character. The verb *l'satan* means to obstruct, thwart, oppose, or challenge. The expression in either form may be applied to supernatural beings or to human beings. The first occurrence of *satan* is in Numbers 22: 22 in connection with a supernatural being, namely an angel of God, with a drawn sword, who confronts Balaam on his ass to obstruct him (*l'satan lo*).[2] In Job 1: 6–12, the satan (*hasatan*, with a definite article) is the challenger, who is said to be one of the 'sons of God' (*bnei elohim*). There is no suggestion whatsoever that he is an incarnation of evil. In Zechariah 3: 1–2 the satan is simply an accuser, who is duly rebuked by God. All other occurrences of the expression and its cognates are applications to human beings. In 1 Samuel 29: 4 David becomes a satan, that is an adversary, to the Philistines in war. Hadad the Edomite (1 Kings 11: 14) and Rezon (1 Kings 11: 23) are said to be satans, that is adversaries, of David and Solomon.

Diabological sources of human evil　　　The more common alternative is to explain the wickedness and evil in man by reference to external sources of evil. The most popular form of

---

save him. The Virgin then descended to hell, seized the contract from the Devil, and ascended to Heaven to appeal to God, who forgave the sinner. Theophilos accordingly changed his abode for the better. This legend was translated into Latin in the nineth century, and the tale of Theophilos was widely circulated throughout Europe thereafter. It also appeared in medieval illuminated manuscripts such as the thirteenth-century Lambeth Apocalypse. The story furthered the cult of the Virgin, and encouraged yet more anti-Semitism. It initiated the idea of making pacts with the Devil (hence is the remote source of the Dr Faustus tale of the sixteenth century). Pacts, it was believed, could be discovered by tests (immersion, holding red hot irons, interrogation under torture). Just how far this canard had reached by the fifteenth century is evident in Chaucer's Prioress's Tale. The idea of a pact with the devil became a cornerstone for the demonization of minorities (e.g. Jews, gypsies, Muslims) and their torture and murder. Burning at the stake became the means of purging the Christian community from the contamination.

[2] The King James translation has 'the angel of the Lord stood in the way for an adversary against him'. This is mistaken, as *l'satan* is a verb.

explanation is diabological. Zoroastrianism (Mazdayasna) was a weighted dualist Persian religion (one of the two powers being greater than the other), acknowledging Ahura Mazda (the Wise Lord Creator) as the principle of goodness, and Angra Mainyu (later personified as Ahriman) as the principle of evil and the spirit of chaos and destruction. Ahura Mazda (like Jehovah) created the universe out of chaos, imposing upon it order and truth. Zoroastrianism insisted that human beings are free to choose between good and evil, right and wrong. It influenced the much later development of Mandaeism (which was more emphatically dualist, the two cosmic powers being equipolent), Christianity,[3] Gnosticism,[4] and subsequently Manichaeism.[5] Christianity externalized evil by reference to the Devil (Satan, Lucifer). In this it was influenced by Zoroastrianism, Gnosticism, and pseudo-epigraphic apocalyptic writings that were prevalent during the period of the Second Temple in the first centuries BC and AD (including the books of Enoch and of Jubilees) which acknowledge a supernatural spirit of evil. These were sacred texts of various little-known Jewish and non-Jewish sects of this period. The book of Revelation, which patently belongs to this tradition, was included in the canon of Christianity. Other apocryphal books such as the Apocalypse of Paul (third century) and the Gospel of Nicodemus (fourth century) were not. They revel in the horrors of hell and the eternal punishment of the damned, and contributed greatly to the Christian conception of hell for the next millennium and more. This is absent from the Hebrew Bible.

Mistranslations of 'satan';
invention of Lucifer

The Hebrew *satan* was incorrectly translated in the Septuagint as *diabolos*, which meant 'slanderer' rather than 'adversary' or

---

[3] Ahriman, in later Zoroastrianism, is roughly the equivalent of the Devil in Christianity. Unlike Christianity, the strife is directly between Ahura Mazda and Ahriman (as opposed to being at God's arm's length, as it were, between Christ and Satan), with mankind playing an important role in this cosmic strife. Zoroastrianism embraced resurrection, the end of days in which good finally triumphs over evil, the last judgement, heaven and hell, and an ultimate saviour, Saoshyant, born of a virgin impregnated by the seed of Zarathustra while bathing in a lake.

[4] Gnosticism is a group of second-century mystical and apocalyptic doctrines. Despite their Judaeo-Christian roots in Alexandria, they were ferociously anti-Semitic. Gershom Scholem, the great scholar of Jewish mysticism, described Gnosticism as 'the greatest case of metaphysical antisemitism'.

[5] Manichaeism is a religious doctrine founded by Mani (216–276) during the Sassanian empire. It flourished from the third to the twelfth centuries, and for a time competed with Christianity (Augustine was a Manichaean before he became a Christian). It is an ancestor of Bogomil and Cathar beliefs.

'obstructer', which led to the Latin *diabolis*, which is the source of 'devil' and 'Teufel' alike. *Satan* was mistakenly assumed to be a proper name of the master evil spirit, and the snake, in the story of Adam and Eve, was, without any textual warrant, taken to be the Devil as tempter.[6] 'Lucifer', allegedly the name of a fallen angel, is derived from a misreading of Isaiah 14: 12, in which Isaiah prophesies the fall of the king of Babylon and compares his predicted fall to the fall of a rebellious god in Babylonian mythology. The tale is perhaps an eastern variant of the Phaeton myth, or of the allocation of the realms of the cosmos, the underworld (akin to Hades) going to Ben Shakhar, who had rebelled against the High God, El. Its appearance in Isaiah is merely as an analogy for the prophesied fall of the king of Babylon. The Hebrew runs *heilel, Ben Shakhar. Ben Shakhar* literally means 'the son of dawn', perhaps a reference to Hesperus, the morning star. It is obscure what *heilel* means. There is no particular reason for supposing it to be a noun at all, but the translators who edited the Septuagint translated the word as *heōsphoros*, meaning literally 'bringer of dawn' (which makes it a redundant redescription of the son of dawn). The Vulgate opted for 'Lucifer' ('bringer of light'). However, there seem no grounds for this. It may well be a verb meaning 'howl', that is 'How you have fallen from the sky! Howl, Son of Dawn. You have been hacked down to earth, defeater of nations', since the god is being hurled down from the heavens. Lucifer, thus misconstrued and misconceived, was fitted into Christian diabology as the prince of the angels who rebelled against God, and was cast out of heaven.

The Devil in the New Testament The New Testament is replete with numerous references to the Devil and Satan: *diabolos* occurs thirty-four times, *satanas* thirty-six times in the Greek, and many other names and descriptions are also used.[7] In due course, between the third and sixth centuries, the Devil, Satan, Lucifer, and the Evil One were gradually united in a single personage.[8] The devil was conceived to be an external personification of evil, constantly

---

[6] The first to suggest that the snake in Genesis chapter 3 is the Devil appears to be Justin Martyr (AD 100–165) in his *Dialogue with Trypho*, chs 45 and 79. This suggestion was supported by Tertullian (AD 155–240) and in due course became orthodox doctrine.

[7] The same figure is also referred to as Abaddon, Apollyon, Beelzebub, Belial, Beliar, Leviathan, the Evil One, the Father of Lies, the Great Dragon, the Ruler of the Power of the Air, the Tempter, Angel of the Bottomless Pit – all of which show the varied sources in the post-Zoroastrian dualist mythologies of Middle Eastern religions on which the authors of the New Testament drew.

[8] See e.g. Origen (third century), *On First Principles*, 1. 5. 5.

at work in the world. He had been created by God as an angel, but rebelled against him out of pride. He was thought to take possession of human beings, to sign pacts with them, to tempt them to sin, and also to punish sinners in the afterlife in hell in the most horrific ways Christians could imagine. The devil headed an army of other fallen angels and minor devils in a constant war against God and the faithful. Like Ahriman, he would be defeated at the end of days.

The temptations of the Devil   Evil having thus been externalized, human wickedness and evil were explained by reference to the temptations the Devil lays before human beings and the pacts he makes with them. But, predestinarianism and demonic possession apart, human beings are held to be free to choose between Christ and the Devil, and are responsible for their choice. In the absence of remorse, confession, and atonement, punishment for evil-doing will ensue after death.[9] Evil-doers will be assigned to hell, where they will burn in flames for eternity, much to the pleasure of those who are saved, who will be able to observe their agonies but without feeling any pity, since what they observe is the justice of God.[10]

Paradoxes in Christian diabology   There were, and remained, paradoxes in Christian diabology. On the one hand, the Devil was God's ultimate adversary; on the other, he was God's agent, being his ordained executioner. Similarly, he was consigned to hell, but roamed the earth tempting humans to do evil. He was characterized as 'the Prince of the Air' (Ephesians 2: 2), dwelling in the sky beneath heaven, but also dwelt beneath the earth as the ruler of hell.

---

[9] The history of the Devil, Satan, Lucifer, and Mephistopheles from antiquity to the twentieth century is traced in painstaking detail by Jeffrey Burton Russell in a four-volume study (Cornell, 1977–90) to which I am much indebted.

[10] This doctrine was celebrated by Tertullian (whom Nietzsche mocked in *The Genealogy of Morals*, section 15) and repeated over the next fifteen centuries or more by numerous worthy Catholic and Protestant theologians alike, who revelled in the thought of watching unspeakable tortures and torments. Tertullian wrote: 'At that greatest of all spectacles, that last and eternal judgement how shall I admire, how laugh, how rejoice, how exult, when I behold so many proud monarchs groaning in the lowest abyss of darkness; so many magistrates liquefying in fiercer flames than they ever kindled against Christians; so many sages philosophers blushing in red hot fires' (*De Spectaculis*, 30). Augustine, unsurprisingly, followed suit. But one cannot but be shocked to discover that Thomas Aquinas of all people succumbed to this morally repulsive and wicked belief: 'In order that the happiness of the saints may be more delightful to them and that they may render more copious thanks to God, they are allowed to see perfectly the suffering of the damned ... So that they may be urged more to praise God' (*Summa theologiae*, suppl. qu. xciv, art. 1).

Original sin   A darker complication was added to the Christian world-view by the doctrine of original sin. Arising from a discussion of the Fall by Irenaeus (ca. 130–ca. 202), it was developed and advanced by Augustine of Hippo (354–430). According to Augustine, human beings, at birth, inherit Adam's original sin of disobedience by the concupiscent form of their creation, hence becoming a *massa damnata*, subject by nature to the sin of sexual desire that is not under the control of reason. Sin is in man from birth.[11] This primal flaw can be mitigated by baptism. Man can overcome the temptations of evil only by the grace of God. Pelagius's (360–418) attempt to overthrow this ugly doctrine and to insist that man has the ability to overcome the temptations of evil by his own free will alone was thwarted by Augustine's counter-polemics and the church councils of Carthage (418) and Ephesus (431).

Little good and much suffering came from this explanation of the existence of evil, even though it has lasted for eighteen centuries. Early diabology in Eastern Christianity was further developed in the West by leading figures such as Gregory the Great (ca. 540–604), Isidora of Seville (ca. 560–636), Bede (ca. 670–735), and Alcuin (ca. 730–804). There were three great conciliar characterizations of the Devil, the Council of Braga (563), the Fourth Lateran Council (1204), and the Council of Trent (1546). Representations of the Devil flourished in medieval village plays, in which the Devil was largely a figure of fun, equipped with horns, a tail, and cloven feet (borrowed from the god Pan), and covered with hair from the waist down, sometimes holding a trident (borrowed from Poseidon). He was farcically ugly in appearance and, to the delight of vulgar audiences (no different in their sense of humour from modern ones), farted a great deal.[12] However, representations of the Last Judgement and hence of hell in frescoes and in sculpture were prolific and anything but funny.

Romanticization   Representations of the Devil became more and *of the Devil*   more horrific and repulsive as the centuries went by. With the rise of witch-hunting in the sixteenth and seventeenth centuries, diabology assumed a seriousness and theoretical

---

[11] A prominent theme in the apocryphal second book of Esdras: 'For the first Adam, burdened with an evil heart, transgressed and was overcome, as were also all who were descended from him. Thus the disease became permanent; the law was in the hearts of the people along with the evil root; but what was good departed and the evil remained' (Esdras 3: 21–2; see also 4: 30).

[12] Apotropaic magic was widely accepted. One could ward off the Devil by wearing a garland of stinking garlic or asafoetida (still known in German as *Teufelsdreck*).

urgency. Heinrich Kramer's *Malleus Maleficarum* (*The Hammer of the Witches*), published in Speyer in 1487, was a best-seller for two centuries. It was rejected by the Catholic Church, but hideous application of it was made in civil courts that tortured and then sentenced innocent citizens to the stake or to the hangman. Luther wrote more on diabology than anyone since the church fathers, and his writings gave support to the witch-hunting craze in Protestant lands. This died out with the Enlightenment, but fascination with the Devil and the personification of evil did not. It became romanticized. This was partly due to the Romantics' reading of Milton's *Paradise Lost* (1667, 1674), in which the rebel angel Lucifer was viewed as a heroic figure[13] (frustrated as they were by the restoration of the *ancien régime* after the fall of Napoleon). It was also partly attributable to Goethe's *Faust*, in which Mephistopheles is a well-dressed gentleman of considerable charm and much wit.[14] The romanticization of the Devil led to the decadent and satanist movements among French authors and poets in the mid- and late nineteenth century (from Baudelaire (*Les Fleurs du mal*), through Lautréamont (*Les Chants de Maldoror*) and Rimbaud (*Une saison en enfer*) to such dandy poseurs as J.-K. Huysmans).

First depiction of a satan/the Devil It is curious to note that the first known depiction of the Devil, or more accurately of a satan, in Christian art is to be found among the mosaics at Sant'Apollinare Nuovo[15] in Ravenna.[16] They were made in 504, almost 200 years after Constantine the Great recognized Christianity as a licit religion and favoured it. Why do depictions of the Devil, so

---

[13] As is evident in Gustav Doré's illustrations for *Paradise Lost* (1866). It is instructive to compare these with pictures of the Devil in late medieval manuscripts and frescoes, in which the Devil is a repulsive monster.

[14] Dostoyevsky's picture of the Devil in *The Brothers Karamazov* presents him as a vulgar, ill-dressed, and uncouth personage, with whom no gentleman would mix. Thomas Mann followed suit in *Doctor Faustus*. Both were intent on distancing themselves from the romanticization of evil.

[15] Sant'Apollinare Nuovo began in 504 as an Arian church dedicated to Christ the Redeemer, built by King Theodoric the Great, the Ostrogoth. It was reconsecrated in the reign of the Byzantine emperor Justinian I in 561, dedicated to St Martin of Tours, an adversary of Arianism, and named Sanctus Martinus in Coelo Aureo. It was again renamed Sant'Apollinare Nuovo in 856, when the bones of Sant'Apollinaris were transferred to it from Sant'Apollinare in Classe.

[16] The first to identify the Devil in this mosaic was E. Kirschbaum, in 'L'angelo rosso e l'angelo turchino', *Rivista di Archeologia Cristiano* 17 (1940), 209–48. For further discussion see Beat Brenk, *Tradition und Neuerung in der christlichen Kunst in der ersten Jahrtausends* (Vienna, 1966).

prominent a figure in Christian scripture, emerge so late in Christian art? In a superstitious society it is likely that pictorial representations were endowed with magical powers and a depiction of the Devil was thought to be dangerous. On the upper band of the Theodorician mosaics on the left lateral wall of the church is a mosaic illustrating Matthew 25: 31–46, in which Christ separates the sheep from the goats, the former to be sent to heaven and the latter to hell. Christ, dressed in a purple toga, is seated on a central throne, flanked by two angels: a red angel on his right above a group of sheep and a blue angel on his left above a group of goats. The representation has occasioned puzzlement and the identification of the blue angel as the Devil has been challenged for two reasons. First, the Devil is supposed to be the recipient of sinners in hell, not their dispatcher from the side of Christ. Secondly, red is supposed to be the colour of the flames of hell and so an identifying mark of the Devil, not of the angel who leads the virtuous to heaven. Both qualms can be answered.

The 'Devil' in this mosaic is a satan in the Old Testament sense, that is an opposer, adversary, antagonist, accuser, challenger, not a monster from hell. His role is to challenge and accuse, and, on Christ's judgement, to dispatch sinners (the goats) to hell. Furthermore, the colour red, at this stage in Christian iconography, was the colour of glory. Only later did it become an identifying mark of the Devil. Confirmation of this comes from an illustration in the sixth century Rabbula Gospels (folio 13v), in which angels carrying the Virgin to Heaven are robed in red. Blue is the colour of the air, the *corpus aereum*, where Lucifer, divine accuser and challenger, is located. The Devil, Satan (as understood by Christianity), takes on the form of a monster coloured red or surrounded by red flames only in the High Middle Ages (he appeared as black in the ninth-century Stuttgart Psalter[17]). Although he originally had feathered wings like angels, in the twelfth century he acquired bat wings. He reached the pinnacle of hideousness and monstrosity in the fifteenth and sixteenth centuries with the rise of widespread witch-hunting throughout Europe.[18]

---

[17] Little black homomorphic figures representing evil spirits or imps, called *eidolons* by Brenk, were common in Christian iconography from the sixth century.

[18] I am grateful to Anthony Kenny and Hans Oberdiek for extensive discussion and cooperative research on this topic, and to Natasha O'Hear for learned and helpful correspondence.

Appendix 3:
Hannah Arendt and
the Banality of Evil

The phrase 'the banality of evil' has become a byword in discussions of evil since its introduction by Hannah Arendt (1906–75) in her *New Yorker* article *Eichmann in Jerusalem: A Report on the Banality of Evil* (1963). I shall argue that it is a dangerous 'ready-made' that obstructs clarity of thought.

Arendt Arendt was a distinguished German Jewish political theorist, who had studied under Heidegger and Karl Jaspers. She escaped from Nazi Europe and became a professor of political philosophy at various leading American universities. She wrote numerous renowned works, such as *The Origins of Totalitarianism* (1951), *The Human Condition* (1958), and *On Revolution* (1963). When the Israeli secret service captured Adolf Eichmann and abducted him to Jerusalem for trial (1961), Arendt arranged to attend the trial as a reporter for the *New Yorker*. The reports she wrote and the subsequent book raised an impassioned debate that continues to this day. Some, such as Daniel Bell, Alfred Kazin, Mary McCarthy, and Bruno Bettelheim, contended that she had deep insights into the nature of evil and of the Nazi genocide, while others, such as David Cesarini, Jacob Robinson, Gershom Scholem, H. G. Adler, Mark Lilla, and Primo Levi held that her claims were tendentious, bigoted, and riddled with error.

Offensiveness and arrogance Arendt's book was indisputably offensive. The offensiveness is patent in the title of the first chapter,

'The House of Justice', and in the opening sentence: '*Beth Hamishpath* [*sic*] – the House of Justice: these words shouted by the court usher at the top of his voice make us jump to our feet as they announce the arrival of the three judges.' But *Beth Hamishpat* means 'house of justice' only to the extent that *Habayit Hasheni* means 'the second house'. In fact, the announcement simply means 'the Court', just as *Habayit Hasheni* means 'the Second Temple'. She asserts that the German translation of the proceedings for the accused (of which she gives no examples) was inadequate and often ludicrous, and ascribes this to characteristic Israeli nepotism in the appointment of translators. She avers that the proceedings were a 'show trial' stage-managed by the Israeli prime minister, Ben Gurion. A show trial, a term introduced in the 1930s to characterize the trials of the fathers of the Russian Revolution in the Soviet Union, is a trial in which the accused is innocent but has been forced under torture to confess to crimes he never committed, the proceedings disregard rules of judicial evidence, and the verdict bears no relation to the validity of the evidence. This was patently not the case in the Eichmann trial. Mossad did not begin its hunt for Eichmann in 1957 in response to instructions from Ben Gurion. The televising of the trial was not Ben Gurion's idea, but Milton Fruchtman's, a young American television producer, who persuaded him to license it with the permission of the court. What is true is that Ben Gurion wanted the events of the Holocaust to be presented in Israel before the court by the personal evidence of Jewish survivors. Their voices had not been heard at Nuremberg.

Arendt's attitudes were prejudiced and arrogant. In a letter written to her erstwhile teacher and friend Karl Jaspers shortly after the trial began, she wrote:

> My first impression: On top, the judges, the best of German Jewry. Below them, the prosecuting attorneys, Galicians, but still Europeans. Everything is organized by a police force that gives me the creeps, speaks only Hebrew, and looks Arabic. Some downright brutal types among them. They would obey any order. And outside the doors, the Oriental mob, as if one were in Istanbul or some other half Asiatic country. In addition, and very visible in Jerusalem the peies [sidelocks] and caftan Jews, who make life impossible for all reasonable people here.

Her judgement was biased by preconceptions concerning the nature of totalitarianism. Her postscript to her book *Eichmann in Jerusalem* (2006) summarised her notion of totalitarianism: 'The essence of totalitarian government, and perhaps the nature of every bureaucracy, is to make functionaries and mere cogs in the administrative machinery out of men, and thus to dehumanize them.'[1] It was through these spectacles that she viewed Eichmann. I shall show that they were warped. The resultant observations were perverse and morally reprehensible.

There were three themes in her *Eichmann in Jerusalem*: Eichmann and the nature of evil; the iniquities of Ben Gurion, the young state of Israel, and the legality of the trial; and the *Judenräte* (the Jewish councils in the ghettos that cooperated with the Nazis in the vain hope of being able to save some of the inmates). The actions of the *Judenräte*, she wrote, were to the Jews 'undoubtedly the darkest chapter in the whole dark story' – so, presumably darker than Eichmann, and darker than the Holocaust itself. This judgement should give one pause.

Four main claims concerning Eichmann Of these three themes, only the first concerns me here. Her salient claims were:

(i)   that only good has any depth, so only good can be radical; evil cannot, and can only be extreme, for it is neither deep nor demonic;
(ii) that Eichmann was simply a prototype of a totalitarian functionary, with no anti-Semitic beliefs, committed only to the supreme value of obeying orders and to the advancement of his career;
(iii) that the source of the evil he did was thoughtlessness;
(iv) that evil is banal.

It is noteworthy that Arendt only watched Eichmann giving evidence for four days, when he was being questioned by his defence attorney, Dr Servatius, after which she left to go on holiday with her husband. She did not see Eichmann antagonistically questioned for three weeks by the prosecuting attorney Gideon Hausner, whom she held in contempt because he was a Galician Jew who, she wrongly hypothesized, 'probably spoke no languages'. Instead she subsequently read the proceedings and pre-trial interrogations by the police. She described her response to her reading: 'I thought Eichmann was a fool', 'Eichmann was a clown', 'I laughed at the transcription'. No one who actually attended Eichmann's cross-examination by Hausner found the proceedings funny.

---

[1] H. Arendt, *Eichmann in Jerusalem: A Report on the Banality of Evil*, revised edition with postscript (Penguin Books, New York, 2006; original publication 1963, revised ed. 1964).

(i) *Radical evil*: the concept of radical evil is Kantian. For Kant radical evil (*radix malorum*) is acting in a manner coextensive with the requirements of the moral law for reasons of self-interest, rather than 'determining the moral law as the highest principle of determination of one's maxims'. This, Kant held, signifies a radical perversity that is symptomatic of a disposition that has been corrupted by an innate propensity to evil. This is to be contrasted with the *devilish*, which consists in adopting the bad as the maxim of one's actions. This is not what Arendt had in mind at all. In her book on totalitarianism she characterized man in a totalitarian regime, who, in obeying orders, acts merely like a cog in the machine, as radically evil. By the time she wrote her book on Eichmann she had changed her mind, as she explained in a letter to Mary McCarthy:

> You are quite right, I changed my mind and do no longer speak of 'radical evil'. ... It is indeed my opinion now that evil is never 'radical', that it is only extreme, and that it possesses neither depth nor any demonic dimension. It can overgrow and lay waste the whole world precisely because it spreads like a fungus on the surface. It is 'thought-defying', as I said, because thought tries to reach some depth, to go to the roots, and the moment it concerns itself with evil, it is frustrated because there is nothing. That is its 'banality'. Only the good has depth that can be radical.

Arendt's misguided Augustinianism   The claim that evil is nothing, that it is merely privative, is Augustinian. It is striking that Arendt's doctoral dissertation had been on Augustine. This supposition was forced on Christian thinkers by the need for a theodicy. For if evil is not privative, if it exists in and of itself, and if God has created all things, then he must have created evil. That would be inconsistent with his infinite goodness. Arendt's thesis that evil is superficial, a mere surface phenomenon, is her variant on this misguided Augustinian doctrine. Darkness is privative, being simply the absence of light. But there is nothing privative, insubstantive, or unreal about being beaten to death, about the death pits of Babi Yar, or about being gassed with Zyklon-B. Does evil lack depth? If this means that evil deeds cannot be supported by evil beliefs and evil ideologies, that is patently false. The Cathar genocide of the Albigensian Crusade, the horrors of the Spanish Inquisition, the Herero–Nama genocide by the Germans in German South West Africa (Namibia), Stalin's 'dekulakization' in the Ukraine, and the Holocaust were unquestionably supported by elaborate evil beliefs and evil ideologies. If Arendt's thesis of the lack of depth of evil means that the evil beliefs that motivate evil deeds are not deeply rooted and exceedingly difficult to forswear, that too seems false.

Evil and the demonic   Can evil not be demonic? It is far from clear what she means by demonic. The only clue she gives us lies in the averral that 'Eichmann was not Iago and not Macbeth, and nothing would have been further from his mind than to determine with Richard III "to prove a villain". ... With the best will in the world one cannot extract any diabolical or demonic profundity from Eichmann.'[2] Is her claim then that in twentieth-century totalitarian regimes there were no demonic figures? That Hitler, Goering, Heydrich, and Himmler were not demonic? That Stalin, Yezhov, Yagoda, and Beria were not diabolical? If so, then it seems that there is no such thing as demonic or diabolical evil short of the imagined superhuman evil of Satan. If not, then evil-doers can be demonic, and her claim is presumably limited to Eichmann and the thoughtless mediocrities of the SS, SA, and the Wehrmacht who were guilty of atrocities. Richard III, Shakespeare suggested, was evil by nature; Iago became evil through resentment and jealousy; Macbeth embraced evil ('Things bad begun make strong themselves by ill' (III. 3)), but Eichmann, according to Arendt 'did not know what he was doing'.

Arendt's misjudgements about Eichmann   Arendt seems to have been over-impressed by the fact that Eichmann, at his trial, did not radiate evil like Lucifer, that he had no aura of evil grandeur. As she wrote: 'Despite all the efforts of the prosecution, everybody could see that this was not a "monster", but it was difficult not to suspect that he was a clown.'[3] Many epithets may occur to one apropos Eichmann, but 'clown' is not one of them. Why did Arendt reach this strange and disturbing conclusion? Apparently because he initially expressed reluctance to give evidence under oath since that would force him to tell the truth with unknown consequences, and subsequently expressed a preference for taking an oath. She also found his attempts to explain Kant's categorical imperative hilarious. They were not hilarious, but chilling, subordinating the individual not to the moral law, but to the Führer's law. Everyone could actually see that he was not a monster, Arendt insisted on the basis of her observations. To be sure, that he was a monster was not to be seen in the glass cage at the trial. It was to be seen fifteen years earlier, when he was in the prime of his power, dressed in gleaming jackboots and starched black SS uniform with silver braid and peaked hat, barking

---

[2] Arendt, *Eichmann in Jerusalem* (Viking Press, New York, 1964), 287–8.

[3] Arendt, *Eichmann in Jerusalem*, 55.

out orders with boundless menace and efficacy. Eichmann changed human history. He wielded the power of death over millions and doubtless felt like a god in exercising it. Fifteen years later, after the fall of the Reich, stripped of uniform and status, kept in gaol, and presented in court under police escort, he may have looked ordinary and mundane, as had all the Nazi leaders at Nuremberg.

Cogs in totalitarian machines

(ii) *Totalitarian functionary*: Eichmann was anything but a mere cog in the bureaucratic machine.[4] *He had designed the machine*. He was a fanatical Nazi, a demonic anti-Semite who was determined (even against Himmler's orders in late 1944 to desist) to make Europe 'Jew free'. He was a highly efficient officer who, despite the chaos of a losing war, was able to organize the transportation of millions of human beings from every corner of the continent to the death camps in Poland. Arendt claimed that Eichmann was not an anti-Semite, and that he had no motives at all except for 'an extraordinary diligence in looking out for his personal advancement. And this diligence was in no ways criminal.' This is demonstrably false. Eichmann had been brought up from the age of five in Austria. He came from a right-wing family, and grew up in a right-wing city (Linz), in a country steeped in anti-Semitism. In one of the numerous interviews he gave to Willem Sassen in 1956 he avowed:

> The cautious bureaucrat, yes, that was me ... But joined to this cautious bureaucrat was a fanatical[5] fighter for the freedom of the blood I descend from.... What is good for my *Volk* is for me a holy command and holy law ... I must honestly tell you that had we killed 10.3 million Jews I would be satisfied and would say, good, we've exterminated the enemy ... We would have completed the task for our blood and our *Volk* and the freedom of nations had we exterminated the most cunning people in the world.[6]

Had he succeeded, he averred, he would have died a happy man.

It is interesting that such a claim of total ideological indifference might have been made of Rudolf Höss, the penultimate commandant

---

[4] The phrase 'cog in the bureaucratic machine' is derived from Max Weber's reflections on the modern bureaucratic state.

[5] The term 'fanatical' in Nazi newspeak was, as Victor Klemperer showed in his *Language of the Third Reich* (Athlone Press, London, 2000), used as a term of the highest praise. The book was originally published as *Lingua Tertii Imperii* in 1947.

[6] Quoted from the Stassen Papers by Bettina Stangneth, *Eichmann before Jerusalem: the Unexamined Life of a Mass Murderer* (Vintage, London, 2014), 392–3.

at Auschwitz who perfected the gas chambers and the use of Zyklon-B. Of him, Primo Levi wrote:

> We can believe him when he claims that he never enjoyed inflicting pain or killing. He was no sadist, he had nothing of the satanist. By contrast satanic features can be found in Höss's portrait of his peer and friend Adolf Eichmann; however, Eichmann was far more intelligent than Höss.

Höss's beliefs, according to Primo Levi, do approximate to the kind of person Arendt had in mind, for he believed that

> Order is necessary in everything; directives have to come from higher up, they are good by definition, and they are to be carried out conscientiously and without discussion; personal initiative is permissible only if it fosters more efficient execution of orders.

He was, Levi continued, one of the worst criminals of all time, but his make-up was not dissimilar from that of any citizen of any country.

(iii) *The source of the evil Eichmann did was thoughtlessness*: the notion of 'thoughtlessness' is used here as a technical term. Arendt derived it from her teacher Heidegger. But it is unclear whether Arendt is using it in the same way as he did. Eichmann, she claimed,

> *merely*, to put the matter colloquially, *never realized what he was doing* … He was not stupid. It was sheer thoughtlessness – something by no means identical with stupidity – that predisposed him to become one of the greatest criminals of that period.… What he said was always the same, expressed in the same words. The longer one listened to him, the more obvious it became that his inability to speak was closely connected with his inability to think, namely to think from the standpoint of somebody else. No communication was possible with him, not because he lied, but because he was surrounded by the most reliable of all safeguards against the words and presence of others, and hence against reality as such. This safeguard is thoughtlessness and consequent remoteness from reality.

Fifteen years later, in an article entitled 'Thinking' written for the *New Yorker*, she had another shot at clarifying what she meant:

> Clichés, stock phrases, adherence to conventional codes of expression and conduct have the socially recognized function of protecting us against reality; that is, against the claim on our thinking attention which all events and facts make by virtue of their existence. If we were

responsive to this claim all the time, we would soon be exhausted; Eichmann differed from the rest of us only in that clearly he knew of no such claim at all.

This account was tailor-made for Arendt's own analysis of modernity, state bureaucracy, and totalitarian regimes. Maybe there is some truth to her analysis, although it is not evident that the human ability to think for oneself was any greater in religion-ridden pre-industrial European society. But it is by no means clear that Eichmann fitted the mould Arendt had prepared for him, among other things because in court he was trying to give the impression that he was indeed but a cog in the machinery of the Nazi state and merely obeying orders. *That was his main defence.* It seems that Arendt fell for this. For it was Eichmann's intention to present himself to the court as an unthinking bureaucrat merely obeying orders. But there is every reason for thinking that he was lying. There is also every reason for thinking that, contrary to Arendt's claim, he had known perfectly well what he was doing. He had been no bureaucratic desk-sitting administrator, and there was nothing 'remote from reality' about the way he had handled his evil tasks.

(iv) *The banality of evil*: the term 'banal' is generally applied to assertions and to thoughts, not to deeds or to agents who perform them. So what Arendt meant by claiming that evil is banal and not radical needs explaining. It is certainly not the evil done, the sufferings of millions of victims, that is said to be banal. It is the evil-doers, in particular Eichmann. But what is meant by saying that they are banal? It is not that they are ordinary people. Nor is it that their thoughts are commonplace. It is simply that they are 'thoughtless', that they shield themselves from their deeds by a wall of platitudes, stock phrases, and banalities. Arendt's conclusion is 'That such remoteness from reality and such thoughtlessness can wreak more havoc than all the evil instincts taken together which, perhaps are inherent in man – that was, in fact, the lesson one could learn in Jerusalem.' This conclusion does indeed fit in well with her analysis of totalitarianism, and with her warranted anxieties about the bureaucratic state. It also fits well Dr Servatius's defence of Eichmann at the trial. However, it does not fit what became evident about Eichmann during the trial, and has become undeniable as further knowledge of Eichmann and his career has come to light since the trial.

Non-banality of Nazi evil-doers That ordinary human beings can do evil is news from nowhere. That exceptionally able human beings, given the opportunity and the right socio-political circumstances can do great evil, is patent throughout history. Doubtless the modern bureaucratic state provides the opportunity, given the corruption of leaders, for relatively unthinking wickedness and evil done by desk-bound officials following orders (state departments and home offices excel at this). But the millions of Nazi troops (not just the SS and SA, but the Wehrmacht too) and their collaborators across the continent of Europe and in conquered Poland and Russia, who actually facilitated and committed genocide, *were not bureaucrats*. They were, for the most part, ordinary human beings, with potentialities for both good and evil, who chose to do evil – to disclose to the Nazis where Jews were hiding, to slaughter innocent people, to burn people alive, to murder children, and to torture victims. There was no case, as far as I know, of a Nazi soldier being punished seriously for refusing to obey such orders. What they did was, in the horrendous context, commonplace, but their activities were anything but 'banal'. They were not pushing pieces of paper around, filling in forms, and passing on instructions by telephone. They were murdering, torturing, and tormenting human beings. Moreover, their generals, officers, commanders, and commandants issued horrendous orders that were neither commonplace nor a matter of mere bureaucratic conformity to orders. They ensured the execution of these orders and looked on at the slaughter and torture with equanimity and satisfaction. There was nothing banal about them or about the evil they did.

Appendix 4:
The Pictorial Representation
of Pleasure in Western Art

As was described in *The Passions: a Study of Human Nature*, the various hedonist vices, such as gluttony, drunkenness, and concupiscence (*luxuria*) have a rich iconographic history in post-medieval Western art. So too does the emotion of love (*amor*). The iconography of pleasure, however, is relatively thin, even though the pictorial representation of pleasure is rich and diverse. Although the latter is a subject for a book in its own right, it is worth surveying briefly, for it is an expression of human attitudes towards pleasure and its significance in human life in Western civilization after the Middle Ages.

In Renaissance and early post-Renaissance art there are three main hedonic themes: (i) bacchanalias; (ii) the choice between a life of pleasure and a life of virtue; and (iii) representations of amorous pleasure.

(i) The first renowned bacchanalia occurs in Giovanni Bellini's well-known *Feast of the Gods* (1514), whch is frankly erotic as well as comic. It was commissioned by Alfonso I d'Este for his *camerino d'alabastro* and depicts an erotic scene from Ovid's *Fasti*. It was subsequently modified by Titian, who later painted two other large paintings for the same room: *The Bacchanal of the Andrians* (1518–23) and *The Worship of Venus* (1518–19). These are similarly representations of pagan divine hedonism, both erotic and alcoholic, this time based on a text by Philostratus (*Imagines*, I. 25, 6). The programme for all was drawn up by Mario Equicola, a humanist scholar in the service of Isabella d'Este in Mantua. These paintings depict idealized

Mediterranean landscapes in which gods (i.e. idealized human beings), satyrs, fauns, naiads, and cupids (*putti*) are celebrating erotic love, wine, and women. The theme was popular throughout the next century. It was a common subject in the classicizing work of Nicolas Poussin. The theme reached the Netherlands in the early seventeenth century.

(ii) A quite different topic in baroque art was the choice between virtue and pleasure, represented by Hercules at the cross-roads. This was a popular theme from the late Renaissance to seventeenth- and eighteenth-century baroque, essayed by such masters as Annibale Carracci, Pompeo Batoni, and Sebastiano Ricci. It enjoyed comparable popularity among printmakers. These representations were didactic, showing Hercules at the cross-roads between the path of pleasure wending its way between groves of fruit trees and delightful streams, and the path of virtue ascending steeply up a rocky and perilous road to the summit of a distant mountain. A scantily dressed voluptuous Venus tempts Hercules to indulgence and a sybaritic life, while a severely dressed stern Minerva, goddess of wisdom, urges him to choose the hard road of virtue. The iconographical markers of Voluptas are the pan pipes or a reed pipe of the satyrs, a tambourine (associated with the ecstatic swirling maenad dances of female devotees of Bacchus), masks symbolizing deceit, playing cards hinting at gambling, a rod and fetters signifying enslavement to the baser passions of *luxuria*.

(iii) Representations of amorous pleasure in Renaissance and Mannerist art are relatively late. Sodoma's *Marriage of Alexander and Roxana* (1517) in the Villa Farnesina, is a celebration of the amorous delights of the marital bed, in which Roxana, seated on the edge of the bed is being disrobed by *putti*. The theme of Leda and the swan provided an acceptable erotic motif precisely because it was mythological. It was painted suggestively by Leonardo in 1508. In 1530 Michelangelo painted Leda in coitus holding the swan between her thighs. Correggio painted a series of highly erotic canvasses of Jupiter's amorous adventures that was commissioned by Federico II Gonzaga for a room designed to display Ovid's *Metamorphoses* in 1530. These large canvases severally depict Jupiter as a cloud embracing Io in orgasmic ecstasy, the abduction of Ganymede, Leda and the swan, and Danaë. It is evident that the painting of erotic pleasure in Renaissance and Cinquecento Italy was respectable only when it was masked by the acceptable veneer of antiquity and tales of Greek and Roman mythology.

An uncommon but widely known theme is an illustration to Macrobius' commentary on Cicero's *Somnium Scipionis* (Scipio Africanus the Younger (236–184 BC)) from Cicero's book *De re publica*. The painting is Raphael's small (17 × 17 cm) allegorical panel (1504–5) in the National Gallery, London, in which Scipio Aemilianus sleeps and dreams he is accosted by two female figures, one offering him a book and a sword, and the other a flower. Unlike the motif of the choice of Hercules, the two figures here do not appear as contestants. Rather, it has been suggested, they are complementary aspects of the perfect life of a nobleman. In Macrobius, Scipio is urged to pursue the active and contemplative life and warned against the voluptuous life. However, that is not the idea conveyed by Raphael's panel, the programme of which is inspired by Marsilio Ficino, who advocated a subtle balance between the active, contemplative, and pleasurable life as the proper road to human felicity. Hence it is this threefold harmony that is the ideal for a young aristocrat.[1]

It was primarily in the art of the Low Countries in the sixteenth and seventeenth centuries that pleasure descended from the gods to mankind. In Antwerp, the greatest painter of vulgar pleasures was Pieter Bruegel the Elder (ca. 1525–69), whose paintings *Children's Games* (1560), *Peasant Wedding* (1566–9), *Peasant Dance* (1568), and *Wedding Dance* (1566) are among his most wonderful portrayals of the social entertainments of the rude peasantry. That tradition was continued by his two sons. He also influenced Karel van Mander, who transmitted the painterly concern with depiction of coarse life to the Dutch, as is patent in his representations of the village Kermis – drunken peasant street celebrations in villages and small towns.

It was in the independent Netherlands, above all, that depiction of human beings taking pleasure became a dominant theme in painting, sometimes didactic, but often just as a record of the pleasures of human life. The range of themes is wide: music and singing are a common motif, for example Gerrit van Honthorst's *Merry Fiddler* (1623) and *The Concert* (1624); Jan Molenaar's *Family Making Music* (1635–6); and Gerard ter Borch's *The Concert* (1657). All such paintings represent the pleasure taken by wealthy families in making music

---

[1] The panel is complementary to a similar-sized painting by Raphael of the Three Graces, Chastity, Beauty, and Pleasure, each holding a golden apple, now in the Musée Condé of Chantilly in France. It has been suggested that they are probably front and back panels for a luxurious binding of a copy of Macrobius' commentary presented to young Scipione Borghese on his birthday in 1505.

together, or by a duo, trio, or quartet of musicians playing for their own delight, as well as the pleasure given by musicians playing in taverns and the enjoyment of solitary players.

The pleasures of food and drink at both ends of the social ladder were often depicted. One social extreme was represented by paintings of burgher companies of civic guards enjoying a banquet, for example Frans Hals's *Officers of the St George Civic-Guard* (1616) and *Officers of the St Adrian Civic-Guard* (1627), both in the Frans Hals Museum in Haarlem. At the other social extreme, Adriaen Brouwer produced numerous paintings of carousing and quarrelling peasants. Adrian van Ostade made somewhat less vicious paintings of boisterous tavern life. Jan Steen painted many humorous and witty pictures of tradesmen's families around a festive table, for example *Merry Company* (1663), *Merry Company on a Terrace* (1673), and *Twelfth Night* (1662). These are always didactic, illustrating one moralizing dictum or another from the rich variety of such pieces of folk wisdom available to the Dutch. Further up the social scale are such depictions of the pleasure of dining in company as Gerrit van Honthorst's wonderful *Supper Party* by candlelight (1620), which vies with the best of Georges de la Tour in its candlelight effects. Merry company paintings of the parties of the wealthy bourgeois and aristocracy were celebratory, with no hint of condemnation.

The pleasures of alcohol are commonly represented, usually didactically. At the lower end of the social scale, the drunkenness and tavern brawling of boors by Brouwer and van Ostade are contemptuous. Jan Steen's depictions of over-indulgent celebrations are affectionate and amused, as is evident in numerous paintings such as *Village Wedding*, *Village Revels*, *Country Wedding*, and *Celebrating a Birth*. Nevertheless, they are didactic, as is often evident from a painted dictum, or by a symbolic representation of a dictum. At the same time, it is difficult to see Frans Hals's *The Merry Drinker* (ca. 1628–30) as anything other than an expression of merriment, and if Rembrandt's painting of himself, raising his glass in a toast with Saskia on his lap is meant to be a prodigal son painting it is of an uncommonly cheerful and as yet unrepentant young man.

Pictures of gallant garden parties were also popular, originating in moralizing paintings of the prodigal son. The moralizing flavour of this theme rapidly disappeared, as in the garden banquet paintings of Esaias van de Velde, in Frans Hals's early (1610) painting of a banquet in a park or Jan Steen's late painting (1677) of an idealized garden party in the grounds of the patrician Paedts family. So such depictions of upper-class pleasures became idealized records of memorable social occasions.

Brothel scenes were common and were dire warnings, for example Vermeer's *The Procuress* (1656); Wouter Crabeth's *Cheating*, depicting a game of cards in a brothel in which a drunken young man is being fleeced; Dirk van Baburen's *The Procuress* (1622); and the famous painting by Gerard ter Borch, misnamed *The Paternal Admonition* (1654–5) by Goethe and admired for its tender sympathy, now in Berlin-Dahlem, which is in fact a brothel scene in which the services of a young woman are being purchased from the madam. Actual portrayals of fornication were restricted to illicit pornographic prints.

The pleasures of games were often depicted, sometimes as warnings against betting or being cheated but often perfectly innocuous representations of amusements. Paintings of card playing abound, some in decorous settings of burghers' houses, others in genteel inns, and yet other in soldiers' guardrooms or in common taverns. Jan Steen's *Bowling Game* (1665), *Skittle Players outside an Inn'*(ca. 1660–3), and *Tric-trac Players* (1667) are examples of outdoor amusements. Ice skating and ice hockey are outdoor sports often depicted by Aert van de Neer, Isaak van Ostade, and Arent Arentsz.

Thenceforth all forms of human pleasure and entertainment were subjects for painting, differently developed in different countries and times. Eighteenth-century French rococo (e.g. Antoine Watteau, Nicholas Lancret, Jean-Baptiste Pater, and François Boucher) presented an erotic fantasy world for the amusement of the French upper classes. The British contributed little to the depiction of pleasure until Hogarth's series of eight paintings *The Rake's Progress* (1732). It is rightly renowned as a fiercely didactic representation of the unrestrained pursuit of pleasure that was subsequently engraved and published as prints (1735). His *Beer Lane* and *Gin Street* (1751) display the pleasures of beer drinking and evils of gin. Depictions of the pleasures of blood sports became common in the late eighteenth and nineteenth century, for example shooting birds and game, as well as bull-fighting in Spain (Goya) and fox-hunting as well as pugilism in Britain.

Index

Pages containing significant discussions appear in **bold**.

ability, 161
  two-way, xxi, 157
abstaining, 189–90, 195
absurd, the, 325
accountability, 199, 200
action, **183–8**
  explanations of, **186–7**, 289
  intentional, 377–8
  reflex, xxi, 161
  responsibility for, 202–6
action potential, 193, 195
aesthetes/aestheticism, 240–2
Ahriman, 392
Ahura Mazda, 392
Akiva, Rabbi, 69, 167
Albigensian Crusade, 72
Allerhand, Maurycy, 129
altruism/selfishness, 36, 272
Améry, Jean (Hans Mayer), 76–7,
  90, 139
analogy, argument from, 290–1
anamnesis, 338
Anglo-Saxon poem

The Wanderer or *The Exile's
  Lament*, 165
animal(s), **361–89**
  classifying, 370–1
  concepts, proto-concepts, 368–9
  cruelty to, 79
  do no evil, 93
  emotions, 384–6
  expectation, 385
  happiness of, 251
  knowledge of other animals,
    378–81
  lack a soul, 135n.
  lack of awareness of mortality,
    336
  love, 385
  and meaning of life, 328
  memory, 373–5
  and mirror recognition, 381
  morality, 18, **361–4**, 382–3
  perception, xx n.
  propositional knowledge, 371–3
  reasoning, 367, 375–6

responsibility of, 376
self-consciousness, 381
thinking, thought, 364–7
Aquinas, xx n., 117, 394n.
Arendt, Hannah, 125, **398–406**
Ariphron of Sikyon
‘Ode to Health’, 273
Aristippus, 237–8
Aristotle, xiii, 5n., 18, 23, 26, 48,
66, 117, 121, 162n., 214, 223,
226–7
on eudaimonia, 274–6
Armenian genocide, 73, 79, 90–1
arm rising, arm raising, 194–5
Arnaud of Amaury, 72
Asch, Solomon, 125n.
attitudes
reactive, 45, 160
Augustine of Hippo, 93, 168,
395, 401
Austen, Jane
Pride and Prejudice, 36, 146
Austin, Jean, 257n.
avoidability v. unavoidability, 194
axiological concepts, need for,
15–16
Ayers, Amanda Conley, 287n.
Aztecs, 27, 28–9

Babel, Isaac
Red Cavalry, 109
‘The Tale of My Dovecot’, 80
badness, varieties of, **65–8**
Baker, G. P., 176n.
‘banal’, 405
banality of evil, *see* evil, banality of,
alleged
Baron-Cohen, Simon, 96–100,
120, 293
Bartholomew’s Day massacre,
113–14
Bates, Timothy, 285n.
Bede, the Venerable, 340–1
Belshazzar’s feast, 163n.

beneficial, the, 12–13, 14
benevolence, 45
Bennett, M. R., 63n., 98n.,
180n., 193n.
Bentham, Jeremy, 26, 27, 160n.,
209–10, 223, 231–5, 270, 303
Ben-Yami, Hanoch, 292n.,
368n., 379n.
Ben-Yami, Jotham, 292n., 379n.
Ben-Yami, Maya, 292n., 379n.
Berridge, Kent, 285, 299–300
Beziers, 72
Blackburn, Simon, 317
Blake, William
‘Auguries of Innocence’, 331
Blakemore, Colin, 179
Bley, Helmut, 73n.
Blixen, Karen, 268
blood libel, 118
body, xvi
we are v. we have, xvi, 131
Boethius, 49
Boly, Melanie, 336n.
Boniwell, Ilona, 287n.
brain state, 180–1
Branaugh, Kenneth, 130n.
Brandt, Sebastian
The Ship of Fools, 239–40
Brenk, Beat, 396n., 397n.
Britten, Benjamin, 105n.
Brontë, Charlotte
Jane Eyre, 146
Brosnan, Sarah F., 383n.
Brown, Dee, 95n.
Browning, Christopher
on Police Battalion 101, 127–8
Browning, Robert
‘The Last Duchess’, 110, 388
‘Rabbi ben Ezra’, 260
Bryant, Peter, 379n.
Büchner, Georg
Danton’s Death, 114
Burgess, Anthony
A Clockwork Orange, 108, 118

Caesar, Julius Gnaius, 72, 94
Calvin, Johannes, 61
Camus, Albert, 318, 324–6
  The Myth of Sisyphus, 317
  The Rebel, 317
Čapek, Karel
  The Makropoulos Case, 347,
    350–2
caring, 21, 39–40
Cartwright, Nancy, 175n.
Castellio, Sebastian, 29, 58n.,
    61–2, 333
Catharism, 72n.
causal explanation, 175
causal powers of man, xxi
causation, 174–5
choice, 158
Christianity, 31, 59–60
  and diabology, 390–7
  and hedonism, 239
  on human freedom, 167–8
Chrysippus's dog, 368
Cicero, Marcus Tullius, 49n.,
    238, 269
Cinderella effect, 93n.
Clayton, Nicola S., 375n.
clockwork, 169, 175–6
cogitative powers, xix
cognitive powers, xviii
cognitive receptivity, xvii
Cole, Philip, 83n., 85n.
Coleman, Martin R., 336n.
commensurability v. incommensura-
    bility, xiii
competence, normal human, 200
concepts / concept possession,
    369–71
Conrad, Joseph
  Heart of Darkness, 75–6, 129–30
conscience, 26, 75, 137–8, 140
consciousness
  intransitive v. transitive, xvii
  transitive perceptual, xvii–xviii
contentment, 248–50

control, 203–6
  lack of, 182–3
cooperation, 45
corpse, xvi
could/would, 158–9
'crimes against humanity', 73n.
Croesus, King of Lydia, 261
cruelty, 78, 109

Dally, Joanna M., 375n.
Darrow, Clarence, 86n.
Darwin, Charles, 21, 383
David, King of Israel, 110, 387–8
David, Susan, 287n.
Davis, Matthew H., 336n.
Dawkins, Richard, 179
Dayan, Peter, 296n., 298–9
death, conceptions of, 335–6;
    see also thanatophobia
decision-making, neural pathways
    for, 376–7
deeds, moral goodness of, 35–7
dehumanization, 112
dekulakization, 113
demonic, 402–3
Descartes, René, 169–70
desire, *see* wanting, kinds of
determinism
  neuroscientific, **179–82**
  nomological, 169–72
  physical, 170–1
  and predictability, 171
  reductive, **173–7**
Devil
  absence of in Hebrew Bible, 89;
    see also satan, in Hebrew Bible
  in Christianity, 393–7
  depiction of in Christian art,
    396–7
d'Holbach, Pierre-Henri Thiry,
    Baron, 172
diabology, 89n., 390–7
Dickens, Charles
  David Copperfield, 349

Great Expectations, 14
  A Tale of Two Cities, 250
Digby, Kenelm, 368n.
dignitas, 48
dignity, 49–50
  formal, 39, 48
Dilman, Ilham, 130n.
Dolan, Raymond J., 294, 299
Donne, John
  'Holy Sonnets', 357n.
Doré, Gustav, 396n.
Dostoyevsky, F., 332, 396n.
  The Brothers Karamazov, 45,
    70, 396
  Crime and Punishment, 119,
    120, 149
  The Possessed, 184, 324
Dretske, Fred, 372n.
Dummett, M. A. E., xviii n.
Dupré, John, 175n.
duress, 160n., 162, 202
duty, path of, 25
duty for duty's sake, 39, 43
duty-imposing rules, 25

Ecclesiastes, 314, 319, 320, 322–3,
  326, 342, 390
economics, marginalist, 282–3
Eichmann, Adolf, 312, **398–406**
Einsatzgruppen, *see* SS
    Einsatzgruppen
Eldar, E., 294, 299
Eliot, George
  Middlemarch, 14, 268
Emery, Nathan J., 375n.
emotions, xxiii, 21, 180, 386–7
empathy, 21, **96–9**
enjoyment, 212–19
Enlightenment, 31–2, 56–7, 62
Enver Pasha, 73, 111
Epicurus, 211–12, 220–1, 238, 355
eschatology, 334, 338
ethics and aesthetics, alleged unity
  of, **122–4**

eudaimonia, 269, **271–6**, 285n.
Euripides
  Iphigenia in Aulis, 271
  Medea, 271
evil, 53–4, **76–82**
  apparent unintelligibility of,
    90–1
  banality of, alleged, 125, **398–406**
  blindness to, 83–4
  and death of the soul, **138–43**
  degrees of, 78
  and the demonic, 402–3
  diabological sources of, 391–2
  disintegrative force of, **143–8**
  -doers, 78, 98–9, 140, **141–3**
  -doers and happiness, 149
  -doers and love, 150
  etymology of, 80–1
  excuses and justifications for, 116
  existence of, **83–9**
  explanations of, 68–9, 91, 94–5,
    101–28
  and explanations of behaviour,
    85–6
  group-, attractions of, 95–6
  as a motive, **115–21**
  motives for, 105, 119–20
  natural, super-natural, human,
    68–76
  as privative, **87–9**
  propensity for, 92
  pure, 106
  radical, 89, 400–1
  rationality/irrationality of, 118
  reasons for non-existence of, 84
  roots of, **65–100**
  science of, 96–9
  understanding of, 103–5
  and the unforgivable,
    148–9
evolution, 94, 321–2, 362
existentialism, 317
experiences, as source of meaning,
  309–10

fact/value gap, 3
Fall, doctrine of the, 91–2, 121,
    167, 395
false belief test, 292–3, 379
Farnell, L., 98n.
fatalism, **157–68**
    global, 162–4
    individual, 166
Faustus, 391n.
Fielding, Jonathan
    Jonathan Wild, the Great, 119
first-/third-person asymmetry,
    293–4
First World War, 63n.
Flavius Josephus, *see* Josephus
    Flavius
Foot, Philippa, 20n., 43, 46
forgiveness, **143–8**
Forster, E. M., 105n.
Frankfurt, Harry, 255n., 376n.
Frankl, Viktor, 308, 330n., 332
Freedland, Mark, 85n.
freedom, 39, 48, 157–8
    of action, 161–2
    of choice, 161
    and divine foreknowledge, 168
friendship, 41–2
Frith, Christopher D., 291
Frith, Uta, 98n., 293

Gaita, Rai, 53n., 77
Gallup, Gordon, 381n.
Gauthier, David, 363n.
Gellhorn, Martha, 129, 133
Genesis, creation myth of, 48,
    321–2
Genghis Khan, 332
genocide, 72–5, 79, 148–9, 401
Gibbon, Edward, 72, 252
Gibson, W. G., 98n.
Gide, André
    The Cellars of the Vatican, 177
Gilgamesh, 237
gladiatorial games, 29n.

Glock, H.-J., 368n., 372n.
goals, regulative v. terminative, 308
God, death of, 317–18
Goethe, Johann Wolfgang von
    Faust, 345, 396
    Roman Elegies no. 19, 260
    'Willkommen und Abschied',
        251–2, 259
Golden Rule, 61, 64
good
    grammar of, 6
    of man, 14–15, 16, 22, 33–5,
        37–8, 41, 45, 59, 64, 236
    as a means, 4
good and evil
    knowledge of, **121–4**
    potentiality for, 92
goodness, varieties of, xiii, 5, 7,
    8–17
    moral, **17–24**, **33–8**, 45
Goya, Francisco
    Black Paintings, 312
Great Chain of Being (*scala
    naturae*), 49–50
Greene, Graham
    Brighton Rock, 106
Greene, Joshua, 377n.
Greenwood, Edward
    'A Meditation', 357
    'The Valuers', xxv
Grossman, Vasily
    Everything Flows, 113, 139, 151
    Life and Fate, 130

Hacker, Jonathan, 118n.
Hacker, P. M. S., 26n., 63n., 176n.,
    180n., 182n., 193n., 286n.,
    336n., 380n.
Haidt, Jonathan, 376n.
Hannibal Barca, 144, 278
happiness, **243–80**
    conceptual multiplicity of, 262–3,
        267–8
    and contentment, 249–50

degrees of, 253–4
and enjoyment, 250
epistemology of, **266–9**
etymology of, 244–5
and eudaimonia, 278
and health, 263–4
kinds of, 257–61, 262–3
measurement of, **294–8**
and morality, 276–80
neuroscientific investigation of, **295–8**
objective v. subjective, 246–8
and pleasure, 250
preconditions of, **263–9**
social, 251, 292
and temperament, 252–3
and unhappiness, 245–6
varieties of, 261–2
vocabulary of, 247
and the wicked, 279–80
happiness, science of, **281–303**
hard data for, 284–5
happy
life, 261–3, 276–9
mood, 252–3
society, 303
Harris, Robert
Cicero Trilogy, 249
Harrison, Bernard, 75n.
Hart, H. L. A., 19n., 86n., 197n.
hate, need for, 112–13
Hatton, Sean N., 98n.
health, 8–9, 66
Hebrew Bible (Old Testament), 85, 89
and fatalism, 166–7
on responsibility for good and evil, 390
hedonic calculus, 231–5
hedonic goodness, 13, 217
hedonic life, **237–42**
hedonic utterances, 235–6
hedonic vocabulary, 213
hedonism, 209–12, 270
psychological, 229

Heinsohn, Gunnar, 75n., 111n.
Hemingway, Ernest
A Farewell to Arms, 349
Henley, W. E.
'Invictus', 162
Henry, O.
'Roads to Destiny', 166
Herero genocide, 72–3
Hermens, Daniel F., 98n.
Hillel, the elder, 61, 64
House of, 333
Hitler, Adolf, 75n., 111n.
Hofman, Mark, 177n.
Hogarth, William
The Rake's Progress, 277
Holocaust, Jewish, 69n., 73–4, 77–8, 90–1, 112, 127–8, 400–6
uniqueness of, 73–4
Holodomor (Ukraine), 73
Holt, Jim, 382n.
Homer
Iliad, 59, 164–5, 355
Homo loquens, xxiii, 22, 131, 328
Homo ludens, 22
Höss, Rudolf, 403–4
Hugo, Victor
Les Misérables, 144
Huizinga, Johannes, 22
human behaviour, explanations of, **182–8**
human beings, xiii, xiv, xvi–xxiii
awareness of death, 336–7
biological nature of, 4–5, 20–21
good human beings, 14–15, 59, 64
human nature
constants in, **61–4**
contingencies of, 19–20
contrasted with machines, 179–80
plasticity of, 23
Hume, David, xv, 18, 38–9, 40, 45, 97, 170, 249n.
humiliation, 56, 112

Huppert, Felicia, 284
Huxley, Aldous
　　Brave New World, 257
Huysmans, J.-K.
　　Against Nature, 277
Hyman, John, 195n.

Ibsen, H.
　　Hedda Gabler, 108
　　The Masterbuilder, 43
immortality, as robbing life of
　　meaning, 346, 352n.
impulse, irresistible, **203–6**
inaction, explanations of, **187–8**,
　　188–90, 290
inevitability, 160, 194
Innocent III, Pope, 72, 95
Inquisition, Spanish, 115
intemperance, 236–7
intentions, 184–5
　　goodness of, 35–6
Irenaeus, 395
Isenberg, Arnold, 387n.
Ish Gamzu, Rabbi Nahum, 69n.
Islam, Sunni, 163

Jacobins, 115
Jesus, 61n.
Jewish genocide, *see* Holocaust,
　　Jewish
Job, 69, 89
Josephus Flavius, 167, 338
Julian of Norwich, 312
Justin Martyr, 393n.

Der Kampf, 72–3
Kanai, Ryota, 285n.
Kant, xi, 36, 39, 43, 50–2, 55, 122,
　　400–1
Kaplan, Chaim, 77
Kekes, John, 76n.
Kenny, A. J. P., 59n., 204–6, 223,
　　263n., 397
Kenny, Charles, 263n.

Kepler, Johannes, 169
Kershaw, Ian, 109, 150
Kierkegaard, Søren, 322, 324
Kirschbaum, E., 396n.
Kirwan, Christopher, 93n.
Klemperer, Victor, 150, 403n.
knowledge
　　ability-like, xviii
　　not a mental or neural state, 191
　　retention of, xviii
Koestler, Arthur
　　Darkness at Noon, 115
Korsgaard, Christine, 3n.
Kramer, Heinrich, 396
Kringelbach, Morten, 285,
　　299–300

Laclos, Pierre Choderlos de
　　Les Liaisons Dangereuses, 107–8,
　　250, 277
Ladurie, Emmanuel Le Roy, 60n., 239
Lagopoulos, Jim, 98n.
Landau, Iddo, 314, 325n.
Laplace, Pierre-Simon, Marquis de,
　　170–1
Larkin, Philip
　　'Aubade', 356
Laureola, Rosanna, 244n., 272n.
Laureys, Steven, 336n.
Lauterpacht, Hersch, 73n., 129
Layard, Richard, 284, 287, 288,
　　294, 295
Le Carré, John
　　The Constant Gardener, 118
Le Guin, Ursula
　　The Left Hand of Darkness,
　　20n.
Lemkin, Rafael, 72n., 129
Leopold and Loeb case, 86n., 120
Lermontov, Mikhael
　　A Hero of our Time, 108
Lesius, Dr Johannes, 111
Lewis, Gary J., 285n.
Libet, Benjamin, 182

life
  eternal, **349–53**; *see also*
    immortality, as robbing life of
    meaning
  meaningful, **307–13**
  meaning of, 326–8
  neither reasonable nor
    unreasonable, 325
  sanctity of, 337
  sources of, **330–2**
  transience of, 330
  valuelessness of, **341–44**
  value of, 337, **344–9**
London, Jack
  Martin Eden, 354
Lovejoy, A. O., 49n.
Lucifer, 120, 393
'Lucifer', mistranslation of Hebrew
  text, 89n., 393
Lucretius, 169, 358–9
Lumet, Sidney
  The Pawnbroker, 139
Luther, Martin, 396
Lykken, David, 295, 299

Macfarlane, Alan, 83, 85n.
machine, xvii, 179–80
  as symbol, 176
Mackie, John, 85
Magdeburg, sack of, 89n.
Malcolm, Norman, xvii, 222n.
Mandelstam, Osip, 151–2
Mann, Thomas
  Doctor Faustus, 396n.
  The Magic Mountain, 278
Margalit, Avishai, 54n., 56n., 112n.
marginalist economics, *see* econom-
  ics, marginalist
Marlowe, Christopher
  Tamburlane, 332n.
meaning
  atemporality of, 318
  field of, 310
  illusory, 312

intransitive v. transitive, 310–11
  of life, *see* life, meaning of
  loss of, **313–16**
  sentimental, 311–12
  subjective v. giving meaning to a
    life, 311
meaningfulness in life, 309–10
meaninglessness, roots of, **316–26**
Medick, Hans, 89n.
Melville, Herman
  Billy Budd, 105–6, 205
memory, xviii
  animal, *see* animal(s), memory
mereological fallacy in neuroscience,
  180
metaphysics, xii
Milgram, Stanley, experiment of, **125–7**
Mill, J. S., 26, 223, 270, 315
Miller, Arthur
  All My Sons, 117
  The Crucible, 36
Milton, John
  Paradise Lost, 120, 396
mind, xx
  acquisition of, 136
  mind–body relationship, 130–1
  -reading, 379
Mirandola, Giovanni Pico della,
  49–50
mirror recognition, 381
misery, science of, 302–3
M'Naghten Rules, 203–6
Moirai, 165
Moore, G. E., 217, 218, 309
moral goodness, *see* goodness,
  varieties of, moral
morality, 23, **24–30**
  animal, *see* animal(s), morality
  composition of, 25
  critical, 31–2
  limits of, 29–30
  naturalist account of, 39
  social, 30–1
  source of, 18

moral maturity, 138
moral progress, 62–3
moral relativism, 63–4
morals, autonomy of, 4–5
moral value, 4–5, 26
  roots of, **38–46**
mortality, 326
  human awareness of, 336
Motzkin, Gabriel, 112n.
movement
  bare bodily, 290
  description of, 185
  neural explanation of, 181–2
Mueller, Anselm, 109n.

Nachev, Parashkev, 182n., 336n.
Nagel, Thomas, 347–9, 382
Nama genocide, 72–3
nature, contingencies of, 19–20
Nazis/Nazism, 27, 28, 69n., 73, 110
needs, xxi–xxii, 13
Neurath, Otto, 284
neuroscience, cognitive
  explanatory limits of, **190–2**,
    195
Newton, I., 169
Newtonian physics, 173
Nietzsche, Friedrich, 5, 59–60,
    176–7, 317, 346, 394n.
Nino, C. S., 90n.
Niv, Yael, 294, 299
Noble, Denis, 175n.
norms, ideal, 25
Norns, 165
Norse, 60
Noyes, Alfred
  'The Highwayman', 275
numbness to value, 313–14

Ogawa, Seiji, 98n.
O'Hear, Anthony, 19n., 397n.
O'Hear, Natasha, 397n.
Omar Khayyam
  Rubáiyát, 163–4

omissions, explanation of, **187–90,**
  195
opportunity, xxi, 157
Origen, 393n.
original sin, *see* sin, original
Orwell, George
  1984, 58, 70, 205
ought to be/ought to do, 25, 38
Ovadia, Rabbi of Bartenora, 167n.
Owen, Adrian M., 336n.
Oz, Amos
  A Perfect Peace, 57

pain, 221–2
Pappenheim, Graf zu, 89n.
Parcae, 165
Pascal, Blaise, 316–17
Pasternak, Boris, 151–2
  Doctor Zhivago, 258, 267
Pater, Walter, 240–2
Pelagius, 395
perceptual powers, xvi–xvii
Perner, Josef, 292, 379n.
person(s), xi–xii
  concept of, 33–4, 47–8, 53
  as ends-in-themselves, 52, 55
  goodness of, 34–5, 38
philia, 41
philosophical anthropology,
    xi–xv, 18
  contrasted with philosophy of
    mind, xiv
  and moral philosophy, xiv–xv
  synthetic nature of, xii
  and teleological explanation, xiii
Piaget, Jean, 379
Pickkard, John D., 336n.
Pindar
  Pythian, 271, 273
Plato, 48
play, 22
pleased, being, 212–19, 230–1
pleasure, **209–42**
  active/passive, 224–6

and awareness, 224
and contentment, 248–9
of desire, 215
and first-person authority, 235
and the good of man, 236
and happiness, 250
of intellect, 216
intentionality of, 224
pictorial representation of,
    407–11
and sensation, 222
of the senses, 214–15
sociable, 216
subjective/objective, 219
as unimpeded activity, **226–7**
of will, 215–16
Plotinus, 49n.
polytheism, 68
Pope, Alexander
    'The Quiet Life', 253n.
possibility of an actuality v. actual-
    ity of a possibility, 159–60,
    193–4
possibility of v. possibility for,
    159–60
powers, cogitative
    presupposed by morality, 24
Premack, David, 292, 378–9
Primo Levi, Michele, 404
Prior, Arthur, 162n.
progress, moral, *see* moral progress
prophets, in ancient Israel, 30
Psalm 8, 48, 50
Pseudo-Dionysius, 87
psuchē, 5, 275
psychopaths, 136–7
purpose, 185
    meaningful, 308–9
    objective v. subjective, 309
Pushkin, Alexander
    Yevgeny Onegin, 108

random, 177
rationality, xiii, xx, xxiii, 184n.

Rausching, Hermann, 75n.
Raz, Joseph, 200n., 347, 351
reactive attitudes, *see* attitudes,
    reactive
reasonableness, xiii, xxiii
reasoning, xix, 180
    theoretical/practical, xix
reasons, xxii, 184
    forms of, 196
    what makes a reason mine, 184–5
reductionism in biology, 176
Rees, Geraint, 285n.
reflexes, 4
refraining, 189, 195
relativism, moral, 19, 63–4
remembering, 180
remorse, 139
Rescorla, Michael, 368n.
respect, **46–57**
respect, formal, 39, 46n., 52–7
    and evil-doers, 55
responsibility, 160
    varieties of, **196–206**
resurrection, 340–1
right and wrong, forgetting the
    difference between, 121–2
rights, 53, 62, 114
Robespierre, Maximilien, 114
robots, xvii
Rorty, Richard, 83
Rossellini, Roberto
    Socrates, 31n.
Rousseau, Jean-Jacques, 92
Rundle, Bede, 373n.
Russell, Jeffrey Burton, 394
Rutledge, Robb B., 294, 298–9
Ryan, Robert
    Billy Budd, 105n.
Ryle, Gilbert, 121

SA (*Sturmabteilung*, Brownshirts), 79
Sackville-West, Vita, 319–20
Saint-Just, Louis Antoine de, 114
sanctity of life, 337

Sands, Philippe, 73n., 129
Santos, Laurie, 301
Satan, in Christianity, 392, 393–7
satan, in Hebrew Bible, 391–3
satisfaction, logical v. psychological,
    230, 324
Saul, envy for David, 387–8
Scheffler, Samuel, 329n., 359
Schiller, Friedrich, 43n., 174
Schlick, M., 341
Schopenhauer, Arthur, 321, 343–4
Schwarz-Bart, André
    The Last of the Just, 310
science, value-free, 9
scrub jays, 375
Sebag-Montefiore, Simon, 142n.
self, 381
self-consciousness, 381
self-interest, 362–3
Seligman, Martin, 300–1
Sellars, Peter
    Dr Strangelove, 183
Sellars, Wilfred, 292n., 378, 380
Selwyn, Pamela, 89
Seneca, Lucius Annaeus, 238
Sereny, Gita, 90, 112n., 133
Shakespeare, William
    Hamlet, 50n., 314
    Julius Caesar, 106–7
    King Lear, 70, 109
    Othello, 107
    Richard III, 109, 110, 140–1
    Sonnet 91, 387
Shark Island, 73
Shaw, George Bernard
    Pygmalion, 117
Shelley, Percy Byshe
    'Ozymandias', 319n.
Shirley, James
    *The Contention of Ajax and
        Ulysses*, 342n.
show trial, 399
Siedentop, L., 51n.
sin, original, 93, 167, 395

Sisyphus, 324–6
Skandali, Nikolina, 295, 296n.,
    298–9
skills, xxi, 9–10, 12
Small, Thomas, 118n.
Smith, Adam, 39, 97
snake as the Devil, in Genesis, 393n.
Socrates, 30–1, 133, 134, 137, 138,
    149, 338, 356–7
solidarity, 43, 45
Solon, 261
Sophocles
    Oedipus at Colonus, 343
    Oedipus Tyrannus, 165, 271
soul, 334, **338–41**
    acquisition of, 136
    death of, **138–43**
    and the flesh, 134–5
    as form of representation,
        135–6, 154
    secular conception of, 130–1,
        132–7, 152–4
    withering of, 150–1
Speer, Albert, 133–4
SS *Einsatzgruppen* ('Task forces':
    murder squads), 79, 116n.
Stalin, 150n., 151
Stangl, Franz, 90, 112
Stangneth, Bettina, 403n.
state, 172–3
    of the brain, 180–1
Steiger, Rod
    The Pawnbroker, 139
Steinbeck, John
    East of Eden, 65, 97, 119
Stevenson, Robert Louis
    Master of Ballantrae, 106
Stoics/Stoicism, 238–9, 269, 272
Strawson, Peter, 45n.
Styron, William
    Sophie's Choice, 139
substances, xiii
suicide, 337
summa anthropologica, xv

summum bonum, 277–8
supervenience, 174
Swift, Jonathan
   Gulliver's Travels, 352n.
Swinburne, Algernon
   'Proserpine', 360
sympathy, 21, 40–2, 43–4

Talaat Bey, 73
teleological explanation, xxii
teleology, rational, 328
Telfer, Elizabeth, 213n.
Tellegen, Auke, 295, 299
Tennyson, Alfred, 320–1
Tertullian, 393n., 394n.
Teskilât-I Mahsusa (Special
   Organization: murder squads), 79
Thackeray, William Makepiece,
   326n.
Thagard, Paul, 287, 298n.
thanatophobia, **353–60**
theodicy, 69, 89
Theophilus of Cilecia, 390
theory of mind theory, 291–4,
   378–81
Thomas, Dylan
   'And death shall have no
     dominion', 345
   'Fern Hill', 258–9
Thomas, Keith, 78n.
Thoreau, Henry David, 295
thoughtfulness, 180
Tichborne, Chidiock
   'My prime of youth', 350
Tillyard, E. M. W., 49n.
Titchener, Edward, 96n.
toleration, 61
Tolstoy, Leo
   Anna Karenina, 14, 246, 315
   A Confession, 315–16
   The Death of Ivan Ilych, 316n.
   The Kreuzer Sonata, 388
   War and Peace, 45, 257–8,
     267, 315

tools, xxi, 11
Trollope, Anthony
   Can You Forgive Her, 309
   He Knew He Was Right, 388
   Phineas Finn, 257
truth, as a variety concept, 7n.
Turgenev, Ivan
   Torrents of Spring, 278, 344

Ulitskaya, Ludmilla
   The Big Green Tent, 136n.
universe, state of the, 172–3
utilitarians/utilitarianism, 211–12,
   240, 270–1
   flaws in classical, 281–2

value
   absolute/relative, 4
value free science, *see* science,
   value-free
values, in a world of facts, 3, 8
   internally related to life, 7
   moral, *see* moral value
Vehib Pasha (Commander of
   Ottoman Third Army), 79n.
virtue, path of, 25–6
virtues, 275–6
   as protective, 274
   relative permanence of, **58–60**
Vlastos, Gregory, 46
voluntary/involuntary, non-voluntary, 160n., 183
   partly, 182, 206
Von Wright, G. H., 5, 6–7, 8, 10–11,
   16–17, 25, 37–8, 66, 217–18

Waal, Frans de, 374n., 376n., 377,
   382–3
Walpole, Hugh
   Judith Paris, 109
wanting, kinds of, 254–5, 257
wants, second order, 255n., 376
   objects of, 256
   terminating v. regulative, 256–7

Wantzel, Pierre Laurent, 153
Warnock, Geoffrey, 37–8
Weil, Simone, 20, 53
Weiskrantz, Larry, 195n.
Welles, Orson
    Citizen Kane, 14
Werfel, Franz
    Forty Days of Musa Dagh, 111
    Jacobowsky and the Colonel, 252
White, Alan, 159n., 160n., 203n.
wickedness, 68
Wiesenthal, Simon, 112n.
Wiggins, David, 20n., 43, 46
Wilde, Oscar
    The Picture of Dorian Gray, 108,
        250, 277
will, xx, xxii
Williams, Bernard, 44, 351

Wimmer, Heinz, 292, 379n.
Wittgenstein, Ludwig, xx, 37, 122,
        124n., 152–4, 176n., 194–5,
        310, 313, 327, 385
Woodruff, Guy, 292, 378–9
Wootton, Baroness Barbara,
        86n.
Wordsworth, William
    'The Prelude', 259

Yeats, William Butler
    'What Then?', 323–4

Zajonc, Robert, 376
Zimbardo, Philip, experiment of,
        126n.
Zoroastrianism, 391–2
Zweig, Stefan, 61n.

Printed and bound by CPI Group (UK) Ltd, Croydon, CR0 4YY

07/07/2026

14916223-0004